Depolarizing Collisions in Nonlinear Electrodynamics

Depolarizing Collisions in Nonlinear Electrodynamics

Igor V. Yevseyev
Valery M. Yermachenko
Vitaly V. Samartsev

CRC PRESS

Boca Raton London New York Washington, D.C.

Library of Congress Cataloging-in-Publication Data

Evseev, I.V. (Igor Victorovich)
Depolarizing collisions in nonlinear electrodynamics / Igor V. Yevseyev, Valery M. Yermachenko, Vitaly V. Samartsev.
p. cm.
Includes bibliographical references and index.
ISBN 0-415-28416-3 (alk. paper)
1. Quantum electronics. 2. Nonlinear optics. 3. Electrodynamics. 4. Collisions (Physics). 5. Gas dynamics. I. Ermachenko, V. M. (Valerii Mikhailovich) II.Samartsev, V.V. (Vitalii Vladimirovich) III. Title.
QC446.2E95 2004
537.5—dc22 2003070022

International Standard Book Number 0-415-28416-3
Library of Congress Card Number 2003070022
Printed in the United States of America 1 2 3 4 5 6 7 8 9 0
Printed on acid-free paper

Preface

In many cases, when one considers the interaction of resonant electromagnetic radiation with a gas medium, it is sufficient to take into account the interaction between atoms (molecules) of the gas in the approximation of pair collisions. As is well known, the cross section of elastic collisions, which do not change the populations of the considered resonant atomic levels, is usually much higher than the cross section of inelastic collisions. Elastic collisions not only result in changes in the velocities of resonant atoms, but also lead to the redistribution of atoms over the Zeeman sublevels of resonant levels, which are usually degenerate. In those cases when we investigate the nonlinear interaction of electromagnetic fields of different polarizations with a gas medium, elastic depolarizing collisions, leading to the redistribution of resonant atoms over the Zeeman sublevels, play an especially important role. The significance of such collisions is associated with the fact that, due to selection rules, the electromagnetic field of certain polarization couples only definite Zeeman sublevels of resonant atomic levels, while depolarizing collisions involve other sublevels in the interaction with the field.

In this monograph, we provide a consistent theory of elastic depolarizing collisions. This theory is then employed for the investigation of some nonlinear electromagnetic phenomena in a gas medium, including different types of the photon echo, double-mode lasing in gas lasers with orthogonal polarizations of laser modes, and the interaction of strong and weak electromagnetic waves passing through a resonant gas medium. It will be demonstrated that the description of several effects involving the interaction of electromagnetic waves with different polarizations even at the qualitative level requires the analysis of elastic depolarizing collisions.

A considerable part of this monograph is devoted to the discussion of the nonlinear electrodynamics of the photon echo in a gas medium. First, we will present the theory of the photon echo ignoring depolarizing collisions and provide a review of experimental data that can be understood in terms of such a theory. Then, we develop the theory of the photon echo and its modifications including elastic depolarizing collisions and discuss the available experimental results that can be interpreted only with allowance for elastic depolarizing collisions. Finally, based on this theoretical background, we propose new experiments that may provide an additional spectroscopic information on atoms (molecules) in a gas medium.

In this monograph, we also present the theory of double-mode lasing in standing-wave gas lasers with allowance for elastic depolarizing collisions. We will demonstrate that, in the case of modes with mutually orthogonal polarizations, the inclusion of such collisions in the analysis of the problem is of fundamental importance. These collisions determine the minimum intermode separation that still provides stable double-mode lasing.

The results of investigations of nonlinear interaction between weak and strong electromagnetic waves passing through a gas medium presented in this monograph reveal a considerable influence of elastic depolarizing collisions on this interaction. Depending on the polarizations of the strong and weak waves, the width of the resonance in the gain of the weak wave is determined by different relaxation parameters introduced in the theory of depolarizing collisions.

Writing this monograph, we employed the results of our investigations that have been carried out for many years in collaboration with our colleagues: Yu.A. Vdovin, V.K. Matskevich, V.A. Reshetov, D.S. Bakaev, V.N. Tsikunov, V.A. Zuikov, V.A. Pirozhkov, I.I. Popov, I.S. Bikbov, R.G. Usmanov, and others. We are deeply grateful to all of them. We also thank V.P. Chebotayev, N.V. Karlov, S.S. Alimpiev, L.S. Vasilenko, and N.N. Rubtsova for useful discussions and support at different stages of our investigations. We also acknowledge a valuable help of Dr. A.M. Zheltikov for translation of this book from Russian into English.

CONTENTS

Chapter 1

INTERACTION OF ATOMS IN THE APPROXIMATION OF DEPOLARIZING COLLISIONS

1.1 The Integral of Elastic Atomic Collisions

The effect of atomic collisions on the optical characteristics of an ensemble of excited atoms is well known [1, 2]. The evolution of laser techniques and the resulting interest in various nonlinear-spectroscopy methods have suggested that atomic collisions have a significant effect on the lasing mode of gas lasers and the shape and width of various nonlinear signals. For a long time attention has been focused on the important role of elastic collisions, among other things, and various methods have been proposed for describing such collisions [3].

The possibility of consistently describing elastic collisions follows from the fact that the systems being investigated (for example, gases used as active media in gas lasers) are characterized by pressures for which the binary approximation is valid. Besides, in such systems the number of atoms excited to active levels is considerably lower than the number of nonexcited atoms, therefore it is sufficient to take into account only collisions of excited atoms with nonexcited atoms.

We will describe the state of atoms excited to levels a and b with angular momenta j_a and j_b using the density matrix technique. We define $\rho^{(bb)}_{mm_1}(\boldsymbol{r},\mathbf{v},t)$ as the density-matrix elements belonging to state b, with m and m_1 the projections of the angular momentum j_b on the quantization axis ($|m| < j_b$) and $\boldsymbol{r}$ and $\mathbf{v}$ the quantities characterizing the motion of an atom as a whole (the atom radius vector and velocity at time t, i.e., in contrast to the internal state, the motion of the atom as a whole is described classically), $\rho^{(aa)}_{mm_1}(\boldsymbol{r},\mathbf{v},t)$ as the same quantity for state a, and $\rho^{(ba)}_{mm_1}(\boldsymbol{r},\mathbf{v},t)$ as the density-matrix elements describing the optical coherence (transitions) between the levels.

It is assumed that the ground state of the nonexcited atoms with which the excited atoms interact (collide) has an angular momentum $j_0 = 0$, with the nonexcited atoms being uniformly distributed in space with a density n and characterized by a Maxwell velocity distribution $F(\mathbf{v})$ normalized to unity.

As a result of an elastic collision, the excited atom changes its velocity and internal state. For a fixed initial state, the final state of the atom after scattering is completely determined by the scattering amplitude [4]. In terms of the amplitudes of atomic elastic scattering in states a and b one can also express the changes in the corresponding elements of the density matrix brought about by such collisions, or the collision integral. The diagrammatic technique developed in [5] has led to the following equations for the density-matrix elements [6]:

$$\left(\frac{\partial}{\partial t} + \frac{\boldsymbol{p}}{M_1}\nabla_r + i\omega_b - i\omega_a + \frac{\gamma_b^{(0)} + \gamma_a^{(0)}}{2}\right)\rho_{mm_1}^{(ba)}(\boldsymbol{r},\boldsymbol{p},t) +$$

$$+ i\Sigma_{mm_2}^{(b)}(\boldsymbol{p})\,\rho_{m_2m_1}^{(ba)}(\boldsymbol{r},\boldsymbol{p},t) - i\,\rho_{mm_2}^{(ba)}(\boldsymbol{r},\boldsymbol{p},t)\,\Sigma_{m_1m_2}^{(a)*}(\boldsymbol{p}) =$$

$$= \frac{n\pi}{2M^2}\int\frac{d\boldsymbol{p}_2}{(2\pi)^3}\int\frac{d\boldsymbol{p}_1}{(2\pi)^3}F(\boldsymbol{p}_1)\,f_{mm_3}^{(b)}(\boldsymbol{p}-\boldsymbol{p}_2, \boldsymbol{p}_2 - M_1(\boldsymbol{p}_2+\boldsymbol{p}_1)/M_0)\times$$

$$\times\,\rho_{m_3m_2}^{(ba)}(\boldsymbol{r},\boldsymbol{p}_2,t)\,f_{m_1m_2}^{(a)*}(\boldsymbol{p}-\boldsymbol{p}_2, \boldsymbol{p}_2 - M_1(\boldsymbol{p}_2+\boldsymbol{p}_1)/M_0)\times$$

$$\times\delta(\varepsilon(\boldsymbol{p}) - \varepsilon(\boldsymbol{p}_2) + \varepsilon^{(0)}(\boldsymbol{p}_2+\boldsymbol{p}_1-\boldsymbol{p}) - \varepsilon^{(0)}(\boldsymbol{p}_1)), \tag{1.1.1}$$

$$\Sigma_{mm_1}^{(b)}(\boldsymbol{p}) = \frac{n}{2M}\int\frac{d\boldsymbol{p}_1}{(2\pi)^3}f_{mm_1}^{(b)}(0, \boldsymbol{p} - M_1(\boldsymbol{p}+\boldsymbol{p}_1)/M_0). \tag{1.1.2}$$

Here the quantity $\Sigma_{mm_1}^{(a)}(\boldsymbol{p})$ is obtained from (1.1.2) by substituting index a for index b, and the following notation is employed. Instead of the velocity $\mathbf{v}$ of the excited atom we introduce the momentum vector $\boldsymbol{p} = M_1\mathbf{v}$, where M_1 is the mass of the excited atom. Since the other atom participating in the collision may be of a different kind (say, helium in the helium–neon laser), by M_2 we denote the mass of the nonexcited atom and introduce the reduced mass of the colliding atoms, $M = M_1M_2/M_0$, with $M_0 = M_1 + M_2$; and

$$\varepsilon(\boldsymbol{p}) = \boldsymbol{p}^2/2M_1;\quad \varepsilon^{(0)}(\boldsymbol{p}) = \boldsymbol{p}^2/2M_2.$$

$F(\boldsymbol{p})$ is the Maxwell distribution over the momenta of the nonexcited atoms (normalized to unity), $\delta(x)$ is the delta-function, $\omega_{b(a)}$ is the transition frequency from the upper (lower) active level to the ground state;

$$f_{mm_1}^{(b)}(\boldsymbol{p}-\boldsymbol{p}_2, \boldsymbol{p}_2 - M_1(\boldsymbol{p}_2+\boldsymbol{p}_1)/M_0)$$

is the scattering amplitude of the atom in state b, and $f^{(a)}_{mm_1}$ is the similar quantity for the atom in state a.

The scattering amplitude depends on the velocity transfer in the collision process, $\mathbf{v} - \mathbf{v}_2 = (\boldsymbol{p} - \boldsymbol{p}_2)/M_1$, and the relative velocity of the colliding atoms $\mathbf{v}_2 - \mathbf{v}_1 = (\boldsymbol{p}_2 - M_1(\boldsymbol{p}_2 + \boldsymbol{p}_1)/M_0)/M$.

The quantities $\gamma^{(0)}_b$ and $\gamma^{(0)}_a$ characterize the rate at which the active levels are depleted by radiative decay and inelastic collisions. The equation for $\rho^{(bb)}_{mm_1}(\boldsymbol{r}, \boldsymbol{p}, t)$ is obtained from (1.1.1) by replacing a with b, and the equation for $\rho^{(aa)}_{mm_1}(\boldsymbol{r}, \boldsymbol{p}, t)$ is obtained in a similar way. Finally, throughout the chapter we employ the summation convention for repeated indices.

Similar equations have been obtained in [7], with the only difference that the scattering amplitudes are replaced there with their Born approximations in explicit form. In [8] a complete quantum mechanical treatment of the problem was carried out, and the motion of an atom as a whole was also considered in the quantum mechanical setting. In [9] the approach to describing collision integrals was similar to our approach. Both [9] and [10] contain large lists of references devoted to the collision-integral problem.

1.2 The Model of Depolarizing Collisions

In what follows it is convenient to separate the collision integral into two parts: one takes into account scattering through a zero angle only, while the other is associated with variations of atomic velocities in the collision process. Shifting the terms associated with the collisions to the right-hand side of equation (1.1.1), we can write them in the following form:

$$\hat{A}(\boldsymbol{p})\,\hat{\rho}(\boldsymbol{p}) + \int d\boldsymbol{p}_1 \hat{B}(\boldsymbol{p}, \boldsymbol{p}_1)\,\hat{\rho}(\boldsymbol{p}_1) = [\,\hat{A}(\boldsymbol{p}) + \\ + \int d\boldsymbol{p}_1 \hat{B}(\boldsymbol{p}, \boldsymbol{p}_1)\,]\,\hat{\rho}(\boldsymbol{p}) + \int d\boldsymbol{p}_1 \hat{B}(\boldsymbol{p}, \boldsymbol{p}_1)\,[\,\hat{\rho}(\boldsymbol{p}_1) - \hat{\rho}(\boldsymbol{p})\,]. \qquad (1.2.1)$$

The first term on the right-hand side of (1.2.1) is determined by the total scattering cross section of the colliding atoms and describes the variations in the polarization of the excited atom in the collision process, and the second term is associated with scattering with velocity variations.

For the interaction of the van der Waals type, $V(R) \sim C/R^6$, which we use in all future calculations of relaxation characteristics, distances of the order of the Weisskopf radius [2], $\rho_0 \sim (C/\mathrm{v})^{1/5}$, play an important role in scattering. Here the velocity of the atom is assumed to be so small that the effective distances (the impact parameter) ρ_0 exceed atomic dimensions and the dipole–dipole approximation in the interaction of the atoms can be used. At the same time, the requirement that the motion of an atom as a whole be quasiclassical imposes restrictions on the velocity of the atom's motion from below: $1/M\mathrm{v} << \rho_0$ (here we use atomic units). This means that the effective scattering angle in the considered approximation is fairly small,

$$\theta_{eff} \sim 1/M\mathrm{v}\rho_0 << 1,$$

and as a whole exceeds scattering through small angles, so that

$$\langle\theta^2\rangle << 1, \tag{1.2.2}$$

where $\langle\theta^2\rangle$ is the mean square of the scattering angle of the atom. Hence, the second term on the right-hand side of (1.2.1) contains a small parameter and in some cases can be dropped.

Thus, we arrive at the model of depolarizing collisions in which the variations of atomic velocities in the collision process are ignored, but the variation of the distribution of these atoms over the Zeeman sublevels is taken into account. In some problems it is necessary to allow for the variations in atomic velocities to obtain a detailed explanation of the experimental data. For example, the experimentally observed Lamb dip in the output power of the single-mode lasing of a gas laser has a more complicated shape than one would expect from the model of depolarizing collisions. In order to explain this, one is forced not only to allow for variations in atomic velocities in the collision process but also for resonance radiation trapping (see, for example, [11] and [12]).

Let us now write equation (1.1.1) and the corresponding equation for $\rho^{(bb)}_{mm_1}$ in the model of depolarizing collisions:

$$\left(\frac{\partial}{\partial t} + \mathbf{v}\nabla_r - i\omega_0 + \frac{\gamma_b^{(0)} + \gamma_a^{(0)}}{2}\right)\rho^{(ab)}_{mm_1}(\boldsymbol{r}, \mathbf{v}, t) =$$

$$= -\Gamma_{ab}(\mathbf{v})^{m_2m_3}_{mm_1}\,\rho^{(ab)}_{m_2m_3}(\boldsymbol{r}, \mathbf{v}, t), \tag{1.2.3}$$

$$\left(\frac{\partial}{\partial t} + \mathbf{v}\nabla_r + \gamma_b^{(0)}\right)\rho_{mm_1}^{(bb)}(\boldsymbol{r}, \mathbf{v}, t) = -\,\Gamma_b(\mathbf{v})_{mm_1}^{m_2m_3}\,\rho_{m_2m_3}^{(bb)}(\boldsymbol{r}, \mathbf{v}, t), \tag{1.2.4}$$

where $\omega_0 = \omega_b - \omega_a$ is the central frequency of the transition. The equation for $\rho_{mm_1}^{(aa)}(\boldsymbol{r}, \mathbf{v}, t)$ is obtained by substituting index a for index b everywhere in equation (1.2.4). The matrices $\Gamma_{ab}(\mathbf{v})$, $\Gamma_a(\mathbf{v})$, and $\Gamma_b(\mathbf{v})$ describe the collisional relaxation of the respective components of the density matrix. In the small-angle scattering approximation, the scattering amplitude can be expressed in terms of the S-matrix [4], with the relaxation matrices assuming the following form

$$\Gamma_{ab}(\mathbf{v})_{mm_1}^{m_2m_3} = n\int d\mathbf{v}_0\, d\vec{\rho}\, F(\mathbf{v}_0)|\mathbf{v} - \mathbf{v}_0|[\,\delta_{mm_2}\,\delta_{m_1m_3} - \\ - S_{mm_2}^{(a)}\,S_{m_1m_3}^{(b)*}\,], \tag{1.2.5}$$

$$\Gamma_b(\mathbf{v})_{mm_1}^{m_2m_3} = n\int d\mathbf{v}_0\, d\vec{\rho}\, F(\mathbf{v}_0)|\mathbf{v} - \mathbf{v}_0|[\,\delta_{mm_2}\,\delta_{m_1m_3} - \\ - S_{mm_2}^{(b)}\,S_{m_1m_3}^{(b)*}\,], \tag{1.2.6}$$

where $d\vec{\rho}$ – integration with respect to the impact parameter $\vec{\rho}$ – is carried out in the plane perpendicular to the relative velocity of the colliding particles, $\mathbf{v} - \mathbf{v}_0$. For $\Gamma_a(\mathbf{v})_{mm_1}^{m_2m_3}$, there is a formula similar to (1.2.6). Here $S_{mm_2}^{(a)}$ and $S_{mm_2}^{(b)}$ are the elements of the S-matrix of elastic scattering of the excited atom in states a and b on the nonexcited atom. They depend on the impact parameter $\vec{\rho}$ and the relative velocity of the colliding atoms. The authors of [13] accounted for collisional relaxation in a similar manner, but the relaxation matrices were assumed to be independent of $\mathbf{v}$. The results of [13] can be obtained if one averages (1.2.5) and (1.2.6) over $\mathbf{v}$ with the Maxwell distribution function.

Generally the S-matrix is not diagonal in the magnetic quantum numbers, which reflects the possibility of transitions between the Zeeman sublevels in the scattering process. It satisfies the equation [14]

$$i(\mathbf{v} - \mathbf{v}_0)\nabla S_{mm_1}^{(b)} = V_{mm_2}^{(b)}(\boldsymbol{R})\, S_{m_2m_1}^{(b)} \tag{1.2.7}$$

together with the condition

$$S^{(b)}_{mm_0}(-\infty) = \delta_{mm_0},$$

where m_0 characterizes the state of the atom before the collision, $V^{(b)}_{mm_2}(\boldsymbol{R})$ are the matrix elements of the interaction operator for the excited and nonexcited atoms, and $\boldsymbol{R}$ is the vector connecting these atoms. There is also a similar equation for $S^{(a)}_{mm_1}$. The relaxation matrices $\Gamma_{ab}(\mathbf{v})$, $\Gamma_a(\mathbf{v})$, and $\Gamma_b(\mathbf{v})$ are the functions of the velocity of the excited atom, so they both depend on the magnitude and direction of the velocity vector.

The common approach in studying the effect of depolarizing collisions on the interaction of electromagnetic radiation with gaseous medium is to ignore the velocity dependence of a relaxation matrices, which is equivalent to averaging over velocity with the Maxwell distribution function. Obviously, averaging over the directions of the velocity vector leads to a loss of effects based, for instance, on the difference in the relaxation rates of the components of the density matrix longitudinal and transverse in relation to $\mathbf{v}$. The dependence of the relaxation matrices on the absolute value of velocity that remains after averaging also leads to observable effects, for example, to a difference in the profile of the spectral line of radiation from the Voigt profile. The size of all such effects depends essentially on the mass ratio of the colliding (excited and nonexcited) atoms.

1.3 Dependence of Relaxation Matrices on Atomic Velocities

To establish the way in which the relaxation matrices depend on the velocity of the atoms, one must solve equation (1.2.7) for a specific interaction between the colliding atoms. Equation (1.2.7) is actually a system of equations for the elements of the S-matrix. The matrix has the simplest form in a system of coordinates in which the z-axis is directed along the vector $\mathbf{v}_{rel} = \mathbf{v} - \mathbf{v}_0$, and the zx-plane coincides with the collision plane. By $\widetilde{S}^{(b)}_{mm_1}$ and $\widetilde{S}^{(a)}_{mm_1}$ we denote the elements of the respective S-matrices in this system of coordinates. The transformation to a system of coordinates in which the polar axis is arbitrarily directed is then carried out in the ordinary way via rotation matrices [15]. In studying specific problems of the interaction of electromagnetic radiation with a medium, it has proved expedient to solve the system of equations for the elements of the density matrix $\rho^{(aa)}_{mm_1}$, $\rho^{(bb)}_{mm_1}$, $\rho^{(ab)}_{mm_1}$ by expanding the density matrix and the relaxation matrices in irreducible tensor operators [16, 17]. Here we use the notation accepted in [13]:

$$\rho_{mm_1}^{(bb)}(\boldsymbol{r},\mathbf{v},t) = (-1)^{j_b - m}(2j_b + 1)^{-1/2}\sum_{\kappa q}(2\kappa + 1)\begin{pmatrix} j_b & j_b & \kappa \\ m & -m_1 & q \end{pmatrix}\times$$
$$\times f_q^{(\kappa)}(\boldsymbol{r},\mathbf{v},t),$$

$$\rho_{mm_1}^{(aa)}(\boldsymbol{r},\mathbf{v},t) = (-1)^{j_a - m}(2j_a + 1)^{-1/2}\sum_{\kappa q}(2\kappa + 1)\begin{pmatrix} j_a & j_a & \kappa \\ m & -m_1 & q \end{pmatrix}\times$$
$$\times \phi_q^{(\kappa)}(\boldsymbol{r},\mathbf{v},t),$$

$$\rho_{mm_1}^{(ab)}(\mathbf{r},\mathbf{v},\mathrm{t}) = (-1)^{j_a - m}(2j_a + 1)^{-1/2}\sum_{\kappa q}(2\kappa + 1)\begin{pmatrix} j_a & j_b & \kappa \\ m & -m_1 & q \end{pmatrix}\times$$
$$\times \psi_q^{(\kappa)}(\boldsymbol{r},\mathbf{v},t), \qquad (1.3.1)$$

where

$$\begin{pmatrix} j_a & j_b & \kappa \\ m & -m_1 & q \end{pmatrix}$$

are $3j$-symbols. If we combine equations (1.2.3) and (1.2.4) with the appropriate equation for $\rho_{mm_1}^{(aa)}$ and employ the properties of $3j$-symbols, we get

$$\left(\frac{\partial}{\partial t} + \mathbf{v}\nabla_r - i\omega_0 + \frac{\gamma_b^{(0)} + \gamma_a^{(0)}}{2}\right)\psi_q^{(\kappa)}(\boldsymbol{r},\mathbf{v},t) =$$
$$= -\sum_{\kappa_1 q_1}\Gamma_{ab}(\mathbf{v})_{qq_1}^{\kappa\kappa_1}\psi_{q_1}^{(\kappa_1)}(\boldsymbol{r},\mathbf{v},t), \qquad (1.3.2)$$

$$\left(\frac{\partial}{\partial t} + \mathbf{v}\nabla_r + \gamma_b^{(0)}\right)f_q^{(\kappa)}(\boldsymbol{r},\mathbf{v},t) =$$
$$= -\sum_{\kappa_1 q_1}\Gamma_b(\mathbf{v})_{qq_1}^{\kappa\kappa_1}f_{q_1}^{(\kappa_1)}(\boldsymbol{r},\mathbf{v},t), \qquad (1.3.3)$$

$$\left(\frac{\partial}{\partial t} + \mathbf{v}\nabla_r + \gamma_a^{(0)}\right)\phi_q^{(\kappa)}(\boldsymbol{r}, \mathbf{v}, t) =$$
$$= - \sum_{\kappa_1 q_1} \Gamma_a(\mathbf{v})_{qq_1}^{\kappa\kappa_1}\, \phi_{q_1}^{(\kappa_1)}(\boldsymbol{r},\mathbf{v},t). \tag{1.3.4}$$

In this representation depolarizing collisions are described in terms of the relaxation matrices $\Gamma_a(\mathbf{v})_{qq_1}^{\kappa\kappa_1}$, $\Gamma_b(\mathbf{v})_{qq_1}^{\kappa\kappa_1}$, $\Gamma_{ab}(\mathbf{v})_{qq_1}^{\kappa\kappa_1}$, which are related to the scattering matrices $\widetilde{S}_{mm_1}^{(b)}$ and $\widetilde{S}_{mm_1}^{(a)}$ by the following formulas:

$$\Gamma_{ab}(\mathbf{v})_{qq_1}^{\kappa\kappa_1} = \sum_{Jq_2} (-1)^{q+q_2}[4\pi(2J+1)]^{1/2}\begin{pmatrix} \kappa & \kappa_1 & J \\ q & -q_1 & p \end{pmatrix} \times$$
$$\times \begin{pmatrix} \kappa & \kappa_1 & J \\ q_2 & -q_2 & p \end{pmatrix} Y_{Jp}(\theta,\phi)\,\Gamma_{ab}(\mathrm{v})_{q_2}^{\kappa\kappa_1}, \tag{1.3.5}$$

where

$$\Gamma_{ab}(\mathrm{v})_{q}^{\kappa\kappa_1} = \sum(-1)^{2j_a+m_2+m_3+q+\kappa}(2\kappa_1+1)(2J+1)\times$$
$$\times \begin{pmatrix} j_a & j_b & \kappa \\ m_2 & -m & k \end{pmatrix}\begin{pmatrix} j_a & j_b & \kappa_1 \\ m_3 & -m_1 & k \end{pmatrix}\begin{pmatrix} \kappa & \kappa_1 & J \\ q & -q & 0 \end{pmatrix}\times$$
$$\times \begin{pmatrix} \kappa & \kappa_1 & J \\ k & -k & 0 \end{pmatrix} n\int d\mathbf{v}_0\, F(\mathbf{v}_0)|\mathbf{v}-\mathbf{v}_0|\,P_J(\cos\theta_1)\int d\vec{\rho}\,[\delta_{m_2m_3}\,\delta_{mm_1} -$$
$$- \widetilde{S}_{m_2m_3}^{(a)}\,\widetilde{S}_{mm_1}^{(b)*}] \tag{1.3.6}$$

is the relaxation matrix in a coordinate system in which the z-axis is directed along the velocity vector $\mathbf{v}$, angles θ and ϕ are the polar and azimuth angles of vector $\mathbf{v}$ with respect to an arbitrarily chosen polar axis, θ_1 is the angle between vectors $\mathbf{v}$ and $\mathbf{v}_{rel}$, $Y_{Jp}(\theta,\phi)$ are spherical functions, and $P_J(\cos\theta_1)$ are Legendre polynomials. The summation over m, m_1, m_2, m_3, k, and J is carried out in (1.3.6).

The matrices $\Gamma_a(\mathbf{v})_{qq_1}^{\kappa\kappa_1}$, $\Gamma_b(\mathbf{v})_{qq_1}^{\kappa\kappa_1}$ are related to the corresponding matrices $\Gamma_{a(b)}(\mathrm{v})_q^{\kappa\kappa_1}$ through (1.3.5), where, for instance, $\Gamma_b(\mathrm{v})_q^{\kappa\kappa_1}$ is determined by relation (1.3.6) in whose right-hand side a is replaced with b, m_2 with m, m_3 with m_1, m_1 with m_3, and m with m_2. The formula for $\Gamma_a(\mathrm{v})_q^{\kappa\kappa_1}$ is obtained from that for $\Gamma_b(\mathrm{v})_q^{\kappa\kappa_1}$ by replacing b with a everywhere. In obtaining the above expressions we used the properties of rotation matrices and 3*j*-symbols. Equations (1.3.5) and (1.3.6) give the explicit dependence of the relaxation matrices on the velocity of the excited atom. Note that in averaging over the velocity directions these matrices are diagonalized in indices κ and q.

1.4 Relaxation Characteristics of an Atomic Transition between Levels with Angular Momenta 0 and 1

Let us now apply the general results just obtained to atomic transitions with level angular momenta $j_a = 0$ and $j_b = 1$. As (1.3.6) implies, only the components with $\kappa = \kappa_1 = 1$ remain nonvanishing. Since in this case q assumes the values – 1, 0, and 1, from (1.3.6) it follows that in a system of coordinates in which the z axis is directed along the velocity vector $\mathbf{v}$, the $\Gamma_{ab}(\mathrm{v})_q^{\kappa\kappa_1}$ matrix has the following nonzero components:

$$\Gamma_{ab}(\mathrm{v})_1^{11} = \Gamma_{ab}(\mathrm{v})_{-1}^{11} = \Gamma_1(\mathrm{v}) - i\Delta_1(\mathrm{v}),$$

$$\Gamma_{ab}(\mathrm{v})_0^{11} = \Gamma_0(\mathrm{v}) - i\Delta_0(\mathrm{v}), \tag{1.4.1}$$

where $\Gamma_q(\mathrm{v})$ and $\Delta_q(\mathrm{v})$ $(q = 0, 1)$ are real-valued functions of velocity. Substituting (1.4.1) into (1.3.5), we find all the nonzero matrix elements:

$$\Gamma_{ab}(\mathbf{v})_{qq_1}^{\kappa\kappa_1} - \Gamma_{ab}(\mathbf{v})_{qq_1}^{11}. \tag{1.4.2}$$

Omitting for the sake of brevity the superscripts in (1.4.2), we arrive at the following expressions for the diagonal (in q) matrix elements:

$$\Gamma_{ab}(\mathbf{v})_{11} = \Gamma_{ab}(\mathbf{v})_{-1-1} = [\Gamma_0(\mathrm{v}) - i\Delta_0(\mathrm{v}) + 2\Gamma_1(\mathrm{v}) - 2i\Delta_1(\mathrm{v})]/3 -$$

$$- (1 - 3\cos^2\theta)[\Gamma_1(\mathrm{v}) - i\Delta_1(\mathrm{v}) - \Gamma_0(\mathrm{v}) + i\Delta_0(\mathrm{v})]/6,$$

$$\Gamma_{ab}(\mathbf{v})_{00} = [\Gamma_0(\mathrm{v}) - i\Delta_0(\mathrm{v}) + 2\Gamma_1(\mathrm{v}) - 2i\Delta_1(\mathrm{v})]/3 +$$

$$+ (1 - 3\cos^2\theta)[\Gamma_1(\mathrm{v}) - i\Delta_1(\mathrm{v}) - \Gamma_0(\mathrm{v}) + i\Delta_0(\mathrm{v})]/3. \qquad (1.4.3)$$

The difference in these diagonal elements is determined by the difference of the quantities characterizing the relaxation rates of longitudinal and transverse components of the density matrix (with respect to the direction of **v**), that is,

$$\Gamma_1(\mathrm{v}) - \Gamma_0(\mathrm{v}) - i(\Delta_1(\mathrm{v}) - \Delta_0(\mathrm{v})).$$

All the off-diagonal elements $\Gamma_{ab}(\mathbf{v})_{qq_1}$ $(q \neq q_1)$ are proportional to this difference, e.g.,

$$\Gamma_{ab}(\mathbf{v})_{1-1} = (1/2)[\Gamma_1(\mathrm{v}) - \Gamma_0(\mathrm{v}) - i(\Delta_1(\mathrm{v}) - \Delta_0(\mathrm{v}))]\sin^2\theta\exp(-2i\phi). \qquad (1.4.4)$$

If we are not interested in effects associated with the fact that the off-diagonal elements of the relaxation matrix are nonzero, then the inaccuracy in describing the collision process by a relaxation matrix averaged over the velocity directions is characterized by the ratio of $\Gamma_1(\mathrm{v}) - \Gamma_0(\mathrm{v})$ and $\Delta_1(\mathrm{v}) - \Delta_0(\mathrm{v})$ to $\Gamma_1(\mathrm{v})$, $\Gamma_0(\mathrm{v})$ and $\Delta_1(\mathrm{v})$, $\Delta_0(\mathrm{v})$, respectively.

To establish the size of this effect, calculations were performed assuming that van der Waals forces act between the colliding atoms. At $j_a = 0$ for level a we obtain from (1.2.7) the following formula:

$$\widetilde{S}^{(a)} = \exp(-3\pi i C^{(a)}/(8\rho^5 \mathrm{v}_{\mathrm{rel}})), \qquad (1.4.5)$$

where $C^{(a)}$ is the van der Waals constant for the interaction of an atom in state a, and ρ is the impact parameter. The solution of the system of equations (1.2.7) for matrix $\widetilde{S}^{(b)}_{mm_1}$ was carried out by a standard method (see [14,

18]). We write the matrix elements of the interaction potential of the colliding atoms, $V_{mm_2}(\boldsymbol{R})$, in the form (see [19])

$$V_{mm_1}(\boldsymbol{R}) = [A\,\delta_{mm_1} + B(-1)^{j+m+1}\begin{pmatrix} j & j & 2 \\ m & -m_1 & k \end{pmatrix} Y_{2k}(\theta,0)]/R^6, \qquad (1.4.6)$$

where θ is the angle between vectors $\boldsymbol{R}$ and $\mathbf{v}_{\rm rel}$ with the z-axis directed along $\mathbf{v}_{\rm rel}$ (for the sake of brevity we have omitted the index b on A, B, j, and $V_{mm_1}(\boldsymbol{R})$). The constants A and B can be expressed in terms of the van der Waals constants C_m, which characterize the interaction of atoms with projection m of the angular momentum of the excited atom on the direction of $\boldsymbol{R}$. At $j = 1$ we have

$$A = (2C_1 + C_0)/3\,, \quad B = (24\pi)^{1/2}(C_0 - C_1)/3. \qquad (1.4.7)$$

Combining (1.4.6) with (1.2.7) yields the following system of equations:

$$i v_{\rm rel}\rho^5 \frac{dS_{mm_1}}{d\xi} = (1-\xi^2)^{3/2}[A\,S_{mm_1} + B(-1)^{j+m+1}\sum_{m_2}\begin{pmatrix} j & j & 2 \\ m & -m_2 & k \end{pmatrix}\times$$

$$\times Y_{2k}(\xi,0)\,S_{m_2m_1}], \qquad (1.4.8)$$

where the variable ξ is defined via the relation

$$\xi = \cos\theta = z/(z^2+\rho^2)^{1/2}, \qquad -1 \le \xi \le 1,$$

and the boundary conditions are

$$S_{mm_1}(\xi=-1) = \delta_{mm_1}\,. \qquad (1.4.9)$$

The elements $\widetilde{S}_{mm_1}$ of the scattering matrix entering into the relaxation matrices (1.3.6) depend on parameters v_{rel} and ρ and are obtained by solving system (1.4.8) with $\xi = 1$:

$$\widetilde{S}_{mm_1} = S_{mm_1}(\xi = 1), \tag{1.4.10}$$

where for the sake of brevity we have omitted parameters v_{rel} and ρ as independent variables.

The system of equations (1.4.8) implies that the matrix elements $S_{mm_1}(\xi)$ satisfy the following relations:

$$S_{mm_1} = (-1)^{m-m_1} S_{-m-m_1}. \tag{1.4.11}$$

At $j = 1$ this yields

$$S_{11} = S_{-1-1}, \quad S_{1-1} = S_{-11}, S_{10} = -S_{-10}, \quad S_{01} = -S_{0-1}. \tag{1.4.12}$$

Expressing the constants A and B in (1.4.8) in terms of C_1 and C_0 of (1.4.7) and writing $S_{mm_1}(\xi)$ in the form

$$S_{mm_1}(\xi) = T_{mm_1}(\xi)\exp\{-i[(C_1 + C_0)/(2\rho^5 v_{\text{rel}})]\int_{-1}^{\xi} dt\,(1 - t^2)^{3/2}\} \tag{1.4.13}$$

we arrive at the following system of equations for the $T_{mm_1}(\xi)$:

$$iG\frac{dT_{11}}{d\xi} = (1 - \xi^2)^{3/2}\{\xi^2 T_{11} + (1 - \xi^2)T_{1-1} + \xi[2(1 - \xi^2)]^{1/2}\, T_{01}\},$$

$$iG\frac{dT_{1-1}}{d\xi} = (1-\xi^2)^{3/2}\{\xi^2 T_{1-1} + (1-\xi^2)T_{11} - \xi[2(1-\xi^2)]^{1/2}\, T_{01}\},$$

$$iG\frac{dT_{01}}{d\xi} = (1-\xi^2)^{3/2}\{(1-2\xi^2)T_{01} + \xi[2(1-\xi^2)]^{1/2}\,(T_{11} - T_{1-1})\}, \qquad (1.4.14)$$

$$iG\frac{dT_{00}}{d\xi} = (1-\xi^2)^{3/2}\{(1-2\xi^2)T_{00} + 2\xi[2(1-\xi^2)]^{1/2}\, T_{10}\},$$

$$iG\frac{dT_{10}}{d\xi} = (1-\xi^2)^{3/2}\{-(1-2\xi^2)T_{10} + \xi[2(1-\xi^2)]^{1/2}\, T_{00}\},$$

where we have allowed for (1.4.11) and

$$G = 2\rho^5 v_{\rm rel}/(C_1 - C_0). \qquad (1.4.15)$$

The boundary conditions for $T_{mm_1}(\xi)$ are obtained from (1.4.9):

$$T_{mm_1}(\xi = -1) = \delta_{mm_1}. \qquad (1.4.16)$$

The form of equations (1.4.14) implies that the first three equations and the last two form independent subsystems. Let us transform the subsystem of the (1.4.10) first three equations. Summing the first two equations and combining the result with (1.4.16) yields

$$T_{11}(\xi) + T_{1-1}(\xi) = \exp[-(i/G)\int_{-1}^{\xi} dt\,(1-t^2)^{3/2}]. \qquad (1.4.17)$$

Subtracting the second equation from the first and introducing the notation

$$T_{11}(\xi) - T_{1-1}(\xi) = w(\xi), \qquad T_{01}(\xi) = -\,u(\xi)/\sqrt{2}\,, \tag{1.4.18}$$

we get

$$iG\frac{dw}{d\xi} = -\,(1-\xi^2)^{3/2}[(1-2\xi^2)w + 2\xi(1-\xi^2)^{1/2}\,u]. \tag{1.4.19}$$

In terms of the quantities introduced in (1.4.18), the third equation in (1.4.14) assumes the form

$$iG\frac{du}{d\xi} = -\,(1-\xi^2)^{3/2}[2\xi(1-2\xi^2)^{1/2}w - (1-2\xi^2)u]. \tag{1.4.20}$$

The boundary conditions for $u(\xi)$ and $w(\xi)$ follow from (1.4.18) and (1.4.16):

$$u(\xi = -\,1) = 0, \qquad w(\xi = -\,1) = 1. \tag{1.4.21}$$

In what follows we use a positive definite parameter

$$g = |G| \tag{1.4.22}$$

instead of parameter G, which means that $g = G$ for $C_1 > C_0$, while $g = -\,G$ for $C_1 < C_0$, and the solution of the equations is expressed in terms of the complex conjugate functions $u(\xi, g)^*$ and $w(\xi, g)^*$.

Thus, the first three equations in (1.4.14) are reduced to a system of equations for functions $u(\xi, g)$ and $w(\xi, g)$,

$$ig\frac{dw}{d\xi} = -(1-\xi^2)^{3/2}[(1-2\xi^2)w + 2\xi(1-\xi^2)^{1/2}u],$$

$$ig\frac{du}{d\xi} = -(1-\xi^2)^{3/2}[2\xi(1-2\xi^2)^{1/2}w - (1-2\xi^2)u] \qquad (1.4.23)$$

with the boundary conditions (1.4.21). Note that this system coincides with equations (10) of [18], which emerged from the studies of quasiresonant collisions of atoms.

Comparing the fourth and fifth equations in (1.4.14) with system (1.4.23), we arrive at the following relations:

$$T_{00}(\xi) = w(\xi, g)^*, \quad T_{10}(\xi) = u(\xi, g)^*/\sqrt{2}. \qquad (1.4.24)$$

Thus, to calculate all the elements of matrix $\widetilde{S}_{mm_1}$ at $j = 1$ it is sufficient to find the solution of the system (1.4.23) with boundary conditions (1.4.21) and g as parameter. The system can be solved analytically only in approximate form for high and low values of parameter g (see [14]). For arbitrary values of g it has been solved in numerical form on computers, where for additional accuracy control relations that follow from the properties of the functions w and u were employed (see [17]):

$$|w(\xi, g)|^2 + |u(\xi, g)|^2 = 1, \quad u(\xi = 1, g) = u(\xi = 1, g)^*. \qquad (1.4.25)$$

Let us now assume that $j_a = 0$, $j_b = 1$ in (1.3.6) and substitute the expression for $\widetilde{S}^{(a)}$ (1.4.5), and the corresponding expressions for the matrix elements $\widetilde{S}^{(b)}_{mm_1}$, written in terms of the functions $u(\xi = 1, g)$ and $w(\xi = 1, g)$. From integration with respect to impact parameter ρ we go over to integration with respect to parameter g and from integration with respect to $\mathbf{v}_0$ to integration with respect to $\mathbf{v}_{rel}$. The result are formulas that express the velocity dependence of the $\Gamma_{ab}(\mathrm{v})^{11}_{q}$ with factors dependent on parameter

$$S = (2C^{(a)} - C^{(b)}_0 - C^{(b)}_1)/(C^{(b)}_1 - C^{(b)}_0) \qquad (1.4.26)$$

and determined through numerical integration with respect to g and preliminary solution of system (1.4.23).

Omitting the above-described transformations, we write the final result:

$$\Gamma_1(\mathrm{v}) - i\Delta_1(\mathrm{v})\mathrm{sgn}(C_1^{(b)} - C_0^{(b)}) = \pi n|C_1^{(b)} - C_0^{(b)}|^{2/5}(2T/M_2)^{3/10}[\theta_1(S)I_1(\alpha x) + \\ + \theta_2(S)I_2(\alpha x)], \tag{1.4.27}$$

$$\Gamma_0(\mathrm{v}) - i\Delta_0(\mathrm{v})\mathrm{sgn}(C_1^{(b)} - C_0^{(b)}) = \pi n|C_1^{(b)} - C_0^{(b)}|^{2/5}(2T/M_2)^{3/10}[\theta_1(S)I_1(\alpha x) - \\ - 2\theta_2(S)I_2(\alpha x)], \tag{1.4.28}$$

where T is the temperature of the gas in energy units, sgn(z) the sign function

$$\alpha = (M_2/M_1)^{1/2},$$

and

$$x = \mathrm{v}(M_1/2T)^{1/2}$$

is the dimensionless velocity of the excited atoms. The functions $I_1(\alpha x)$ and $I_2(\alpha x)$, which determine the dependence of the relaxation characteristics on the velocity of the atoms, are expressed in terms of the hypergeometric functions $\Phi(\gamma, \beta, x)$ [20] as follows:

$$I_1(\alpha x) = \exp(-\alpha^2x^2)\Phi(9/5, 3/2, \alpha^2x^2),$$

$$I_2(\alpha x) = \exp(-\alpha^2x^2)[\Phi(9/5, 3/2, \alpha^2x^2) - \Phi(9/5, 5/2, \alpha^2x^2)]. \tag{1.4.29}$$

The complex-valued functions $\theta_1(S)$ and $\theta_2(S)$ can be found by numerically integrating the appropriate expressions with respect to parameter g with preliminary solution of system (1.4.23). Figures 1.1 and 1.2 depict $\theta_1(S)$ and $\theta_2(S)$ as functions of parameter S. The reader can see that $\mathrm{Re}\theta_1(S)$ considerably exceeds $\mathrm{Re}\theta_2(S)$ in the entire range of S, while $\mathrm{Im}\theta_1(S)$ exceeds $\mathrm{Im}\theta_2(S)$ in absolute value everywhere except the region $|S| \leq 1$. The functions $\mathrm{Re}\theta_2(S)$ and $\mathrm{Im}\theta_2(S)$ attain their maxima in absolute value at $S = -1$, which corresponds to $C_0^{(b)} = C^{(a)}$.

For sufficiently large values of parameter $|S|$ asymptotic expressions for $\theta_1(S)$ and $\theta_2(S)$ have been obtained. For instance,

$$\theta_1(S) = 2(3\pi|S|/16)^{2/5}\Gamma(9/5)\Gamma(3/5)[\cos(\pi/5) + i(S/|S|)\sin(\pi/5)]/\sqrt{\pi} \quad (1.4.30)$$

with $\Gamma(x)$ – the gamma function. This yields the following approximate formulas:

$$\mathrm{Re}\theta_1(S) \approx 1.03|S|^{2/5}, \qquad \mathrm{Im}\theta_1(S) \approx 0.746(S/|S|)|S|^{2/5}.$$

Similarly, we find that

$$\mathrm{Re}\theta_2(S) \approx 0.07(S/|S|)|S|^{-3/5}, \qquad \mathrm{Im}\theta_2(S) \approx 0.05|S|^{-3/5}.$$

Note that for $|S| \geq 5$ the difference between these asymptotic expressions and the results of numerical calculations does not exceed 10%.

Combining (1.4.27) with (1.4.28), we arrive at an expression that enables us to estimate the effect of the direction of velocity **v** on the relaxation characteristics:

$$\Gamma_1(\mathrm{v}) - \Gamma_0(\mathrm{v}) - i\,\mathrm{sgn}(C_1^{(b)} - C_0^{(b)})[\Delta_1(\mathrm{v}) - \Delta_0(\mathrm{v})] =$$

$$= 3\pi n|C_1^{(b)} - C_0^{(b)}|^{2/5}(2T/M_2)^{3/10}\theta_2(S)I_2(\alpha x). \quad (1.4.31)$$

This implies that the difference between $\Gamma_1(\mathrm{v})$ and $\Gamma_0(\mathrm{v})$ becomes significant in the neighborhood of $S = -1$ and $S = 1$, where $|\mathrm{Re}\theta_2(S)|$ is of the order of $(0.1 - 0.2)\mathrm{Re}\theta_1(S)$. Here parameter α plays an important role because $\Gamma_1(\mathrm{v}) - \Gamma_0(\mathrm{v})$ must be compared with $\Gamma_1(\mathrm{v})$ and $\Gamma_0(\mathrm{v})$ for values of v of the order of the mean thermal velocity of the atoms, that is, at $x = 1$. From (1.4.29) it follows that $I_2(\alpha)$ is of considerable magnitude only when $\alpha \approx 1$, that is, when the mass of the nonexcited atom participating in the collision is

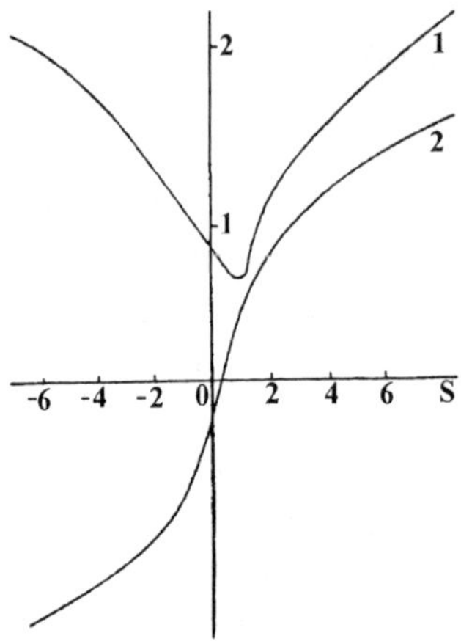

Fig. 1.1 $\mathrm{Re}\theta_1$ (curve 1) and $\mathrm{Im}\theta_1$ (curve 2) as functions of parameter S.

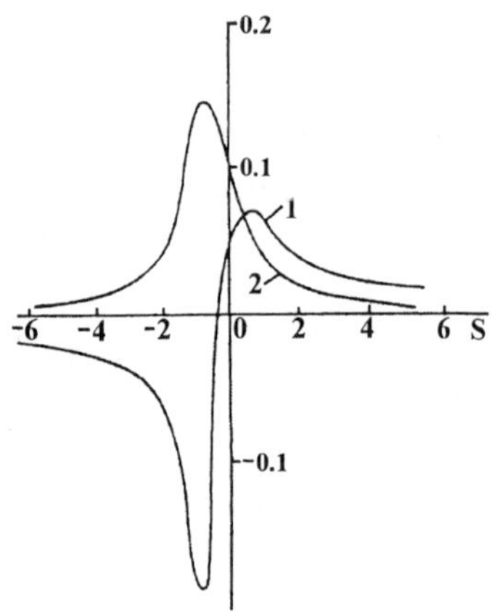

Fig. 1.2 $\mathrm{Re}\theta_2$ (curve 1) and $\mathrm{Im}\theta_2$ (curve 2) as functions of parameter S.

considerably greater than that of the excited atom. Hence, for example, in describing the lasing of helium–neon lasers, where this parameter does not exceed unity, the use of relaxation matrices averaged over the directions of the atomic velocities proves to be a good approximation.

Here are some effects in which the dependence of the relaxation characteristics on the directions of atomic velocities may manifest itself. For example, we can write the profile of the spectral line that represents spontaneous emission in the process of the transition being considered and allowing for depolarizing collisions in the following form:

$$I(\omega) \sim \int d\mathrm{v}\, F(\mathrm{v})\{[1 - (\mathbf{kv})^2/(k\mathrm{v})^2] \times$$

$$\times \frac{\gamma^{(0)} + \Gamma_1(\mathrm{v})}{(\omega - \omega_0 - \Delta_1(\mathrm{v}) - \mathbf{kv})^2 + (\gamma^{(0)} + \Gamma_1(\mathrm{v}))^2} +$$

$$+\frac{(\mathbf{kv})^2}{(kv)^2}\frac{\gamma^{(0)}+\Gamma_0(v)}{(\omega-\omega_0-\Delta_0(v)-\mathbf{kv})^2+(\gamma^{(0)}+\Gamma_0(v))^2}\},$$

where $\mathbf{k} = \mathbf{n}\omega/c$, with $\mathbf{n}$ being the unit vector pointing in the direction of propagation of the radiation, and $\gamma^{(0)} = (\gamma_b^{(0)}+\gamma_a^{(0)})/2$. The difference of $I(\omega)$ and the Voigt profile is due to the fact that in the former we have allowed for the dependence of the relaxation characteristics on both the direction and magnitude of atomic velocities. The fact that we have allowed for directions manifests itself in the appearance of two functions (instead of one), $\Gamma_1(v)$ and $\Gamma_0(v)$, and of two functions $\Delta_1(v)$ and $\Delta_0(v)$. If we were to use relaxation matrices averaged over the velocity directions, we still would not get the Voigt profile because of the dependence of these quantities on the magnitude of the velocity. This, apparently, means that the data on $I(\omega)$ is insufficient to isolate the effect of collisional anisotropy.

Effects related to the dependence of the relaxation characteristics on the directions of atomic velocities have been considered, for example, in [21–23]. The results obtained in this section will be employed in what follows for the investigation of the photon echo in a gas medium.

1.5 Relaxation Characteristics Averaged over the Directions of Atomic Velocities

The investigation performed in the previous section suggests that in many problems of laser physics the averaging of relaxation characteristics over the directions of atomic velocities introduces no significant error in the final results. At the same time this approach considerably simplifies the solution of many problems since it diagonalizes the matrices

$$\Gamma_a(\mathbf{v})_{qq_1}^{\kappa\kappa_1},\ \Gamma_b(\mathbf{v})_{qq_1}^{\kappa\kappa_1},\ \Gamma_{ab}(\mathbf{v})_{qq_1}^{\kappa\kappa_1}$$

in indices κ and q.

We introduce the following notation:

$$\langle \Gamma_{ab}(\mathbf{v})^{\kappa\kappa_1}_{qq_1} \rangle = \Gamma(\mathrm{v})^{(\kappa)}\, \delta_{\kappa\kappa_1}\, \delta_{qq_1}, \tag{1.5.1}$$

$$\langle \Gamma_{a}(\mathbf{v})^{\kappa\kappa_1}_{qq_1} \rangle = \Gamma_a(\mathrm{v})^{(\kappa)}\, \delta_{\kappa\kappa_1}\, \delta_{qq_1},$$

$$\langle \Gamma_{b}(\mathbf{v})^{\kappa\kappa_1}_{qq_1} \rangle = \Gamma_b(\mathrm{v})^{(\kappa)}\, \delta_{\kappa\kappa_1}\, \delta_{qq_1}, \tag{1.5.2}$$

where the $\langle ... \rangle$ stands for averaging over the directions of the velocity vector. Combining (1.3.5) with (1.3.6), we arrive at the following formulas for the quantities $\Gamma(\mathrm{v})^{(\kappa)}$, $\Gamma_a(\mathrm{v})^{(\kappa)}$, $\Gamma_b(\mathrm{v})^{(\kappa)}$, which depend on the length of the velocity vector $\mathbf{v}$:

$$\Gamma(\mathrm{v})^{(\kappa)} = n\sum(-1)^{2j_a+m_2+m_3}\begin{pmatrix} j_a & j_b & \kappa \\ m_2 & -m & k \end{pmatrix}\begin{pmatrix} j_a & j_b & \kappa \\ m_3 & -m_1 & k \end{pmatrix}\times$$

$$\times\int d\mathbf{v}_0\, F(\mathbf{v}_0)|\mathbf{v}-\mathbf{v}_0| \int d\vec{\rho}\, [\delta_{m_2m_3}\,\delta_{mm_1} - \tilde{S}^{(a)}_{m_2m_3}\,\tilde{S}^{(b)*}_{mm_1}], \tag{1.5.3}$$

$$\Gamma_b(\mathrm{v})^{(\kappa)} = n\sum(-1)^{2j_b+m_2+m_3}\begin{pmatrix} j_b & j_b & \kappa \\ m_2 & -m & k \end{pmatrix}\begin{pmatrix} j_b & j_b & \kappa \\ m_3 & -m_1 & k \end{pmatrix}\times$$

$$\times\int d\mathbf{v}_0\, F(\mathbf{v}_0)|\mathbf{v}-\mathbf{v}_0| \int d\vec{\rho}\, [\delta_{m_2m_3}\,\delta_{mm_1} - \tilde{S}^{(b)}_{m_2m_3}\,\tilde{S}^{(b)*}_{mm_1}], \tag{1.5.4}$$

where summation is performed over the indices m, m_1, m_2, and m_3, and the expression for $\Gamma_a(\mathrm{v})^{(\kappa)}$ obtained from (1.5.4) by replacing b with a. In this approximation each irreducible component of the density matrix, $\psi_q^{(\kappa)}(\boldsymbol{r},\mathbf{v},t)$, $f_q^{(\kappa)}(\boldsymbol{r},\mathbf{v},t)$, and $\phi_q^{(\kappa)}(\boldsymbol{r},\mathbf{v},t)$ relaxes independently under collisions, just as in the approximation accepted in [13], and equations (1.3.2)–(1.3.4) take the form

$$\left(\frac{\partial}{\partial t} + \mathbf{v}\nabla_r - i\omega_0 + \frac{\gamma_b^{(0)}+\gamma_a^{(0)}}{2}\right)\psi_q^{(\kappa)}(\boldsymbol{r},\mathbf{v},t) =$$

$$= -\Gamma(\mathrm{v})^{(\kappa)}\,\psi_q^{(\kappa)}(\boldsymbol{r},\mathbf{v},t), \tag{1.5.5}$$

$$\left(\frac{\partial}{\partial t} + \mathbf{v}\nabla_r + \gamma_b^{(0)}\right) f_q^{(\kappa)}(\boldsymbol{r},\mathbf{v},t) =$$
$$= -\Gamma_b(\mathrm{v})^{(\kappa)} f_q^{(\kappa)}(\boldsymbol{r},\mathbf{v},t), \quad (1.5.6)$$

$$\left(\frac{\partial}{\partial t} + \mathbf{v}\nabla_r + \gamma_a^{(0)}\right) \phi_q^{(\kappa)}(\boldsymbol{r},\mathbf{v},t) =$$
$$= -\Gamma_a(\mathrm{v})^{(\kappa)} \phi_q^{(\kappa)}(\boldsymbol{r},\mathbf{v},t). \quad (1.5.7)$$

Specific calculations of relaxation characteristics have been carried out for the van der Waals interaction (1.4.6) between colliding atoms for the following three types of optical transitions:

$$a)\ j_a = 0, j_b = 1; \quad b)\ j_a = j_b = 1; \quad c)\ j_a = 2, j_b = 1.$$

Below we give the results of numerical calculations for each type.

a) The transition with $j_a = 0, j_b = 1$

In this case the only $\Gamma(\mathrm{v})^{(\kappa)}$ that is nonzero is the one with $\kappa = 1$, which we denote by $\Gamma(\mathrm{v})$. Using the results obtained in Section 1.4, we find that

$$\Gamma(\mathrm{v}) = [\Gamma_0(\mathrm{v}) + 2\Gamma_1(\mathrm{v}) - i\Delta_0(\mathrm{v}) - 2i\Delta_1(\mathrm{v})]/3 =$$
$$= \gamma(\mathrm{v})[\mathrm{Re}\theta_1(S) + i\,\mathrm{sgn}(C_1^{(b)} - C_0^{(b)})\mathrm{Im}\theta_1(S)] \quad (1.5.8)$$

with

$$\gamma(\mathrm{v}) = n\pi|C_1^{(b)} - C_0^{(b)}|^{2/5}(2T/M_2)^{3/10} I_1(\alpha x). \quad (1.5.9)$$

The quantity $\Gamma(\mathrm{v})$ determines, for example, the width and shift of the spectral line of spontaneous emission corresponding to the transition and is dependent on the parameter S. The results presented in Section 1.4 of numerical calculations of the function $\theta_1(S)$ agree with the similar results of

[24]. Level a in this case is nondegenerate, so that only one value $\kappa = 0$ is possible for it, with $\Gamma_a(v)^{(0)} = 0$, so that elastic collisions do not change the level population. In this connection it must be noted that for all values of level angular momentum j_a, j_b

$$\Gamma_a(v)^{(0)} - \Gamma_b(v)^{(0)} = 0. \tag{1.5.10}$$

The polarization characteristics of the b level, $\Gamma_b(v)^{(1)}$ and $\Gamma_b(v)^{(2)}$, which reflect the decay of orientation and alignment (the magnetic and quadrupole moments of the level) [17], were calculated using equation (1.5.4) and the expressions for the elements of $\widetilde{S}^{(b)}_{mm_1}$ obtained in Section 1.4. As noted earlier, after we have switched to integration with respect to parameter g, the velocity dependence of $\Gamma_b(v)^{(\kappa)}$ becomes isolated in the form of a factor and the integral determines only numerical coefficients depending on κ. As a result we get

$$\Gamma_b(v)^{(\kappa)} = a_\kappa \gamma(v), \quad (\kappa = 1,2), \tag{1.5.11}$$

where $\gamma(v)$ is defined in (1.5.9), and the numerical coefficients were found to be $a_1 = 0.860$ and $a_2 = 0.761$.

Thus, for a level with angular momentum $j_b = 1$ the ratio of the rates of relaxation of orientation and alignment under elastic depolarizing collisions in the event of van der Waals interaction is independent of all parameters and has the following approximate value:

$$\Gamma_b(v)^{(1)} / \Gamma_b(v)^{(2)} \approx 1.13. \tag{1.5.12}$$

This value coincides with the results arrived at in [25, 26] and differs from the value of 5/3 obtained in [27].

The ratios of $\mathrm{Re}\Gamma(v)$ and $\mathrm{Im}\Gamma(v)$, with the first quantity determining the broadening and the second the shift of the spectral line emitted in the transition, to $\Gamma_b(v)^{(2)}$ are independent of velocity but are dependent on parameter S. This means, among other things, that, as pressure increases, the ratio of spectral-line broadening to the broadening of the polarization characteristics of a level may take on different values depending on the value of S:

$$\frac{d\mathrm{Re}\Gamma(\mathrm{v})}{d\rho} \Big/ \frac{d\,\mathrm{Re}\,\Gamma_b(\mathrm{v})^{(2)}}{d\rho} \approx 1.3\mathrm{Re}\theta_1(S).$$

Figure 1.1 shows that with the exception of the neighborhood of the values $|S| \leq 1$ the ratio is high, in other words, under elastic collisions the line broadens faster than the polarization characteristics of levels. The ratio of the shift $\mathrm{Im}\Gamma(\mathrm{v})$ to the line width $\mathrm{Re}\Gamma(\mathrm{v})$ is also independent of velocity but is dependent on S. For $|S| \approx 1$ the ratio tends toward the classical limit corresponding to transitions between non-degenerate levels [2].

b) The transition with $j_a = j_b = 1$

Formula (1.5.3) implies that for the given transition the values of $\Gamma(\mathrm{v})^{(\kappa)}$ that are nonzero have $\kappa = 0$, 1, and 2, namely, $\Gamma(\mathrm{v})^{(0)}$, $\Gamma(\mathrm{v})^{(1)}$, and $\Gamma(\mathrm{v})^{(2)}$. The most studied, both theoretically and experimentally, is $\Gamma(\mathrm{v})^{(1)}$, which characterizes the polarization vector of the medium, the macroscopic dipole moment. Various problems of nonlinear electrodynamics require data on the other quantities $\Gamma(\mathrm{v})^{(\kappa)}$ ($\kappa \neq 1$).

The formulas obtained in Section 1.4 that express the elements of the scattering matrix $\tilde{S}^{(b)}_{mm_1}$ in terms of the functions $w(\xi = 1,g)$ and $u(\xi = 1,g)$ remain valid for $\tilde{S}^{(a)}_{mm_1}$, the only difference being that here

$$g = 2\mathrm{v}_{rel}\rho^5/|C_1^{(a)} - C_0^{(a)}|.$$

As a result, when from integration with respect to ρ we go over to integration with respect to g, the functions w and u related to state b become dependent on parameter g/t, with

$$t = (C_1^{(b)} - C_0^{(b)})/(C_1^{(a)} - C_0^{(a)}) \tag{1.5.13}$$

(in comparison to [28], we have reversed the notation of the indices: b is replaced with a and vice versa). Omitting intermediate calculations, we write the final expressions for $\Gamma(\mathrm{v})^{(\kappa)}$:

$$\Gamma(v)^{(\kappa)} = \gamma(v)[\mathrm{ReF}(S,t)^{(\kappa)} + i\,\mathrm{sgn}(C_1^{(a)} - C_0^{(a)})\mathrm{ImF}(S,t)^{(\kappa)}],$$

$$(\kappa = 0, 1, 2) \tag{1.5.14}$$

with

$$\gamma(v) = n\pi |C_1^{(a)} - C_0^{(a)}|^{2/5}(2T/M_2)^{3/10} I_1(\alpha x), \tag{1.5.9'}$$

which is obtained from (1.5.9) by replacing index b with index a. The dimensionless complex-valued functions $\mathrm{F}(S,t)^{(\kappa)}$, which depend on parameters t and S

$$S = \frac{C_1^{(a)} + C_0^{(a)} - C_1^{(b)} - C_0^{(b)}}{C_1^{(a)} - C_0^{(a)}} \tag{1.5.15}$$

are defined in terms of the following integrals:

$$\mathrm{F}(S,t)^{(0)} = B\int dg\ \{3 - [2\mathrm{Re}w(g)w(g/t)^* + 2u(g)u(g/t)^* +$$
$$+ \exp(-3i\pi(1-t)/8g)]\exp(-3i\pi S/8g)\}/g^{3/5}, \tag{1.5.16}$$

$$\mathrm{F}(S,t)^{(1)} = B\int dg\ \{3 - [\mathrm{Re}w(g)w(g/t) + u(g)u(g/t)^* + \mathrm{Re}w(g)\exp(3i\pi t/8g) +$$
$$+ \mathrm{Re}w(g/t)\exp(-3i\pi/8g)]\exp(-3i\pi S/8g)\}/g^{3/5}, \tag{1.5.17}$$

$$\mathrm{F}(S,t)^{(2)} = 2\mathrm{F}(S,t)^{(0)}/5 + 3\mathrm{F}(S,t)^{(1)}/5 +$$
$$+ (12B/5)\int dg\ u(g)u(g/t)^*\exp(-3i\pi S/8g)\}/g^{3/5} \tag{1.5.18}$$

where $B = 2^{8/5}\Gamma(9/5)/15\sqrt{\pi}$. Here w and u are the solutions of the set (1.4.23) with $\xi = 1$.

We should note some general properties of the functions $\mathrm{F}(S,t)^{(\kappa)}$. At $t = 1$, $S = 0$ the interaction potentials of the atom in states a and b coincide. In this case the matrix elements of $\widetilde{S}_{mm_1}^{(b)}$ and $\widetilde{S}_{mm_1}^{(a)}$ also coincide and expressions (1.5.14)–(1.5.18) simultaneously determine the polarization characteristics (mentioned earlier) of the level with angular momentum j = 1. It follows particularly from (1.5.16) and (1.4.25) that $F(S = 0, t = 1)^{(0)} = 0$. Formulas (1.5.17) and (1.5.18) then determine the values of the numerical factors a_1 and a_2 entering into (1.5.11):

$$a_1 = F(S = 0, t = 1)^{(1)} = B \int dg \, \{3 - [\mathrm{Re}w(g)^2 + |u(g)|^2 +$$

$$+ \, 2\mathrm{Re}w(g)\cos(3\pi/8g)]\}/g^{3/5}, \tag{1.5.19}$$

$$a_2 = 3a_1/5 + (12B/5) \int dg \, |u(g)|^2/g^{3/5}. \tag{1.5.20}$$

Thus, from (1.5.20) it follows that the ratio a_1/a_2 is lower than 5/3, and the result obtained in [27] corresponds to (1.5.20) without the second term on the right hand side.

Let us now examine the behavior of the function $F(S,t)^{(\kappa)}$ for $|t| >> 1$. To simplify matters, we assume that $t > 0$ and let $t \to \infty$, which corresponds to

$$C_1^{(a)} - C_0^{(a)} = 0. \tag{1.5.21}$$

In this case $\widetilde{S}^{(a)}_{mm_1}$ is equal to the unity matrix multiplied by (1.4.5). For this reason expressions (1.5.14)–(1.5.18) transform into (1.5.8) obtained for the transition with $j_a = 0$, $j_b = 1$. To show this, in (1.5.16)–(1.5.18) we change the variable of integration: $g = g't$. As a result the integrand acquires the factor $|t|^{2/5}$, which together with $\gamma(v) \sim |C_1^{(a)} - C_0^{(a)}|^{2/5}$ of (1.5.9′) leads to formula (1.5.9) for $\gamma(v)$. Allowing also for the fact that

$$S/t = \frac{C_1^{(a)} + C_0^{(a)} - C_1^{(b)} - C_0^{(b)}}{C_1^{(b)} - C_0^{(b)}} = S'$$

coincides with the definition of S in (1.4.26) if one takes into account (1.5.21), for the integrals that remain after we have let $t \to \infty$ we arrive at an expression coinciding with the definition of $\theta_1(S')$:

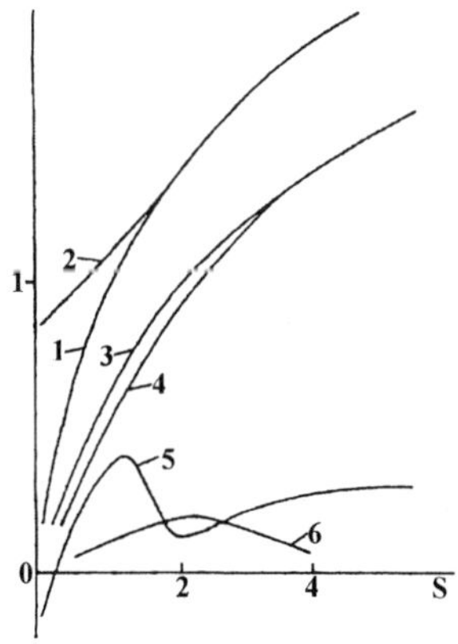

Fig. 1.3 $\mathrm{ReF}^{(0)}$ (curve 1), $\mathrm{ReF}^{(1)}$ (curve 2), $\mathrm{ImF}^{(0)}$ (curve 3), $\mathrm{ImF}^{(1)}$ (curve 4), $2\mathrm{Re}(\mathrm{F}^{(2)} - \mathrm{F}^{(1)})$ (curve 5), and $2\mathrm{Im}(\mathrm{F}^{(2)} - \mathrm{F}^{(1)})$ (curve 6) as functions of parameter S.

$$\theta_1(S') = B \int dg\ \{3 - [2\mathrm{Re}w(g) + \exp(3i\pi/8g)]\exp(-3i\pi S'/8g)\}/g^{3/5}. \quad (1.5.22)$$

Thus, for $|t| >> 1$ the $\Gamma(\mathrm{v})^{(\kappa)}$ are weakly dependent on index κ and parameter t. In a similar manner it can be shown that for $|t| << 1$ the $\Gamma(\mathrm{v})^{(\kappa)}$ are also weakly dependent on κ and t. Let us now discuss the behavior of the $\mathrm{F}(S, t)^{(\kappa)}$ for $|S| >> 1$, $|t| << |S|$. For fairly high values of $|S|$ (formally, as $|S| \to \infty$), these functions are weakly dependent on κ and t and tend to their asymptotic limit given by formula (1.4.30) for function $\theta_1(S)$. From (1.5.16)–(1.5.18) it follows (say, for $t > 0$) that

$$\mathrm{F}(-S/t, 1/t)^{(\kappa)} = t^{-2/5}\mathrm{F}(S, t)^{(\kappa)*}.$$

Hence, we find that at $t = 1$

$$\mathrm{ReF}(-S, 1)^{(\kappa)} = \mathrm{ReF}(S, 1)^{(\kappa)}, \qquad \mathrm{ImF}(-S, 1)^{(\kappa)} = -\,\mathrm{ImF}(S, 1)^{(\kappa)}.$$

These relations and the behavior of the $F(S, t)^{(\kappa)}$ established above have been used to determine the accuracy of numerical calculations for two values of parameter t: 0.1 and 1. Parameter S assumed the values $0, \pm 0.2, \pm 0.4, \pm 0.6, \ldots, \pm 1, \pm 2, \pm 3, \ldots, \pm 10$. Figure 1.3 depicts the results of numerical calculations of the functions $F(S, 1)^{(0)}$, $F(S, 1)^{(1)}$, $F(S, 1)^{(2)}$. The dependence on κ is essential for $\mathrm{Re}F(S, 1)^{(\kappa)}$ only when $|S| < 1.5$, while for $\mathrm{Im}F(S, 1)^{(\kappa)}$ it is essential when $|S| < 3$. In order to determine a more detailed description of the disparity between $\mathrm{Re}F(S, 1)^{(2)}$ and $\mathrm{Re}F(S, 1)^{(1)}$ and between $\mathrm{Im}F(S, 1)^{(2)}$ and $\mathrm{Im}F(S, 1)^{(1)}$, the difference of such quantities has been calculated separately with enhanced accuracy. The results of these calculations are represented by curves 5 and 6 in Fig. 1.3.

Figure 1.4 shows the results of calculations of the $F(S, 0.1)^{(\kappa)}$. Within the accuracy of calculations, the quantities $F(S, 0.1)^{(0)}$, $F(S, 0.1)^{(1)}$, and $F(S, 0.1)^{(2)}$ practically coincide for all values of S. In order to determine the extent to which $F(S, 0.1)^{(2)}$ differs from $F(S, 0.1)^{(1)}$, their difference has been calculated separately with enhanced accuracy. The results of calculations of $\mathrm{Re}(F(S, 0.1)^{(2)} - F(S, 0.1)^{(1)})$ are represented by curve 3 in Fig. 1.4. In absolute value the difference of the imaginary parts of these functions does not exceed 0.05 and, therefore, is not shown in Fig. 1.4. Note that as a rule, for the cases that have been calculated so far

$$|\mathrm{Re}(F(S, 0.1)^{(2)} - F(S, 0.1)^{(1)})| >> |\mathrm{Im}(F(S, 0.1)^{(2)} - F(S, 0.1)^{(1)})|.$$

Thus, the results of numerical calculations demonstrate that with the exception of the region where $|S| < 3$ the quantities $\Gamma(v)^{(\kappa)}$ ($\kappa = 0, 1, 2$) are approximately equal for all values of parameter t:

$$\Gamma(v)^{(0)} \approx \Gamma(v)^{(1)} \approx \Gamma(v)^{(2)}. \tag{1.5.23}$$

For $|t| << 1$ and $|t| >> 1$ this is true for all values of S. Also, in the region where $|S| > 3$ the functions $F(S, t)^{(\kappa)}$ are weakly dependent on parameter t.

The behavior of the functions $\mathrm{Re}F(S, t)^{(0)}$, $\mathrm{Re}F(S, t)^{(1)}$, and $\mathrm{Re}F(S, t)^{(2)}$ differs drastically for small values of S and for values of t on the order of unity. From Fig. 1.3 it follows that at $t = 1$ the function $\mathrm{Re}F(S, t)^{(0)}$ tends to zero as $S \to 0$, while $\mathrm{Re}F(S, t)^{(1)}$ and $\mathrm{Re}F(S, t)^{(2)}$ tend to values on the order of unity. Note that, just as in the case of the transition with $j_a = 0, j_b = 1$, the broadening of the spectral line corresponding to the transition in question

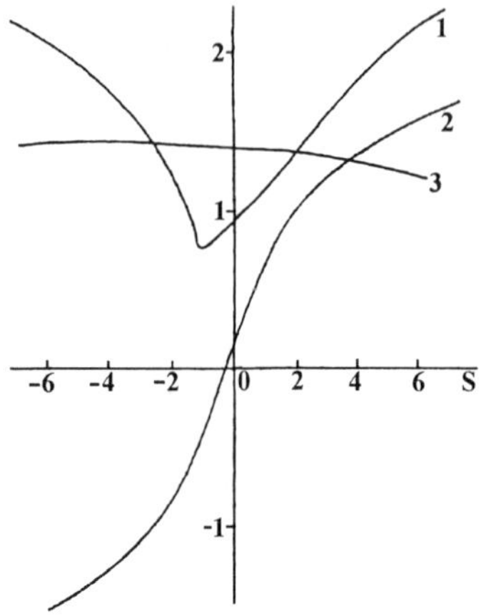

Fig. 1.4. $\mathrm{Re}F^{(0)}$ (curve 1), $\mathrm{Im}F^{(0)}$ (curve 2), and $10\mathrm{Re}(F^{(1)} - F^{(2)})$ (curve 3) as functions of parameter S.

occurs faster than the broadening of the polarization characteristics of the levels for all values of parameters S and t, except in the region where $|S| < 1$.

c) *The transition with* $j_a = 2, j_b = 1$

Here one must calculate the relaxation characteristics of the optical coherence matrix $\Gamma(v)^{(\kappa)}$ ($\kappa = 1, 2, 3$) and the polarization characteristics of level a, $\Gamma_a(v)^{(\kappa)}$ ($\kappa = 1, 2, 3, 4$). The solution of this problem requires a preliminary calculation of all the elements of the scattering matrix $\widetilde{S}^{(a)}_{mm_1}$ which can be found by solving equation (1.4.8) at $j = 2$. There exist the following relations between the constants A and B and the constants of the van der Waals interaction potential, $C_0^{(a)}$, $C_1^{(a)}$, and $C_2^{(a)}$ at $j = 2$:

$$A = (C_0^{(a)} + C_2^{(a)})/2, \qquad B = 2(14\pi)^{1/2}(C_0^{(a)} - C_1^{(a)}),$$

$$C_2^{(a)} = 4C_1^{(a)} - 3C_0^{(a)}. \tag{1.5.24}$$

We now take equation (1.4.8) at $j = 2$, drop index a for the sake of brevity, and introduce matrix T_{mm_1} instead of matrix S_{mm_1} by the relation

$$S_{mm_1}(\xi) = T_{mm_1}(\xi)\exp\{-i[(C_1 + C_2)/(2\rho^5 v_{rel})]\int_{-1}^{\xi} dt\,(1 - t^2)^{3/2}\}. \qquad (1.5.25)$$

This leads us to a system of equations for the elements of T_{mm_1},

$$i v_{rel}\rho^5 \frac{dT_{mm_1}}{d\xi} = (1 - \xi^2)^{3/2}[(C_0 - C_1)\,T_{mm_1}(\xi)/2 + (-1)^{m+1}2(14\pi)^{1/2}(C_0 - C_1)\times$$
$$\times\sum_{m_2}\begin{pmatrix} 2 & 2 & 2 \\ m & -m_2 & k \end{pmatrix} Y_{2k}(\xi,0)\,T_{m_2m_1}(\xi)], \qquad (1.5.26)$$

with the boundary conditions (1.4.16). Formula (1.4.11) makes it possible to consider as independent only those elements of T_{mm_1} that have $m \geq 0$. If we write the equations for T_{mm_1} with $m \geq 0$, we find that the system breaks down into three subsystems. The first couples the matrix elements T_{22}, T_{2-2}, T_{12}, T_{1-2}, T_{02}, the second involves T_{21}, T_{2-1}, T_{11}, T_{1-1}, T_{01}, and the third couples T_{20}, T_{10}, T_{00}. In turn, the first two subsets break down into systems of three and two equations. For example, writing equation (1.5.26) for the differences $(T_{22} - T_{2-2})$ and $(T_{1-2} - T_{12})$, we obtain

$$iG_1\frac{d(T_{22} - T_{2-2})}{d\xi} = -(1 - \xi^2)^{3/2}[(1 - 2\xi^2)(T_{22} - T_{2-2}) +$$
$$+ 2\xi(1 - \xi^2)^{1/2}(T_{1-2} - T_{12})],$$

$$iG_1\frac{d(T_{1-2} - T_{12})}{d\xi} = -(1 - \xi^2)^{3/2}[2\xi(1 - \xi^2)^{1/2}(T_{22} - T_{2-2}) -$$
$$-(1 - 2\xi^2)(T_{1-2} - T_{12})], \qquad (1.5.27)$$

where $G_1 = G/3$, with parameter G defined in (1.4.15). Introducing the notation

$$g = |G|/3, \tag{1.5.28}$$

we see that allowing for the boundary conditions the system of equations (1.5.27) is equivalent to (1.4.23), where

$$w(\xi, g) = T_{22}(\xi, g) - T_{2-2}(\xi, g), \quad u(\xi, g) = T_{1-2}(\xi, g) - T_{12}(\xi, g). \tag{1.5.29}$$

For T_{02}, $(T_{22} + T_{2-2})$, $(T_{12} + T_{1-2})$ we get a set of three equations. Introducing the notation

$$T_{02} = \phi_1/(3/8)^{1/2}, \quad \psi_1 = -(T_{12} + T_{1-2}), \quad \chi_1 = T_{22} + T_{2-2} - \phi_1/2, \tag{1.5.30}$$

we arrive at a set of equations for the functions ϕ_1, ψ_1, χ_1:

$$ig\frac{d\chi_1}{d\xi} = (1-\xi^2)^{3/2}[\,(8\xi^2/3 - 5/3)\chi_1 + 4\xi^2\phi_1/3 - 4\xi(1-\xi^2)^{1/2}\psi_1/3],$$

$$ig\frac{d\phi_1}{d\xi} = (1-\xi^2)^{3/2}[\,(1 - 8\xi^2/3)\phi_1 + 4(1-\xi^2)\chi_1/3 - 4\xi(1-\xi^2)^{1/2}\psi_1/3],$$

$$ig\frac{d\psi_1}{d\xi} = -(1-\xi^2)^{3/2}[\psi_1 + 2\xi(1-\xi^2)^{1/2}(\phi_1 + \chi_1)] \tag{1.5.31}$$

with the boundary conditions

$$\chi_1(\xi = -1) = 1, \; \phi_1(\xi = -1) = \psi_1(\xi = -1) = 0. \tag{1.5.32}$$

Employing (1.5.29) and (1.5.30), we can express the matrix elements T_{mm_1} in terms of the functions w, u, ϕ_1, ψ_1, χ_1:

$$T_{12} = (u + \psi_1)/2, \quad T_{1-2} = (u - \psi_1), \quad T_{22} = (w + \phi_1/2 + \chi_1)/2,$$

$$T_{2-2} = (\phi_1/2 + \chi_1 - w)/2.$$

Reasoning in a similar way, we find that the five matrix elements T_{21}, T_{2-1}, T_{11}, T_{1-1}, T_{01} are expressed in terms of the functions w^* and u^* and the functions ϕ_3, ψ_3, and χ_3 satisfying the same set of equations (1.5.31) but with the boundary conditions

$$\phi_3(\xi = -1) = \chi_3(\xi = -1) = 0, \qquad \psi_3(\xi = -1) = 1. \tag{1.5.33}$$

After simple transformations we obtain

$$T_{11} = (\psi_3 + w^*)/2, \quad T_{1-1} = w^* - \psi_3, \quad T_{01} = -(3/8)^{1/2}\phi_3,$$

$$T_{21} = (u^* - \chi_3 - \phi_3/2)/2, \quad T_{2-1} = (u^* + \chi_3 + \phi_3/2)/2.$$

The set of equations for the remaining matrix elements T_{20}, T_{10}, T_{00} can again be reduced to the set (1.5.31) for the functions ϕ_2, ψ_2, χ_2 with the boundary conditions

$$\psi_2(\xi = -1) = \chi_2(\xi = -1) = 0, \qquad \phi_2(\xi = -1) = 1. \tag{1.5.34}$$

After simple transformations we obtain

$$T_{00} = \phi_2 - \phi_1/2, \quad T_{10} = (\psi_1 - 2\psi_2)/\sqrt{6}, \quad T_{20} = (2\chi_2 - \chi_1 + \phi_2 - \phi_1/2)/\sqrt{6}.$$

Thus, to calculate the elements of $\widetilde{S}^{(a)}_{mm_1}$ one must solve the system of equations (1.4.23) with the boundary condition (1.4.21), and then solve the system (1.5.31) three times with the boundary conditions (1.5.32), (1.5.33), and (1.5.34), respectively. These elements are expressed in terms of the functions w, u, ϕ_i, ψ_i, and χ_i ($i = 1, 2, 3$) taken at $\xi = 1$. The functions are dependent on parameter g defined in (1.5.28). If $C_1^{(a)} < C_0^{(a)}$, we must switch to complex-conjugate functions. Let us now turn to formula (1.5.3) for calculating $\Gamma(\nu)^{(\kappa)}$. We go from integration with respect to ρ to integration with respect to parameter g. Then the functions w and u, in terms of which the elements of $\widetilde{S}^{(b)}_{mm_1}$ for the level with $j_b = 1$ are defined, depend on $3g/t$, where t is defined in (1.5.13).

As a result we can once more represent the $\Gamma(\nu)^{(\kappa)}$. ($\kappa = 1, 2, 3$) in the form of (1.5.14):

$$\Gamma(\nu)^{(\kappa)} = \gamma(\nu)[\mathrm{Re}F(S,t)^{(\kappa)} + i\,\mathrm{sgn}(C_1^{(a)} - C_0^{(a)})\mathrm{Im}F(S,t)^{(\kappa)}],$$
$$(\kappa = 1, 2, 3), \qquad (1.5.35)$$

where $\gamma(\nu)$ is defined in (1.5.9′), and the functions $F(S, t)^{(1)}$, $F(S, t)^{(2)}$, and $F(S, t)^{(3)}$ which are dependent on parameters t and

$$S = \frac{5C_1^{(a)} - 3C_0^{(a)} - C_1^{(b)} - C_0^{(b)}}{C_1^{(a)} - C_0^{(a)}} \qquad (1.5.36)$$

are defined in terms of the following integrals:

$$F(S, t)^{(1)} = D\int dg\,\{\exp(-i\pi S/8g)[w(3g/t)^*(6\psi_3 + 6w + 8\chi_1 - 4\chi_2) +$$
$$+ w(3g/t)(6\psi_3 + 6w^* + 8\phi_2 - 4\phi_1) + \exp(i\pi t/8g)(4\phi_1 + 4\phi_2 + 4\chi_1 + 4\chi_2 +$$
$$+ 6w + 6w^*) + u(3g/t)(6\phi_3 - 6\chi_3 + 12\psi_1 -$$
$$- 12\psi_2 + 12u)]/60 - 1\}/g^{3/5}, \qquad (1.5.37)$$

$$F(S, t)^{(2)} = D\int dg\,\{\exp(-i\pi S/8g)[w(3g/t)^*(4\phi_2 + 2\chi_2 + \psi_3 + w + 4w^*) +$$
$$+ w(3g/t)(4\chi_1 + 2\phi_1 + 4w + \psi_3 + w^*) + \exp(i\pi t/8g)(w + 4\psi_3 + w^* +$$

$$+ 2\phi_2 - 2\chi_2 + 2\chi_1 - 2\phi_1) + u(3g/t)^*(\phi_3 - \chi_3 + 2u + 2\psi_1 - - 2\psi_2)]/30 - 1\}/g^{3/5}, \tag{1.5.38}$$

$$\mathrm{F}(S, t)^{(3)} = D \int dg \, \{\exp(- i\pi S/8g)[w(3g/t)^*(8\psi_3 + 8w + 9\chi_1 - 2\chi_2 + + 5w^* + 5\phi_2) + w(3g/t)(5\chi_1 - 2\phi_1 + 5w + 8\psi_3 + 8w^* + 9\phi_2) + \exp(i\pi t/8g)(7\chi_1 + + 2\phi_1 + 8w + 5\psi_3 + 8w^* + 7\phi_2 + 2\chi_2) + 7u(3g/t)^*(\chi_3 - \phi_3 + 2\psi_2 - 2u - - 2\psi_1)]/105 - 1\}/g^{3/5}, \tag{1.5.39}$$

where $D = - 4(3/2)^{2/5}\Gamma(9/5)/(5\sqrt{\pi})$, and $\Gamma(9/5)$ is the gamma function.

For $|t| << 1$ the dependence of the $\mathrm{F}(S,t)^{(\kappa)}$ on κ is weak, and in the limit as $|t| \to 0$ we obtain

$$\mathrm{F}(S,t = 0)^{(1)} = \mathrm{F}(S,t = 0)^{(2)} = \mathrm{F}(S,t = 0)^{(3)} = \mathrm{F}(S)^{(2)} = = D \int dg \, [\exp(- i\pi S/8g)(w + w^* + \chi_1 + \phi_2 + \psi_3)/5 - 1]. \tag{1.5.40}$$

Note that the value of $t = 0$ corresponds to the case where the degeneracy of level b is ignored and the limit (1.5.40) determines the value of the only nonzero relaxation characteristic $\Gamma(\mathrm{v})^{(2)}$ for the optically forbidden transition with $j_b = 0, j_a = 2$:

$$\Gamma(\mathrm{v})^{(2)} = \gamma(\mathrm{v})\mathrm{F}(S)^{(2)}, \tag{1.5.41}$$

where $\gamma(\mathrm{v})$ is defined in (1.5.9′), parameter S is obtained from (1.5.36) at $C_1^{(b)} = C_0^{(b)}$, and the function $\mathrm{F}(S)^{(2)}$ is defined by (1.5.40).

Let us now discuss the behavior of $\mathrm{F}(S,t)^{(\kappa)}$ ($\kappa = 1, 2, 3$) at large values of parameter t, or $|t| >>1$. In formulas (1.5.37)–(1.5.39) we switch to integration with respect to parameter $g' = 3g/t$, assuming for simplicity that $t > 0$, and introduce instead of S the parameter

$$S' = S/t = \frac{5C_1^{(a)} - 3C_0^{(a)} - C_1^{(b)} - C_0^{(b)}}{C_1^{(a)} - C_0^{(a)}}.$$

Then there appears a factor $(t/3)^{2/5}$ in front of the integrals (1.5.37)–(1.5.39). Allowing for the fact that the product of $\gamma(v)$ defined in (1.5.9′) by $t^{2/5}$ is equal to $\gamma(v)$ from formula (1.5.9) and that S' transforms into (1.4.26) at $C_1^{(a)} = C_0^{(a)}$, we see that as $t \to \infty$, the functions $F(S, t)^{(\kappa)}$ cease to be dependent on κ and the $\Gamma(v)^{(\kappa)}$ transform into the function $\Gamma(v)$ defined in (1.5.8) and corresponding to the transition with $j_b = 1, j_a = 0$.

Let us also study the behavior of the $F(S, t)^{(\kappa)}$ ($\kappa = 1, 2, 3$) for $|S| >> 1$, $|t| << |S|$. The principal term of such an asymptotic series has the same form for all three functions and coincides with formula (1.4.30) for $\theta_1(S)$. The following terms are weakly dependent on parameter t and index κ:

$$F(S, t)^{(\kappa)} = 2(3\pi|S|/16)^{2/5}\Gamma(9/5)\Gamma(3/5)[\cos(\pi/5) + i(S/|S|)\sin(\pi/5)][1 - \\ - 2/(5S) - 2t/(15S) - 113/(125S^2) - 17(t/15S)^2 + (68t/375S^2)(\delta_{\kappa 1} - \\ - 32\delta_{\kappa 2}/17 - \delta_{\kappa 3}/34)]/\sqrt{\pi}, \quad (1.5.42)$$

where $\Gamma(x)$ is the gamma function.

Thus, the t-dependence of $F(S, t)^{(\kappa)}$ is determined by the small parameter t/S, while the dependence on κ is determined by the product of the small parameters t/S and $1/S$. For this reason the fractional difference in the functions, say, $(F(S, t)^{(1)} - F(S, t)^{(2)})/F(S, t)^{(1)}$, is proportional to t/S^2 in this range of values of the parameters.

Figure 1.5 demonstrates the results of numerical calculations of $F(S, t)^{(\kappa)}$ ($\kappa = 1, 2, 3$) carried out for $t = \pm 0.1, \pm 1.5$. Parameter S took on the values $0, \pm 1, \pm 2, \dots, \pm 10$.

Within the accuracy of calculations (≈ 20%) these functions coincide and are weakly dependent on parameter t. For the values of t considered and for $|S| > 3$, the values of the functions calculated by computer agree well with those following from (1.5.42).

To establish the dependence of the $F(S, t)^{(\kappa)}$ on the index κ the differences $\mathrm{Re}(F(S, t)^{(3)} - F(S, t)^{(1)})$ and $\mathrm{Im}(F(S, t)^{(3)} - F(S, t)^{(1)})$ were calculated with enhanced accuracy. Analysis of the results of numerical calculations shows these differences assume their maximum values at values of S such that the asymptotic expression (1.5.42) is valid. The results obtained from computer calculations and from formula (1.5.42) agree fairly well. The calculations suggest that the fractional difference of $\Gamma(v)^{(1)}$, $\Gamma(v)^{(2)}$, $\Gamma(v)^{(3)}$ for an atomic transition with angular momenta $j_b = 1$, $j_a = 2$ does not exceed 13%. Hence, if we are not interested in effects associated with the discrepancy between these relaxation characteristics, we can assume that

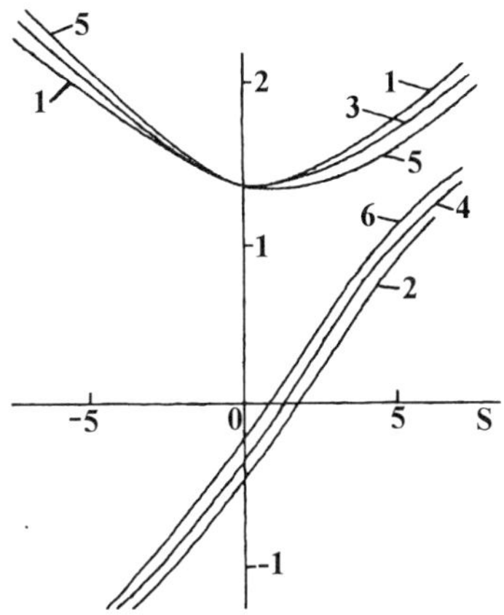

Fig. 1.5 $\mathrm{ReF}(S, t)^{(1)}$ (curve 1) and $\mathrm{ImF}(S, t)^{(1)}$ (curve 2) as functions of parameter S at $t = -1.5$, $\mathrm{ReF}(S, t)^{(1)}$ (curve 3) and $\mathrm{ImF}(S, t)^{(1)}$ (curve 4) as functions of parameter S at $t = -0.1$, and $\mathrm{ReF}(S, t)^{(1)}$ (curve 5) and $\mathrm{ImF}(S,t)^{(1)}$ (curve 6) as functions of parameter S at $t = 1.5$.

$$\Gamma(\mathrm{v})^{(1)} \approx \Gamma(\mathrm{v})^{(2)} \approx \Gamma(\mathrm{v})^{(3)}. \qquad (1.5.43)$$

Let us now discuss the relaxation characteristics $\Gamma_a(\mathrm{v})^{(\kappa)}$ ($\kappa = 0, 1, 2, 3, 4$) of the level with angular momentum $j_a = 2$. To calculate them we use (1.5.4) with a substituted for b. The same line of reasoning as in the investigation of the polarization characteristics of the level with the angular momentum $j = 1$ leads to

$$\Gamma_a(\mathrm{v})^{(\kappa)} = a_\kappa \gamma(\mathrm{v}), \ (\kappa = 0, 1, 2, 3, 4), \qquad (1.5.44)$$

where $\gamma(\mathrm{v})$ is defined in (1.5.9′), and the numerical factors were found to have the following values:

$$a_0 = 0, \ \ a_1 = 1.11, \ \ a_2 = 1.42, \ \ a_3 = 1.64, \ \ a_4 = 1.22. \qquad (1.5.45)$$

These values agree well with the results obtained in [29]. Hence, for the ratio of the quantities characterizing the collisional relaxation of orientation and alignment we have

$$\Gamma_a(\mathrm{v})^{(1)}/\Gamma_a(\mathrm{v})^{(2)} \approx 0.78. \tag{1.5.46}$$

The relaxation characteristics of the transitions considered here exhibit the following general features. The dependence of all these characteristics on atomic velocities is determined by the same function,

$$I_1(\alpha x) = \exp(-\alpha^2 x^2)\Phi(9/5,\, 3/2,\, \alpha^2 x^2),\ \alpha = (M_2/M_1)^{1/2},$$

$$x = \mathrm{v}(M_1/2T)^{1/2}\,. \tag{1.5.47}$$

This dependence follows from the fact that the velocity distribution of the nonexcited atoms is assumed to be Maxwellian, while the interaction of these atoms with the excited atoms is assumed to take place through van der Waals forces. The polarization characteristics of the levels, $\Gamma_a(\mathrm{v})^{(\kappa)}$, $\Gamma_b(\mathrm{v})^{(\kappa)}$, depend on the difference of van der Waals constants pertaining to the given level, $|C_1 - C_0|^{2/5}$, and take into account the redistribution of atoms over the Zeeman sublevels and the disruption of coherence between the sublevels under elastic collisions. Passing to the limit $C_1 = C_0$ is equivalent to ignoring degeneracy, and all the polarization characteristics of the levels vanish as a result. The relaxation characteristics of the optical coherence matrix for atomic transitions with angular momenta j_b and j_a, $\Gamma(\mathrm{v})^{(\kappa)}$ ($|j_b - j_a| \le \kappa \le j_b + j_a$), are dependent on the difference of the constants of one of the levels $|C_1 - C_0|^{2/5}$ and two parameters, t and S (for $j_b, j_a \ge 1$). Parameter t describes the relative contribution to $\Gamma(\mathrm{v})^{(\kappa)}$ of the processes of redistribution of atoms between the sublevels of each of the levels. The values $t = 0$ and $t \to \infty$ correspond to a situation where the degeneracy of one of the levels is ignored. For example, if we go to the limit $C_1^{(a)} = C_0^{(a)}$, we obtain $\Gamma(\mathrm{v})$ corresponding to the transition with angular momenta j_b and $j_a = 0$. This has been demonstrated to be true for both the atomic transition with angular momenta $j_b = j_a = 1$ and the atomic transition with angular momenta $j_b = 1, j_a = 2$, where the limit $C_1^{(a)} = C_0^{(a)}$ led to results corresponding to the transition with $j_b = 1, j_a = 0$, and the limit $C_1^{(b)} = C_0^{(b)}$ to the results corresponding to the transition with $j_b = 0, j_a = 2$.

Parameter S characterizes the interference of phase errors in the upper and lower states of the transition in question. Ignoring the degeneracy of both levels leads to the classical expression for $\Gamma(\mathrm{v})$, which depends on the difference in the constants of the van der Waals interaction on the upper and lower levels. Let us illustrate this using the example of an atomic transition with the angular momenta $j_b = 1$ and $j_a = 0$. For the sake of definiteness we assume that the transition to the limit $C_1^{(b)} = C_0^{(b)} = C_b$ takes place from positive values of $C_1^{(b)} - C_0^{(b)}$. Combining (1.5.8) with (1.5.22) and introducing a new integration variable $g' = g(C_1^{(b)} - C_0^{(b)})$ we obtain

$$\Gamma(\mathrm{v}) = n\pi(2T/M_2)^{3/10} I_1(\alpha x) 2^{8/5}\Gamma(9/5)/(5\sqrt{\pi}) \int dg' \,[1 - \cos(3\pi S'/g') - i\sin(3\pi S'/g')]/(g')^{3/5} \tag{1.5.48}$$

with $S' = 2(C_a - C_b)$. Integration yields the well-known expression [2]

$$\Gamma(\mathrm{v}) = 2n\sqrt{\pi}\,\Gamma(9/5)\Gamma(3/5)|3\pi(C_a - C_b)|^{2/5}(2T/M_2)^{3/10} I_1(\alpha x)[\cos(\pi/5) - i\,\mathrm{sgn}(C_a - C_b)\sin(\pi/5)]. \tag{1.5.49}$$

Note that if we do not allow for degeneracy, the fact that the van der Waals constants of the levels are equal makes $\Gamma(\mathrm{v})$ vanish. If we do allow for degeneracy, the relaxation characteristics $\Gamma(\mathrm{v})^{(\kappa)}$ are generally nonzero at $S = 0$ because of relaxation over the sublevels of the upper and lower levels. This can easily be illustrated by the example of the atomic transitions with angular momenta $j_b = j_a = 1$. Assuming that the interaction potential in states a and b is the same in such transitions, that is,

$$C_1^{(a)} = C_1^{(b)} = C_1,\; C_0^{(a)} = C_0^{(b)} = C_0,$$

and employing equations (1.5.14)–(1.5.18), we see that both $\Gamma(\mathrm{v})^{(1)}$ and $\Gamma(\mathrm{v})^{(2)}$ do not vanish and are proportional to $|C_1 - C_0|^{2/5}$

In conclusion of this chapter we will discuss the dependence of the relaxation characteristics on the velocity v of the atom (formula (1.5.47)). The results obtained in this chapter are used in solving several problems of laser physics. In these the final results are written in the form of integrals with respect to the velocities of the resonant (excited) atoms considered. The velocity dependence of the respective integrands is related to allowing for the Doppler effect in the interaction of the atoms with the electromagnetic field, to the velocity dependence of the relaxation characteristics of the resonance levels and transitions, and to the velocity distribution function of the atoms. In the problems considered in this book the velocity distribution of the atomic motion is everywhere assumed to be Maxwellian. Hence, only velocities up to approximately the thermal velocity of motion are essential in integration, that is, only x values up to the order of unity in (1.5.47) are important. When $\alpha << 1$, the function $I_1(\alpha x)$ varies little as x changes from zero to unity. Then, in integrating with respect to velocity we can assume $I_1(\alpha x)$ constant and equal to $I_1(0)$. Note that this approach is equivalent to averaging over velocity with a Maxwellian distribution function of the relaxation characteristics of the resonance levels and transitions. This averaging formally reduces to substituting for the function $\gamma(\mathrm{v})$, which enters into all relaxation characteristics, a constant obtained from $\gamma(\mathrm{v})$ by setting $\mathrm{v} = 0$ and to replacing M_2 with the reduced mass of the colliding atoms, M.

REFERENCES

1. A.C.G. Mitchell, M.W. Zemansky, *Resonance Radiation and Excited Atoms* (Cambridge: Cambridge Univ. Press, 1971).
2. L.A. Vainshtein, I.I. Sobel'man, E.A. Yukov, *Vozbuzhdenie atomov i ushirenie spetral'nykh linii* (*Excitation of Atoms and Broadening of Spectral Lines*) (Moscow: Nauka, 1979) (in Russian).
3. S.G. Rautian, *Trudy FIAN*, **43**: 3–115 (1968).
4. L.D. Landau, E.M. Lifshitz, *Quantum Mechanics*: *Nonrelativistic Theory* (Oxford: Pergamon, 1977).
5. Yu.A. Vdovin, V.M. Yermachenko, *Zh. Eksp. Teor. Fiz.* **54**: 148–158 (1968).
6. Yu.A. Vdovin, V.M. Galitskii, V.M. Yermachenko, *Proc. All-Union Symp. on Physics of Gas Lasers* (Novosibirsk, 1969), p. 49.
7. E.G. Pestov, S.G. Rautian, *Zh. Eksp. Teor. Fiz.* **56**: 902–913 (1969).
8. V.A. Alekseyev, T.L. Andreeva, I.I. Sobel'man, *Zh. Eksp. Teor. Fiz.* **62**: 614–626 (1972).
9. P.R. Berman, *Phys. Rev.* **A 5**: 927–939 (1972).
10. S.G. Rautian, *Issledovanie stolknovenii metodami nelineinoi spektroskopii* (*Investigation of Collisions by Means of Nonlinear Spectroscopy*) (Novosibirsk: Inst. Automation and Electrometry Sib. Div. USSR Acad. Sci., Preprint No. 82, 1978) (in Russian).
11. Yu.A. Vdovin, V.M. Galitskii, A.I. Gurevich, et al. *O vliyanii atomnykh stolknovenii na kharakteristiki lembovskogo provala* (*On the Influence of Atomic Collisions on the Characteristics of the Lamb Dip*) (Moscow: I.V. Kurchatov Inst. Atom. Energy, Preprint No. 2256, 1973) (in Russian).
12. V.M. Yermachenko, *Kvantovaya Elektron.* No. 1 (13): 134–135 (1973).

13. M.I. D'yakonov, V. I. Perel', *Zh. Eksp. Teor. Fiz.* **48**: 345–352 (1965).
14. A.I. Vainshtein, V.M. Galitskii, *Resonant Excitation Transfer in Collisions of Slow Atoms, in Voprosy teorii atomnykh stolknovenii (Problems of the Theory of Atomic Collisions)* (Moscow: Atomizdat, 1970) (in Russian), p. 39–49.
15. A.S. Davydov, *Quantum Mechanics* (Oxford: Pergamon, 1976).
16. I.I. Sobel'man, *Atomic Spectra and Radiative Transitions* (Berlin: Springer, 1979).
17. A. Omont, *Prog. Quantum Electron.* **5**: 69–138 (1977).
18. Yu.A. Vdovin, V.M. Galitskii, V.M. Yermachenko, *Quasiresonant Collisions of Atoms, in Voprosy teorii atomnykh stolknovenii (Problems of the Theory of Atomic Collisions)* (Moscow: Atomizdat, 1970) (in Russian), p. 78–82.
19. V.K. Matskevich, *Opt. Spektrosk.* **37**: 411–417 (1974).
20. I.S. Gradshtein, I.M. Ryzhik, *Tables of Integrals, Series, and Products* (New York: Academic Press, 1966).
21. V.A. Alekseyev, A.V. Malyugin, *Zh. Eksp. Teor. Fiz.* **74**: 911–923 (1978).
22. T. Manabe, S. Yabuzaki, T. Ogawa, *Phys. Rev.* **A 20**: 1946–1957 (1979).
23. S.G. Rautian, A.G. Rudavets, A.M. Shalagin, *Zh. Eksp. Teor. Fiz.* **78**: 545–560 (1980).
24. V.N. Rebane, *Opt. Spektrosk.* **26**: 673–681 (1969).
25. J.P. Faroux, J. Brossel, *C.R. Acad. Sci.* **B 263**: 612–621 (1966).
26. V.N. Rebane, *Opt. Spektrosk.* **24**: 296–305 (1968).
27. C.H. Wang, W.J. Tomlinson, *Phys. Rev.* **181**: 115–136 (1969).
28. D.S. Bakaev, I.V. Yevseyev, V.M. Yermachenko, *Zh. Eksp. Teor. Fiz.* **76**: 1212–1225 (1979).
29. A.G. Petrashen', V.N. Rebane, T.K. Rebane, *Opt. Spektrosk.* **35**: 408–417 (1973).

Chapter 2

METHODS OF THEORETICAL DESCRIPTION OF THE FORMATION OF PHOTON ECHO AND STIMULATED PHOTON ECHO SIGNALS IN GASES

In this chapter, we provide a review of early theoretical studies on the photon echo in gases. We derive the basic equations, which will be employed in this monograph for the theoretical description of electromagnetic processes in gas media. For the elementary case when the angular momenta J_b and J_a of the upper (b) and lower (a) resonant levels are equal to 1 and 0, respectively, we analyze the formation of photon echo and stimulated photon echo signals in the case of an inhomogeneously broadened spectral line corresponding to an optically allowed $b \to a$ transition. We discuss characteristic parameters involved in the theory of photon echo in gas media.

2.1 Early Theoretical Studies on the Photon Echo in Gases

Photon echo is, in fact, an optical analog of the phenomenon of spin echo, which is well known in nuclear magnetic resonance (NMR). The spin echo in NMR arises in an ensemble of nuclear spins in the presence of a permanent magnetic field under the action of pulses of a weak alternating magnetic field with a resonant frequency. The carrier frequency of the pulses of the alternating field in the case of NMR falls within the radio-frequency range (e.g., see [1–3]).

A phenomenon similar to the spin echo in NMR is also observed in an ensemble of electron spins and is called the electron spin echo (e.g., see [4]). The carrier frequency of pump pulses in experiments on electron spin echo lies within the microwave range.

Beginning with the 1950s, the spin echo in NMR became a powerful spectroscopic tool [1–3]. The use of the spin echo method in the spectroscopy of electron paramagnetic resonance (EPR spectroscopy) became possible somewhat later, because the development of the adequate equipment for electron spin echo encountered several serious technical difficulties [4].

Two modifications of spin echo are usually employed for spectroscopic purposes – primary echo (or simply the spin echo) and stimulated echo. The signal of spin echo is produced when a medium is subject to the influence of two pump pulses separated by a time interval τ. The signal of spin echo arises within the time interval $\approx \tau$ after the second pulse. For the generation of the stimulated spin echo, the medium is irradiated with three rather than two pump pulses. The signal under these conditions arises within the time interval $\approx \tau_1$ after the third pulse. Here, τ_1 is the time interval between the first and second pump pulses. Much experimental data have been obtained in NMR and EPR spectroscopy with the use of the above-described modifications of spin echo.

The second area of practical applications of spin echo is associated with the use of this phenomenon in data writing and processing systems. The possibility of using the spin echo in such systems is based on the paper by Fernbach and Proctor [5], who demonstrated that the signals of spin echo and stimulated spin echo may reproduce under definite conditions the waveform of one of the pump pulses (directly or with time reversal).

We emphasize that, similar to the spin echo, the main areas of practical applications of the photon echo are associated with spectroscopic applications and applications in data writing, storing, and processing. Applications of the photon echo for the spectroscopy of atoms (molecules) in gas media are considered in Chapter 4. The possibilities of using this phenomenon for data writing, storing, and processing are discussed in Chapter 5.

The phenomenon of photon echo was predicted by Soviet scientists Kopvillem and Nagibarov [6]. The subsequent theoretical and experimental study confirmed the possibility of photon echo formation and developed various modifications of the photon echo. Specifically, the primary photon echo was observed for the first time in a ruby sample doped with Cr^{3+} paramagnetic ions by American physicists Kurnit, Abella, and Hartmann [7]. Stimulated photon echo was also observed for the first time with this material [8].

The first experiment on the photon echo in a gas medium was performed in 1968 by Patel and Slusher [9], who employed CO_2-laser pulses for pumping and used an SF_6 molecular gas as a resonant medium. Primary photon echo signals were observed under these conditions.

We should emphasize that a large number of experiments on the photon echo in gases have been performed by now. Not only photon echo signals, but also various modifications of the photon echo were observed in these experiments. The review of these studies is provided in Chapter 3.

Abella, Kurnit, and Hartmann have developed an elementary theory of the formation of a primary photon echo signal in solids [10]. Note that the energy levels of an optically allowed resonant transition involved in the formation of a photon echo signal were assumed to be nondegenerate in this theory.

Two new specific features, which are missing in the case of solids, should be taken into consideration when we describe the photon echo in gas media: first, the thermal motion of gas atoms or molecules and, second, the degeneracy of the energy levels involved in the resonant transition. We should note that the degeneracy is a crucial issue in the investigation of polarization properties of the photon echo and its modifications. This is due to the fact that the generation of such signals involves several transitions between the sublevels of upper and lower degenerate resonant states. The calculation of the collision integral becomes much more complicated in this case. Note that the studies on the photon echo in gases that provide a background for Chapters 2, 4, and 5 of this monograph employed the collision integral calculated in the model of elastic depolarizing collisions described in Chapter 1.

The paper by Scully, Stephen, and Burnham [11] and the paper by Samartsev [12] were the first theoretical papers that took into account the motion of resonant atoms (molecules) in a gas and that, in fact, predicted the possibility of the formation of the photon echo in gas media. The degeneracy of resonant levels was included for the first time in 1969 in the independent papers by Alekseyev and Yevseyev [13] and Gordon *et al.* [14].

The inclusion of the degeneracy of levels involved in a resonant transition in [13, 14] made it possible to carry out the first theoretical investigation of polarization properties of the photon echo in a gas medium. Thus, the studies [13, 14] initiated theoretical investigations of polarization properties of the photon echo and its modifications in gas media. These investigations resulted in the development of the theory of polarization properties of the photon echo and its modifications for gas media.

We should note that several research groups in Russia and abroad were involved in the development of the theory of polarization properties of the photon echo and its modifications for gas media. The studies in this direction were carried out during almost two decades. The studies devoted to the theory of polarization properties of the photon echo and its modifications gradually led to the development of the theory of polarization echo spectroscopy of gas media. This theory is focused on the conditions that allow different spectroscopic characteristics of resonant atoms or molecules to be measured. The theory of polarization echo spectroscopy of gas media is presented in Chapter 4.

Now, let us provide a review of early theoretical papers [13–53] devoted to the photon echo in gases giving a special attention to the development of the theory of polarization properties of the photon echo and its modifications. Note that references [15–53] are arranged in the chronological order in the list of references in accordance with the dates these papers were received by the journals. The paper [53], which is the last reference in this list, in fact, provided a background for the general theory of polarization echo spectroscopy of gas media.

Let us consider first the papers [13, 14], which were mentioned above. The study [13] was devoted to the formation of a photon echo in the case of both narrow [see inequality (2.3.24)] and broad [see (2.3.25)] inhomogeneously broadened [for both pump pulses, see (2.3.29)] spectral lines of a resonant transition. However, analysis performed in [13] was restricted to optically allowed transitions with small angular momenta of the resonant levels. This paper was concentrated on the formation of a photon echo signal by a sequence of two pump pulses linearly polarized in different planes and by sequences of two pulses where the first pulse had a linear polarization, while the second pulse was circularly polarized, or the first pulse was circularly polarized, while the second pulse had a linear polarization, or both pulses were circularly polarized.

Analysis performed in [14] was restricted to a photon echo in the case of an inhomogeneously broadened spectral line, which was narrow for both pump pulses, for arbitrary angular momenta of the levels involved in a resonant transition. The formation of photon echo signals by pump pulses linearly polarized in different planes was also investigated in this paper. We should note (see Chapter 3 for details) that most of the experiments on the photon echo and its modifications in gas media were performed with broad spectral lines.

The authors of [14] derived analytical expressions for the polarization vector of the photon echo signal produced through transitions involving resonant levels with small angular momenta and obtained an approximate expression for this vector for $J \to J\,(J >> 1)$ transitions. Note that the results of this study make it possible to perform numerical simulations for the polarization properties of photon echo signals in the case of arbitrary angular momenta of levels involved in optically allowed resonant transitions. All the numerical simulations in [14] were performed for the optimal areas of the pump pulses. Recall that the area of a pump pulse (see the definition of a pulse area in Section 2.4) is one of the key parameters in the theory of photon echo. This quantity characterizes both a pump pulse and an optically allowed resonant transition involved in the formation of a photon echo or some of its modifications. For linearly polarized pump pulses, the pulse area is defined by formulas (2.4.1)–(2.4.3). The optimal areas of pump pulses are usually defined (e.g., see [54]) as the areas ensuring the maximum intensity of the photon echo or some of its modifications. The wish to obtain results corresponding to the optimal areas of pump pulses is characteristic not only of the study [14], but also of some other early papers on the theory of polarization properties of the photon echo in gas media.

The authors of [14] presented the plots of the angle φ between the polarization vector of the photon echo signal and the polarization vector of the second pump pulse as a function of the angle ψ between the polarization vectors of the first and second pump pulses for the optimal areas of the pump pulses. Unfortunately, the authors of [14] did not analyze the behavior of these dependences in the case when the areas of the pump pulses differ from the optimal ones. Such an analysis seems to be very important because the area of a pump pulse, as will be shown below, is a parameter that is very difficult to measure experimentally.

Both papers [13, 14] speculate on the possibility of determining the angular momenta of resonant levels from polarizations properties of the emerging photon echo signal, i.e., on the possibility of identifying resonant transitions by means of a photon echo. However, such measurements would require expressions describing polarization properties of a photon echo for arbitrary angular momenta of levels involved in optically allowed resonant transitions. The results of [13] cannot be used in such an identification, because only some specific values of the angular momenta of resonant levels are considered in this paper. On the other hand, the results of this paper may be useful to test the results of other experiments on the identification of resonant transitions. Although the results of [14] were obtained for arbitrary values of the angular momenta of levels involved in optically allowed resonant transitions, it would be rather difficult to use these results for the above-described identification procedure, since the general expressions are rather cumbersome and the applicability ranges of both general expressions and the results of numerical simulations are rather narrow.

The authors of [14] also proposed an idea of identifying the type ($J \to J$ or $J \rightleftarrows J+1$) of a resonant transition in the case when $J >> 1$. This idea was based on the following result of numerical simulations carried out in [14]. In the case when pump pulses have optimal areas, the polarization vector of the photon echo signal arising due to $J \to J$ ($J >> 1$) transitions lies within the sector determined by the angle ψ, while for $J \rightleftarrows J+1$ ($J >> 1$) transitions, the polarization vector falls outside this angle. We emphasize that, similar to all the other results presented in [14], this consideration assumes that the photon echo signal corresponds to a narrow spectral line and the carrier frequency ω of the pump pulses is exactly resonant to the frequency ω_0 of the resonant transition, while elastic depolarizing collisions are ignored. In addition, to identify the type of a resonant transition using the method proposed in [14], we need to know *a priori* that the angular momenta of the levels involved in the resonant transition are high.

The idea of identifying the type of a resonant transition proposed in [14] was employed by Alimpiev and Karlov [55], who identified the type of a vibrational–rotational transition in SF_6 molecules resonant to the $P(16)$ oscillation line of a CO_2-laser. Since experiments in [55] were performed with a spectral line that was broad for both pump pulses, the authors of this paper assumed that the polarization vector of a photon echo signal displays the above-mentioned property also in the case of a broad spectral line. Then, *a priori* assuming also that the rotational quantum numbers of the resonant transitions are high, the authors of [55] were able to identify the type of the resonant transition as a transition belonging to the Q branch.

Closing the discussion of the studies [13, 14], we should note that these papers did not provide a consistent analysis of irreversible relaxation, which was included through the transverse relaxation time T_2 [14] or through a single relaxation characteristic of a resonant transition [13], which is equivalent to the introduction of the time T_2. Recall that the inclusion of irreversible relaxation through the time T_2 is characteristic of the spin echo [1–4]. In the case of the photon echo in gas media, this approach, as will be shown below, can be considered only as a very rough approximation.

Now, let us discuss some of the later papers, by considering [15–53]. Dienes [16] has noticed that, in the case of small areas of pump pulses, the formulas derived in [14] are reduced to the expressions for the polarization vector of a medium under conditions when combination modes are generated.

Alekseyev [17] predicted a new effect – non-Faraday rotation of the polarization vector of the photon echo – in the case when a longitudinal magnetic field is applied to a gas medium. The author of [17] proposed to use this effect to determine the g-factor of a degenerate resonant level. Note that irreversible relaxation was ignored in [17].

Alekseyev and Yevseyev [20] were the first to propose to use the non-Faraday rotation of the polarization vector of the photon echo signal to identify resonant transitions. However, since calculations in [17, 20] were performed only for resonant levels with low angular momenta, it is impossible to implement such an identification procedure using exclusively the results of [17, 20]. We should note that it was only the studies [56, 57], where the approximation of small areas of pump pulses was used to derive expressions for the angle of the non-Faraday rotation of the polarization vector of the photon echo signal in a longitudinal magnetic field for arbitrary angular momenta of the levels involved in an optically allowed resonant transition, that the identification of resonant transitions with the use of the above-considered effect in the presence of a longitudinal magnetic field became possible. Irreversible relaxation was included in the analysis performed in [20] through a single relaxation characteristic of a resonant transition, which is equivalent, as mentioned above, to the description of irreversible relaxation in terms of the transverse relaxation time T_2.

In [21], Alekseyev and Yevseyev assigned a specific relaxation characteristic to each resonant level. For gas media, such a model is closer to the reality, but still this approach ignores considerable complications of the collision integral due to the degeneracy of resonant levels.

Wang [25] employed the collision integral calculated in the model of elastic depolarizing collisions and averaged over the modulus and the direction of the velocity **v** of motion of resonant atoms (molecules) to determine the strength of the electric field in the photon echo signal. As mentioned in Chapter 1, such a collision integral can be calculated by averaging the right-hand sides of equations (1.5.5)–(1.5.7) over the modulus of **v**. The author of [25] has also demonstrated that the strength of the electric field in photon echo signals corresponding to narrow spectral lines of $J \rightarrow J$ and $J \rightleftarrows J+1$ transitions depends on the relaxation characteristics $\gamma_{ab}^{(\kappa)}$ ($|J_a - J_b| \leq \kappa \leq$ $\leq J_a + J_b$) with odd κ. Here $\gamma_{ab}^{(\kappa)}$ are the relaxation characteristics of multipole moments of an optically allowed resonant transition $b \rightarrow a$. Therefore, the intensity of the photon echo signal corresponding to a narrow spectral line decays as a function of the time interval τ between the pump pulses in accordance with formula (2.4.9) only for the $0 \rightleftarrows 1$, $1 \rightarrow 1$, $1/2 \rightarrow 1/2$, and $1/2 \rightleftarrows 3/2$ transitions. Note that this property of the above-specified transitions, as independently demonstrated by the authors of [58] and [59], is observed also when the generation of a photon echo signal is associated with a broad spectral line. Thus, one does not need to employ polarization echo spectroscopy when performing optical echo-spectroscopy measurements for the above-specified transitions with the use of a primary photon echo signal. To obtain the experimental information concerning the relaxation characteristic $\gamma_{ab}^{(1)}$ of the dipole moment of a resonant transition in this case, it is sufficient to employ conventional echo spectroscopy. For all the other resonant transitions, the formulas derived in [25] were rather complicated, as they involved a large number of relaxation characteristics $\gamma_{ab}^{(\kappa)}$. We should emphasize that the paper [25] does not comment on how these relaxation characteristics can be measured separately of each other. Another important issue that was not addressed in [25] is how to extract the experimental information on the homogeneous width $\gamma_{ab}^{(1)}$ of a resonant spectral line. Only the authors of [53, 58, 60] determined conditions allowing the quantity $\gamma_{ab}^{(1)}$ and the relaxation characteristics $\gamma_{ab}^{(3)}$ and $\Delta_{ab}^{(3)} - \Delta_{ab}^{(1)}$ to be measured. Note that, in contrast to [25], calculations in [53, 58, 60] were performed for the cases when the photon-echo generation is associated with both narrow and broad spectral lines.

The authors of [36] proposed to identify the type of resonant transitions ($J \to J$ or $J \rightleftarrows J+1$, $J >> 1$) using the difference in the decay of the intensity of the photon echo signal with the growth in the areas of pump pulses in experiments of two types. In experiments of the first type, the photon echo signal is produced by pump pulses linearly polarized in the same plane. In experiments of the second type, the photon echo signal is produced by pump pulses circularly polarized in the same direction. The results of [36] are applicable only for the case when the spectral line involved in the formation of a photon echo signal is narrow for both pump pulses. In addition, the method proposed in this paper required, similar to the approach discussed in [14], an *a priori* knowledge that the levels involved in the resonant transition have large angular momenta. Note that irreversible relaxation was ignored in [36].

Sirasiev and Samartsev [39] predicted a new scheme of photon echo – a modified stimulated photon echo (MSPE). Similar to the stimulated photon echo, this scheme can be implemented when a medium is excited by three pulses. The first two pump pulses in this case still have the carrier frequency ω_1 resonant to the frequency ω_0 of an optically allowed transition $b \to a$. However, in contrast to the stimulated photon echo, the third pump pulse in the modification photon echo has a carrier frequency ω_2 resonant to the frequency $\overline{\omega}_0$ of an optically allowed transition $c \to b$ ($E_c > E_b > E_a$). The modified photon echo signal is produced in a medium at the carrier frequency ω_2 at the moment of time approximately equal to $\omega_1\tau_1/\omega_2$ (after the excitation of the medium by the third pump pulse, Fig. 2.1). Similar to the schemes considered above, τ_1 is the time interval between the first and second pump pulses. Note that the degeneracy of levels involved in the resonant transition was ignored in [39].

The authors of [41] have shown that, for most resonant transitions, polarization properties of the photon echo signal depend on whether the line involved in the formation of the signal is narrow or broad.

The authors of [42] derived formulas for the strength of the electric field in the photon echo signal corresponding to both narrow and broad spectral lines. However, irreversible relaxation was not taken into account, while the angular momenta of the levels involved in the optically allowed resonant transition were assumed to be arbitrary. Using some assumptions that can not be rigorously substantiated for $J \to J$ ($J >> 1$) and $J \rightleftarrows J+1$ ($J >> 1$)

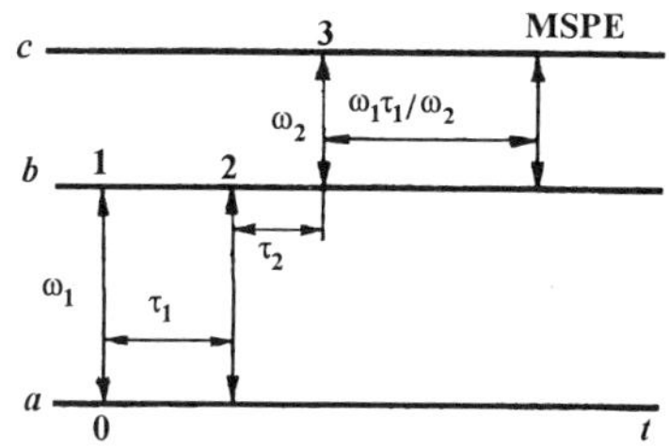

Fig. 2.1. Diagram of the formation of a modified stimulated photon echo signal: 1–3, pump pulses; MSPE, modified stimulated photon echo signal.

transitions, the authors of [42] derived a formula that describes the dependence of the angle φ between the polarization vector of the photon echo signal and the polarization vector of the second pump pulse on the angle ψ between the polarization vectors of the first and second pump pulses:

$$\tan\varphi = \pm\,(1/2)\tan\psi. \tag{2.1.1}$$

The plus sign in this formula corresponds to $J \to J$ ($J >> 1$) transitions, while the minus sign corresponds to $J \rightleftarrows J+1$ ($J >> 1$) transitions. The angles φ and ψ in (2.1.1) are measured from the polarization vector of the second pump pulse and are considered positive for clockwise rotation and negative for counterclockwise rotation. As can be seen from formula (2.1.1), the polarization vector of a photon echo signal produced through $J \to J$ ($J >> 1$) transitions lies within the sector determined by the angle ψ, while the polarization vector of a photon echo signal produced through $J \rightleftarrows J+1$ ($J >> 1$) transitions falls outside this sector.

The authors of [42] employed formula (2.1.1) to interpret the results of polarization experiments [42, 61–63] carried out by their research group for pump pulses with optimal areas. Therefore, we may assume (following, for example, the authors of the monograph [64]) that formula (2.1.1) was derived for pump pulses with optimal areas.

Since the parameter of the relevant expansion was not specified in the derivation of the approximate formula (2.1.1), it was unclear which angular momenta J can be considered as meeting the condition $J >> 1$ and how accurate this formula is. To determine the accuracy of this formula, Yevseyev and Ivliev [65] performed numerical simulations for the ratio $\tan\varphi/\tan\psi$ with different values of J in the case when photon echo signals correspond to a narrow spectral line. These simulations demonstrated that, for example, with $J = 10$, the relative deviation of the results of numerical simulation from the predictions of the approximate formula (2.1.1) is about 14% for $10 \to 10$ transitions and 4% for $10 \rightleftarrows 11$ transitions. In the case when a broad spectral line is involved in the formation of the photon echo signal, such calculations were not performed, and the validity of formula (2.1.1) and its accuracy were not analyzed. However, even if we assume that this expression holds true in the case when a broad spectral line is involved in the formation of the photon echo signal, an *a priori* knowledge that the angular momenta of the levels involved in resonant transitions are high is required for practical applications of this formula. In particular, assuming that expression (2.1.1) holds true in the case when the formation of a photon echo signal involves a broad spectral line and that the resonant levels in vibrational–rotational transitions in SF_6 molecules that are resonant to the $P(14)$ oscillation line of a CO_2-laser are characterized by large rotational quantum numbers, the authors of [63] identified the type of these transitions as transitions that belong to P or R branches. The use of additional experimental data (obtained by means other than the photon echo) allowed the authors of this paper to infer that these transitions belong to the R branch.

The authors of [44] were the first to apply equations of the form (2.3.8)–(2.3.11), which describe the amplitudes of expansion of matrix-density components for resonant levels and transitions in irreducible tensor operators including the interaction of atoms (molecules) in a gas with the electromagnetic field, radiative decay, and other depolarizing and inelastic gas-kinetic collisions, for the description of the photon echo and its modifications. In the subsequent paper [50], these equations were extended to include the terms governing the interaction of atoms (molecules) in a gas with an external permanent magnetic field and the term responsible for the radiative population of the lower resonant level due to spontaneous emission from the upper level. Subsequent studies demonstrated that it is rather convenient to apply equations (2.3.8)–(2.3.11) for the amplitudes of expansion of matrix-density components in irreducible tensor operators for the theoretical investigation of the signals of photon echo and its modifications. This is due to the fact that, when the collision integral is averaged over the direction of motion of resonant atoms (molecules) in the model of elastic depolarizing collisions, each amplitude in the expansion of density-matrix components of resonant levels and transitions of rank κ decays with its own rate (see

Chapter 1). Note also that the authors of [44], who analyzed the example of $1 \rightleftarrows 2$ transitions, were the first to propose to measure the difference in relaxation characteristics $\Delta_{ab}^{(3)} - \Delta_{ab}^{(1)}$ from the beats of the intensity of the photon echo signal as a function of the time interval τ between the pump pulses. This result will be discussed in Section 4.3.

The authors of [45, 47] were the first to take into consideration that the dependence of the collision integral on the direction of the velocity of resonant atoms (molecules) in the model of elastic depolarizing collisions has an influence on the amplitude and polarization of the photon echo signal. Analysis performed in [45, 47], which was restricted to the case of $0 \rightleftarrows 1$ transitions, allowed the authors of these papers to predict a qualitatively new effect. This effect is described in Section 4.6 and is manifested as the appearance of a photon echo signal produced for the considered transitions by pump pulses with orthogonal polarizations. This effect is totally due to the dependence of the collision integral on the direction of the velocity of resonant atoms (molecules). In addition, it was demonstrated in [45] that, by varying the detuning of the carrier frequency ω of the pump pulses from the frequency ω_0 of the resonant transition in experiments, one can perform spectroscopic studies of the dependence of the collision integral on the modulus of the velocity of resonant atoms (molecules) in the model of elastic depolarizing collisions. This result will be discussed in Section 4.5. Note that such measurements have been already performed by Vasilenko, Rubtsova, and Chebotayev [66].

Baer and Abella [46], in fact, repeated calculations of [14] for the polarization vector of a medium induced under the action of two linearly polarized pump pulses. Similar to [14], the resonant spectral line was assumed to be narrow for both pump pulses, and irreversible relaxation was taken into account in an inconsistent way through the introduction of a single relaxation characteristic of a resonant transition, which is equivalent to the description of this relaxation in terms of the transverse relaxation time T_2. The authors of [46] employed the expression for the polarization vector of a medium to find the component of this vector orthogonal to the polarization vectors of the pump pulses. Analyzing this component of the polarization vector of a medium, the authors of [46] considered the limiting case of pump pulses with small areas, which allowed them to determine the component of the electric field strength in the photon echo signal orthogonal to the polarization plane of the pump pulses in the case when the signal is produced through the $7^2P_{1/2} \rightarrow 6^2S_{1/2}$ transition of ^{133}Cs atoms in the presence of a longitudinal magnetic field. Thus, the study [46] solved a particular problem of the formation of a photon echo signal by small-area pump pulses linearly polarized in the same plane for a narrow spectral line corresponding to transitions with electron angular momenta of resonant levels $J_b = 1/2$ and $J_a = 1/2$ in the presence of a longitudinal magnetic field. Note that the authors of [46] were the first to experimentally demonstrate the possibility of detecting a photon echo signal produced by small-area pump pulses.

In [49], Baer proposed to perform experiments with pump pulses of certain (not small) areas to determine relaxation characteristics $\gamma_{ab}^{(\kappa)}$ for odd κ in the case of resonant transitions with some particular values of angular momenta. First, such an approach is not universal, as it can be applied only to specific resonant transitions. Second, even for these transitions, it is difficult to implement this technique, because it is not easy to ensure definite areas of pump pulses with a required accuracy experimentally. The latter difficulty can be illustrated by the following experimental data. In experiments [9], the photon echo signal was produced through transitions of SF_6 molecules resonant to the $P(20)$ oscillation line of a CO_2-laser. The optimal intensity of the first pump pulse, corresponding to the maximum intensity of the photon echo signal, was estimated under these conditions, according to the authors of this paper, as $I_1' \approx 1$ W/cm^2 for a duration of this pulse equal to $\widetilde{T}_1{}' = 200$ ns. However, in experiments [61] performed for the same transitions of SF_6 molecules, the optimal intensity of the first pump pulse was estimated as $I_1'' \approx 0.2$ W/cm^2 for a duration of this pulse equal to $\widetilde{T}_1{}'' =$ $= 300$ ns. Since the product $(I_1)^{1/2}\widetilde{T}_1$ is proportional to the area of the first pump pulse [see formulas (2.4.1)–(2.4.3)], the equality $(I_1')^{1/2}\widetilde{T}_1{}' =$ $= (I_1'')^{1/2}\widetilde{T}_1{}''$ should be satisfied. However, the experimental results discussed above do not meet this equality. This does not seem surprising, as even the experimenters themselves (e.g., see [62]) consider their estimates on the intensities of pump pulses as some approximations that should be treated very carefully.

The authors of [50] were the first to propose to apply the method of photon echo to measure the relaxation characteristics $\gamma_{ab}^{(\kappa)}$ with even κ. Recall that all the earlier papers [44, 49] considered the possibility of measuring the relaxation characteristics $\gamma_{ab}^{(\kappa)}$ and $\Delta_{ab}^{(\kappa)} - \Delta_{ab}^{(1)}$ only with odd κ. To obtain the experimental information on $\gamma_{ab}^{(\kappa)}$ with even κ, the authors of [50] proposed to apply an external longitudinal permanent uniform magnetic field to a gas medium under study. For example, a longitudinal magnetic field applied to a gas medium, as demonstrated in [50], allows the relaxation characteristic $\gamma_{ab}^{(2)}$ of the quadrupole moment of the polarization of the medium to be measured by means of the photon echo for $1/2 \rightleftarrows 3/2$ and $1 \rightarrow 1$ transitions.

We will close our discussion of early theoretical studies on the photon echo and its modifications by considering papers [51, 53]. These papers, in fact, provided a background for the theory of polarization echo spectroscopy. We can make such a conclusion even although the influence of elastic depolarizing collisions on the polarization properties of the photon echo signal, as it was mentioned above, was considered earlier in [25, 44, 45, 47, 49, 50].

Indeed, it is important to keep in mind that the study [25] considered the formation of a photon echo signal only in the case of a narrow spectral line, giving no explanation of how the relaxation characteristics $\gamma_{ab}^{(\kappa)}$ and $\Delta_{ab}^{(\kappa)} - \Delta_{ab}^{(1)}$ with $\kappa = 1, 3, \ldots$ for $J \to J$ ($J \geq 3/2$) and $J \rightleftarrows J+1$ ($J \geq 1$) transitions can be measured. As for the studies [44, 45, 47, 50], the authors of these papers have proposed several methods of such measurements only for some particular transitions.

In [51, 53], the intensity and polarization of the photon echo signal were determined for optically allowed resonant transitions involving levels with arbitrary angular momenta in the cases of both narrow and broad spectral lines. Expressions derived in these papers involve comparatively simple dependences on several parameters of a medium, which makes these expressions very convenient for the processing of the experimental data. Such expressions were derived because the authors of [51, 53] have radically revised the approach to the analysis of the polarization properties of the photon echo in gas media. Instead of investigations for arbitrary areas of pump pulses, it was proposed to perform theoretical studies with small areas of pump pulses and to apply the formulas thus derived for the analysis of the experimental data.

To find the intensity and polarization of the photon echo signal, the authors of [51, 53] solved equations (2.3.8)–(2.3.11) for the amplitudes of expansion of density-matrix components of resonant levels and transitions in irreducible tensor operators in the approximation of small areas of pump pulses. Such an approach made it possible also to perform a theoretical analysis of numerous modifications of the photon echo in gas media. The approach developed in these papers was extended to the case of pump pulses of an arbitrary waveform [67], which allowed, in particular, several new effects associated with the sensitivity of the waveform of some of photon-echo modifications to the waveform of pump pulses to be predicted. These effects will be discussed in Section 5.1.

Note that the approach to the investigation of polarization properties of the photon echo in a gas proposed in [51, 53] was subsequently used by other authors for the solution of similar problems. Specifically, the authors of [68] applied this approach to determine polarization properties of the photon echo signal in solids.

Closing this section, we should emphasize that, before the papers [51, 53] were published, only some particular problems related to the formation of the photon echo through resonant transitions of atoms (molecules) in gases with low angular momenta of resonant levels were studied. The solutions to these problems for resonant transitions with arbitrary angular momenta of the levels and for arbitrary areas of pump pulses were represented by very complex formulas [25], involving a large number of parameters of a gas medium, which made these formulas impractical for extracting the spectroscopic information from the relevant experimental data. In other words, these formulas were unsuitable for polarization echo-spectroscopy. Furthermore, these formulas were derived for a photon echo produced in the case of a narrow spectral line, which is very rarely realized in experiments. Theoretical analysis of such modifications of the photon echo as stimulated and modified stimulated photon echo was restricted to two-level systems, ignoring the degeneracy of resonant levels. Collisional relaxation was included in the analysis of the stimulated photon echo, modified stimulated photon echo, and three-level and modified three-level photon echo (see Section 4.7) within the framework of the scalar model. Thus, the approximations used to analyze the above-specified modifications of the photon echo in a gas medium were very far from reality. This issue is of special importance, because, as will be shown below, these modifications of the photon echo may provide a much richer spectroscopic information on a medium than the photon echo itself.

2.2 The Basic Equations for the Description of Electromagnetic Processes in a Gas Medium

To describe the behavior of atoms (molecules) in a gas in the presence of external resonant fields of pump pulses, we will use the d'Alembert equation,

$$\Box \boldsymbol{E} = \frac{4\pi}{c^2}\frac{\partial^2}{\partial t^2}\int \boldsymbol{P} d\mathrm{v} \tag{2.2.1}$$

and the quantum-mechanical equation for the density matrix $\hat{\rho} = \hat{\rho}(\boldsymbol{r},\mathbf{v},t)$ of resonant atoms (molecules),

$$(\frac{\partial}{\partial t} + \mathbf{v}\nabla + \hat{\Gamma})\hat{\rho} = (i/\hbar)[\hat{\rho}, \hat{H}_0 - \hat{\boldsymbol{d}}\boldsymbol{E}]. \tag{2.2.2}$$

Here $\boldsymbol{E}$ is the strength of the electric field in the medium, $\hat{H}_0$ is the Hamiltonian of an atom (molecule) in the frame of reference related to the center of inertia of this atom (molecule), $\hat{\boldsymbol{d}}$ is the operator of the atomic (molecular) dipole moment, $\boldsymbol{r}$ and $\mathbf{v}$ are the position and the velocity of an atom (molecule) as a whole at the moment of time t, and $\hat{\Gamma}$ is the relaxation matrix, which includes the radiative decay of resonant levels and elastic and inelastic collisions. In what follows, we will describe elastic collisions restricting our consideration to the model of elastic depolarizing collisions discussed in Chapter 1.

The polarization vector of the medium $\boldsymbol{P}$ involved in (2.2.1) is associated with a group of atoms (molecules) moving with a velocity $\mathbf{v}$ and can be expressed in terms of the density matrix $\hat{\rho}$ as

$$\boldsymbol{P} = \mathrm{Sp}(\hat{\rho}\hat{\boldsymbol{d}}). \tag{2.2.3}$$

Let us determine the initial conditions for equations (2.2.1)–(2.2.3). Let the carrier frequency ω of the first pump pulse be close to the frequency ω_0 of an optically allowed resonant transition $b \rightarrow a$ $(E_b > E_a)$. We will restrict our analysis to the consideration of density-matrix elements related only to two resonant levels b and a and to the resonant transition $b \rightarrow a$ between these levels. Along with the angular momenta J_b and J_a, we will employ the projections m_b and m_a of the total angular momentum on the quantization axis to characterize the Zeeman sublevels of resonant levels b and a. For definiteness, the Y axis is chosen along the direction of propagation of the first pump pulse through the gas medium.

Since we are interested in a situation when the first pump pulse incident on the $y = 0$ boundary of a gas medium at the moment of time $t = 0$ propagates in the positive direction of the Y axis, $\xi = t - y/c = 0$ will be chosen the initial moment of time for each point y of the gas medium. Below, we assume that, before the action of the first pump pulse, resonant atoms (molecules) in the gas medium are characterized by a Maxwellian velocity distribution, uniformly distributed in space, and distributed in the Zeeman sublevels of the resonant levels with equal probabilities. Then, at the initial moment of time $\xi = 0$, before the first pump pulse reaches the point y of the gas medium, the density matrix of resonant atoms (molecules) is written as

$$\hat{\rho}\,(\xi = 0) = f(\mathbf{v})[n_b \Sigma\, |J_b m_b\rangle\langle J_b m_b| + n_a \Sigma\, |J_a m_a\rangle\langle J_a m_a|], \tag{2.2.4}$$

where n_b and n_a are the population densities in the Zeeman sublevels of the resonant levels b and a for $\xi \leq 0$ and $f(\mathbf{v})$ describes the Maxwellian distribution of resonant atoms (molecules) in the gas in velocities $\mathbf{v}$. Summation in (2.2.4) is performed over all possible values of projections m_b and m_a.

In considering the interaction of resonant electromagnetic radiation with atoms (molecules) in a gas, it is convenient to expand the components of the density matrix related to resonant levels,

$$\langle J_b m_b|\,\hat{\rho}\,|J_b\, m_b'\rangle = \rho_{mm'}^{(bb)}, \qquad \langle J_a m_a|\,\hat{\rho}\,|J_a\, m_a'\rangle = \rho_{mm'}^{(aa)},$$

and the corresponding transitions,

$$\langle J_a m_a|\,\hat{\rho}\,|J_b\, m_b'\rangle = \rho_{mm'}^{(ab)},$$

in irreducible tensor operators, which is due to the fact that, for many situations considered below, a satisfactory accuracy of calculations can be achieved when the collision integral taken in the model of elastic depolarizing collisions is diagonalized with the use of this expansion. The components of the density matrix of the resonant transition, $\rho_{mm'}^{(ab)}$, and resonant levels, $\rho_{mm'}^{(bb)}$ and $\rho_{mm'}^{(aa)}$, are related to the amplitudes of the expansion in the irreducible tensor operators $\psi_q^{(k)}$, $f_q^{(k)}$, and $\phi_q^{(k)}$ by formulas (1.3.1).

Using formulas (1.3.1) and equation (2.2.2), we derive the following set of equations:

$$[\frac{\partial}{\partial t} + \mathbf{v}\nabla_r + ½(\gamma_a^{(0)} + \gamma_b^{(0)}) - i\omega_0]\,\psi_q^{(k)}(\boldsymbol{r},\mathbf{v},t) =$$
$$= (i/\hbar)(2J_b+1)^{-1/2} d^* \sum E_{-q_1}\,[S_{qq_2q_1}^{\kappa\kappa_1} f_{q_1}^{(k_1)} +$$
$$+ (-1)^{\kappa+\kappa_1} R_{qq_2q_1}^{\kappa\kappa_1}\,\phi_{q_1}^{(k_1)}] + St(\psi_q^{(k)}), \tag{2.2.5}$$

$$(\frac{\partial}{\partial t} + \mathbf{v}\nabla_r + \gamma_b^{(0)})\,f_q^{(k)}(\boldsymbol{r},\mathbf{v},t) =$$
$$= (i/\hbar)(2J_a+1)^{-1/2} \sum E_{-q_1}\,C_{qq_2q_1}^{\kappa\kappa_1}\,[d\psi_{q_1}^{(k_1)} +$$
$$+ d^*\psi_{-q_1}^{(k_1)*}(-1)^{\kappa+\kappa_1+q_2}] + St(f_q^{(k)}), \tag{2.2.6}$$

$$(\frac{\partial}{\partial t} + \mathbf{v}\nabla_r + \gamma_a^{(0)})\,\phi_q^{(k)}(\boldsymbol{r},\mathbf{v},t) =$$
$$= (i/\hbar)(2J_a+1)^{-1/2} \sum E_{-q_1}\,B_{qq_2q_1}^{\kappa\kappa_1}\,[(-1)^{\kappa+\kappa_1} d\psi_{q_1}^{(k_1)} +$$
$$+ d^*\psi_{-q_1}^{(k_1)*}(-1)^{q_2}] + \Gamma_\kappa^{ab}\,f_q^{(k)} + St(\phi_q^{(k)}). \tag{2.2.7}$$

Here, d is the reduced matrix element of the dipole moment operator of the resonant transition $b \to a$, $1/\gamma_a^{(0)}$ and $1/\gamma_b^{(0)}$ are the relaxation times of populations in the resonant levels associated with radiative decay and inelastic gas-kinetic collisions, E_q is the circular component of the vector $\boldsymbol{E}$, and summation is implied over the indices κ_1, q_1, and q_2. Then, as it follows from (2.2.3), the circular component P_q of the vector $\boldsymbol{P}$ is related to $\psi_q^{(1)}$ by the expression

$$P_q = (-1)^{J_b - J_a + q}(2J_a+1)^{-1/2} d\psi_q^{(1)}. \tag{2.2.8}$$

The quantity S in equation (2.2.5) can be determined from the formula

$$C_{qq_2q_1}^{\kappa\kappa_1} = (-1)^{2J_a+q_2}(2J_b+1)^{1/2}(2\kappa_1+1)\times$$

$$\times \begin{pmatrix} \kappa_1 & 1 & \kappa \\ -q_2 & q_1 & q \end{pmatrix} \begin{Bmatrix} \kappa_1 & 1 & \kappa \\ J_b & J_b & J_a \end{Bmatrix} \tag{2.2.9}$$

by the replacement of indices $a \leftrightarrow b$ everywhere except for the second column in the 6j-symbol. The quantities B and R, in their turn, can be found from C and S by renaming $a \leftrightarrow b$.

The second term on the right-hand side of equation (2.2.7) includes the radiative population of the lower resonant level a due to spontaneous emission from the upper level b, which gives

$$\Gamma_{\kappa}^{ab} = (-1)^{1+\kappa+J_a+J_b} \gamma_{ab}[(2J_a + 1)(2J_b + 1)]^{1/2} \times$$
$$\times \begin{Bmatrix} J_b & J_b & \kappa \\ J_a & J_a & 1 \end{Bmatrix}. \tag{2.2.10}$$

The quantity $1/\gamma_{ab}$ is defined as the lifetime of the state b with respect to spontaneous decay to the state a.

Finally, the terms $St(\psi_q^{(k)})$, $St(f_q^{(k)})$, and $St(\phi_q^{(k)})$ on the right-hand sides of equations (2.2.5)–(2.2.7) are the collision integrals, which describe relaxation due to elastic collisions of resonant atoms (molecules) with non-resonant atoms or molecules of buffer gases. Within the framework of the model of elastic depolarizing collisions, these collision integrals are determined either by the right-hand sides of equations (1.3.2)–(1.3.4) or by the right-hand sides of equations (1.5.5)–(1.5.7).

As can be seen from (1.3.1) and (2.2.4), at the initial moment of time ($\xi = 0$), when the first pump pulse reaches the point y of the gas medium, we have the following initial conditions for the amplitudes of expansion of the density-matrix components of resonant levels and transition in irreducible tensor operators:

$$\psi_q^{(k)}(\xi = 0) = 0, \tag{2.2.11}$$

$$f_q^{(k)}(\xi = 0) = (2J_b + 1)n_b\, f(\mathbf{v})\delta_{\kappa 0}\delta_{q0}, \tag{2.2.12}$$

$$\phi_q^{(k)}(\xi = 0) = (2J_a + 1)n_a f(\mathbf{v})\delta_{\kappa 0}\delta_{q0}. \tag{2.2.13}$$

Note that the collision integral determined from (1.5.5)–(1.5.7) by averaging over the modulus of the velocity **v** of resonant atoms (molecules) was employed for the first time by D'yakonov [69] in the problem of resonant light scattering in the presence of an external magnetic field. The set of equations (2.2.5)–(2.2.7) for the amplitudes of expansion of density-matrix components of resonant levels and transitions in irreducible tensor operators involving such a collision integral was presented for the first time in [70, 71], where the theory of a single-mode gas laser was developed. The set of equations (2.2.5)–(2.2.7) was applied for the first time for the description of the photon echo by the authors of [44]. The use of equations (2.2.5)–(2.2.7) with the collision integral determined by (1.5.5)–(1.5.7) allows one [44] to describe the photon echo and its modifications both during the propagation of pump pulses and within the intervals between these pulses by solving the same equations, which is especially important in the case when the durations of the pump pulses are comparable with the times of irreversible relaxation.

2.3 Specific Features of the Formation of Photon Echo Signals in Gases

In this section, we provide a detailed analysis of the formation of the photon echo in a gas medium through an optically allowed transition with angular momenta of resonant levels equal to $J_b = 1$ and $J_a = 0$. Such a transition was chosen because intermediate formulas and the final expression for the strength of the electric field in the photon echo signal are comparatively simple in this case. At the same time, by analyzing this transition, we can understand the main specific features of photon-echo formation in a gas medium. In addition, the photon echo for this transition was experimentally investigated, for example, in ^{174}Yb vapor (see Chapter 3).

We will also assume for simplicity that the mass of a resonant atom (molecule) is greater than or on the order of the mass of a buffer-gas atom (molecule). Such an assumption also allows us to simplify the relevant formulas, because, as demonstrated in Chapter 1, we can employ the collision integral in the form (1.5.5)–(1.5.7) for the description of elastic depolarizing collisions in this case. Finally, we assume that the pump pulses have a rectangular shape and propagate through the gas medium in the same direction, which also simplifies our calculations.

Suppose that the photon echo signal is produced under the action of two pump pulses linearly polarized at an angle ψ with respect to each other. The carrier frequency ω of the pump pulses is assumed to be resonant to the frequency ω_0 of an atomic (molecular) transition $b \to a$. In addition to the energy, quantum-mechanical states of a resonant atom (molecule) will be characterized by the total angular momentum J and its projection m on the quantization axis. The electric-field strengths of the pump pulses can be written as

$$\boldsymbol{E}_1 = e^{(1)}\boldsymbol{l}_1 g_1(\xi)\exp[i(\omega t - ky + \Phi_1)] + \text{c.c.}, \tag{2.3.1}$$

$$\boldsymbol{E}_2 = e^{(2)}\boldsymbol{l}_2 g_2(\xi - \tau - \widetilde{T}_1)\exp[i(\omega t - ky + \Phi_2)] + \text{c.c.}, \tag{2.3.2}$$

where $e^{(n)}$, Φ_n, $\widetilde{T}_n$, and $\boldsymbol{l}_n$ ($n = 1, 2$) are the constant amplitude, phase, duration, and polarization vector of the n-th pump pulse, respectively; $k = \omega/c$; and the function g_n describes the shape of the n-th pump pulse. In the considered case of rectangular pump pulses, the functions g_1 and g_2 can be represented as

$$g_1(\xi) = \theta(\xi) - \theta(\xi - \widetilde{T}_1), \tag{2.3.3}$$

$$g_2(\xi - \tau - \widetilde{T}_1) = \theta(\xi - \tau - \widetilde{T}_1) - \theta(\xi - \tau - \widetilde{T}_1 - \widetilde{T}_2), \tag{2.3.4}$$

where

$$\theta(x) = \begin{cases} 1, & x \geq 0, \\ 0, & x < 0, \end{cases} \tag{2.3.5}$$

and τ is the time interval between the pump pulses.

In what follows, we assume for definiteness that the Z axis is directed along the polarization vector of the second pump pulse ($\boldsymbol{l}_2 = (0, 0, 1)$), and the angle ψ is measured from the Z axis clockwise if we look at this axis along the Y axis ($\boldsymbol{l}_1 = (\sin\psi, 0, \cos\psi)$).

Let us separate fast-oscillating factors of the sought-for functions:

$$\boldsymbol{E} = \boldsymbol{e}\exp[i(\omega t - ky + \Phi)] + \text{c.c.}, \tag{2.3.6}$$

$$\psi_q^{(k)} = \widetilde{\psi}_q^{(k)} \exp[i(\omega t - ky + \Phi)], \tag{2.3.7}$$

where $\boldsymbol{e}$ and $\widetilde{\psi}_q^{(k)}$ are the slowly varying amplitudes, and Φ is the constant phase shift. In the resonant approximation, formulas (2.2.1) and (2.2.5)–(2.2.8) involving the collision integral given by (1.5.5)–(1.5.7) yield the following equations for the slowly varying functions:

$$\left(\frac{\partial}{\partial t} + c\frac{\partial}{\partial y}\right)e_q = -2i\pi\omega \int d\mathbf{v}\, p_q, \tag{2.3.8}$$

$$\begin{aligned}[t]&\left[\frac{\partial}{\partial t} - i(kv_y - \Delta\omega(v) + \Delta_{ab}^{(\kappa)}(v) - \Delta_{ab}^{(1)}(v)) + \gamma_{ab}^{(\kappa)}(v)\right]\widetilde{\psi}_q^{(k)} = \\ &= (i/\hbar)(2J_b + 1)^{-1/2} d^* \sum e_{-q_1} \left[S_{qq_2q_1}^{\kappa\kappa_1} f_{q_1}^{(k_1)} + \right. \\ &\left. + (-1)^{\kappa+\kappa_1} R_{qq_2q_1}^{\kappa\kappa_1} \phi_{q_1}^{(k_1)}\right],\end{aligned} \tag{2.3.9}$$

$$\begin{aligned}[t]&\left(\frac{\partial}{\partial t} + \gamma_b^{(\kappa)}(v)\right) f_q^{(k)} = (i/\hbar)(2J_a + 1)^{-1/2} \sum C_{qq_2q_1}^{\kappa\kappa_1} \left[d\widetilde{\psi}_{q_2}^{(k_1)} (e^*)_{-q_1} + \right. \\ &\left. + (-1)^{\kappa+\kappa_1+q_2} e_{-q_1} d^* \widetilde{\psi}_{-q_2}^{(k_1)*}\right],\end{aligned} \tag{2.3.10}$$

$$\begin{aligned}[t]&\left(\frac{\partial}{\partial t} + \gamma_a^{(\kappa)}(v)\right)\phi_q^{(k)} = (i/\hbar)(2J_a + 1)^{-1/2} \sum (-1)^{\kappa+\kappa_1} B_{qq_2q_1}^{\kappa\kappa_1} \times \\ &\times \left[d\widetilde{\psi}_{q_2}^{(k_1)} (e^*)_{-q_1} + (-1)^{\kappa+\kappa_1+q_2} e_{-q_1} d^* \widetilde{\psi}_{-q_2}^{(k_1)*}\right] + \Gamma_\kappa^{ab} f_q^{(k)},\end{aligned} \tag{2.3.11}$$

where

$$p_q = (-1)^{J_b - J_a + q} (2J_a + 1)^{-1/2} d\widetilde{\psi}_q^{(1)}, \tag{2.3.12}$$

$$\Delta\omega(v) = \omega - \omega_0 - \Delta_{ab}^{(1)}(v), \tag{2.3.13}$$

$$\gamma_{ab}^{(\kappa)}(v) = ½(\gamma_a^{(0)} + \gamma_b^{(0)}) + \Gamma_{ab}^{(\kappa)}(v), \tag{2.3.14}$$

$$\gamma_{a,b}^{(\kappa)}(v) = \gamma_{a,b}^{(0)} + \Gamma_{a,b}^{(\kappa)}(v), \tag{2.3.15}$$

e_q and p_q are the circular components of the corresponding vectors, v_y stands for the projection of the velocity **v** on the Y axis, and $\Gamma_{ab}^{(\kappa)}(v)$ and $-\Delta_{ab}^{(\kappa)}(v)$ are the real and imaginary parts of the quantity $\Gamma(\mathrm{v})^{(\kappa)}$ involved in the right-hand side of equation (1.5.5). The formulas presented above imply summation over κ_1, q_1, and q_2.

Solving the set of equations (2.3.8)–(2.3.11) with initial conditions (2.2.11)–(2.2.13), we assume that the durations $\widetilde{T}_1$ and $\widetilde{T}_2$ of the pump pulses are small as compared with the time interval τ between these pulses. We assume also that the inequalities

$$\gamma_{ab}^{(1)}(v)\,\widetilde{T}_n << 1,\ \Delta_{ab}^{(1)}(v)\,\widetilde{T}_n << 1,\ \gamma_{a,b}^{(0)}(v)\,\widetilde{T}_n << 1,$$

$$\gamma_b^{(1)}(v)\,\widetilde{T}_n << 1,\ \gamma_b^{(2)}(v)\,\widetilde{T}_n << 1 \tag{2.3.16}$$

are satisfied for the characteristic values of v. These inequalities allow us to neglect irreversible relaxation during the time interval when the medium is subject to the influence of the pump pulses. In addition, we will also ignore the reverse effect of the medium on the pump pulses (2.3.1)–(2.3.4). The latter assumption allows us to linearize equations (2.3.8)–(2.3.11).

Suppose that the $y = 0$ boundary of a gas medium is irradiated with the first pump pulse (2.3.1) and (2.3.3) at the moment of time $t = 0$. Then, equations (2.3.8)–(2.3.11) with initial conditions (2.2.11)–(2.2.13) yield the following expression for $\widetilde{\psi}_q^{(\kappa)}$ at the moment of time $t = \widetilde{T}_1 + y/c$, when the first pump pulse emerges from the gas medium at the point y:

$$\widetilde{\psi}_q^{(\kappa)}(\widetilde{T}_1) = (-1)^{q-1} p^{(1)} d^{-1} (l_1)_q \delta_{\kappa,1}. \tag{2.3.17}$$

Here,

$$p^{(1)} = i(3\hbar)^{-1}|d|^2 e^{(1)}\, \widetilde{T}_1 N_0 f(\mathbf{v}) p_1^*,$$

$$(l_1)_q = \delta_{q,0}\cos\psi - 2^{-1/2}\sin\psi(\delta_{q,1} - \delta_{q,-1}),\ N_0 = n_b - n_a, \qquad (2.3.18)$$

where p_1 can be found from

$$p_n = 2(\Omega_n \widetilde{T}_n)^{-1}\sin(\tfrac{1}{2}\Omega_n \widetilde{T}_n)[\cos(\tfrac{1}{2}\Omega_n \widetilde{T}_n) -$$

$$- i(kv_y - \omega + \omega_0)\Omega_n^{-1}\sin(\tfrac{1}{2}\Omega_n \widetilde{T}_n)] \qquad (2.3.19)$$

with $n = 1$, and the quantity Ω_n involved in (2.3.19) is given by

$$(\Omega_n \widetilde{T}_n)^2 = [(kv_y - \omega + \omega_0)\widetilde{T}_n]^2 + \theta_n^2, \qquad (2.3.20)$$

$$\theta_n = \sqrt{4/3}\, e^{(n)}\, |d|\, \widetilde{T}_n \hbar^{-1}. \qquad (2.3.21)$$

Note that the quantity θ_1 determined from (2.3.21) with $n = 1$ is usually called the area of the first pump pulse in the literature. As mentioned above, this quantity characterizes both the pulse itself and the resonant gas medium, being one of the three characteristic parameters in the theory of photon echo.

We emphasize that, as it follows from (2.3.17), in the case of transition with angular momenta of resonant levels equal to $J_a = 0$ and $J_b = 1$, the first pump pulse (2.3.1) and (2.3.3) induces only the polarization dipole moment in a medium.

In accordance with (2.3.8) and (2.3.12), the quantity (2.3.17) induces the field $e_q^{(in)}$ in a medium. This field reaches its maximum near the $y = L$ boundary of the considered gas medium:

$$e_q^{(in)} = -2\pi i\omega L c^{-1} \int d\mathbf{v}\, p_q(\widetilde{T}_1),$$

where

$$p_q(\widetilde{T}_1) = p^{(1)}(l_1)_q \tag{2.3.22}$$

is the circular component of the vector $\boldsymbol{p}(\widetilde{T}_1)$.

To be able to ignore the reverse effect of the medium on the pulse propagating through this medium, we have to assume that

$$|e_q^{(in)}|/e^{(1)} \ll 1. \tag{2.3.23}$$

This inequality imposes certain restrictions on the parameters of a gas medium and a pump pulse. To perform the relevant estimates, we will use the notions of narrow and broad spectral lines of a resonant transition for a given pump pulse, which were introduced in [41]. Indeed, the relation between the Doppler width k_0u of a resonant spectral line and the spectral bandwidth $1/\widetilde{T}_n$ of the n-th pump pulse plays an important role (along with the area of the pump pulse) in the theoretical description of the photon echo. Here, we use $k_0 = \omega_0/c$, where ω_0 is the frequency of the resonant transition $b \to a$, $u = (2T_0/m)^{1/2}$ is the root-mean-square velocity of resonant atoms (molecules) in a gas, m is the mass of these species, and T_0 is the gas temperature measured in energy units. If the inequality

$$k_0u \ll 1/\widetilde{T}_n \tag{2.3.24}$$

is satisfied, then we deal with a photon echo produced in the case of a spectral line that is narrow for a given pump pulse [41]. In such a situation, the pulse with $\omega = \omega_0$ propagating through the medium excites the entire Doppler contour of the spectral line corresponding to the resonant transition. In the opposite case,

$$k_0u \gg 1/\widetilde{T}_n \tag{2.3.25}$$

we deal with a photon echo produced in the case of a spectral line that is broad for a given pump pulse [41]. Under these conditions, the pulse with $\omega = \omega_0$ propagating through the medium excites only a small part of the

Doppler contour of the spectral line corresponding to the resonant transition.

As it follows from (2.3.23), we can neglect the reverse effect of the medium on the pump pulses if the inequality

$$\omega|d|^2 L|N_0|\widetilde{T}_n\,(\hbar c)^{-1} << 1 \qquad (2.3.26)$$

is satisfied with $n = 1, 2$ in the case when the spectral line of the resonant transition is narrow for both pump pulses or if the inequality

$$\omega|d|^2 L|N_0|T_2^*(\hbar c)^{-1} << 1 \qquad (2.3.27)$$

is satisfied in the case of a broad spectral line. Here, $\widetilde{T}_n$ is the duration of the n-th pump pulse, and the time $T_2^* = (k_0 u)^{-1}$ is usually referred to (e.g., see [1–4]) as the time of reversible transverse relaxation. Note that one of the inequalities (2.3.26) and (2.3.27) is usually satisfied in photon-echo experiments in gases.

As can be seen from (2.3.17) and (2.3.22), at the moment of time $t = \widetilde{T}_1 + y/c$, the polarization vector

$$\boldsymbol{P}(\widetilde{T}_1) = \boldsymbol{p}(\widetilde{T}_1)\exp[i(\omega\widetilde{T}_1 + \Phi_1)] + \text{c.c.}$$

of a group of atoms (molecules) having a projection v_y of the velocity on the Y axis is directed along the polarization vector of the first pump pulse, which is a consequence of the constant-field approximation.

Within the interval $0 \le \xi - \widetilde{T}_1 \le \tau$ after the propagation of the first pump pulse, the set of equations (2.3.8)–(2.3.11) can be split into independent equations. Since only the solution to equation (2.3.9) contributes to the photon echo signal, we can employ (2.3.17) to find that

$$\widetilde{\psi}_q^{(\kappa)}(\xi - \widetilde{T}_1) = \widetilde{\psi}_q^{(\kappa)}(\widetilde{T}_1)\exp[-\gamma_{ab}^{(1)}(v)(\xi - \widetilde{T}_1)]\times$$
$$\times\exp[i(kv_y - \Delta\omega(v))(\xi - \widetilde{T}_1)]\delta_{\kappa,1}. \qquad (2.3.28)$$

Thus, the polarization vector

$$\boldsymbol{P}(\xi - \tilde{T}_1) = \boldsymbol{p}(\tilde{T}_1)\exp[-\gamma_{ab}^{(1)}(v)(\xi - \tilde{T}_1)]\exp[i(\omega\xi + \Phi_1) + \\ + i(kv_y - \Delta\omega(v))(\xi - \tilde{T}_1)] + \text{c.c.}$$

oscillates after the propagation of the first pump pulse and differs from zero at all the subsequent moments of time that are less than or on the order of the time of irreversible transverse relaxation $T_2 = 1/\gamma_{ab}^{(1)}(0)$. Note that this relation sets a correspondence between the homogeneous half-width $\gamma_{ab}^{(1)}(0) = \gamma_{ab}^{(1)}$ of the spectral line corresponding to the resonant transition $b \rightarrow a$ and the time T_2 of irreversible transverse relaxation, which is usually introduced in the literature in a phenomenological way (e.g., see [1–4]). The vector of macroscopic polarization of a medium

$$\int dv\, \boldsymbol{P}(\xi - \tilde{T}_1)$$

decays within the time interval $T_2{}^* = (k_0u)^{-1}$ in the case of a narrow spectral line and within the time interval $\tilde{T}_1$ in the case of a broad spectral line.

Note that, in accordance with (2.2.1), this macroscopic polarization of a medium is responsible for a partial coherent emission of atoms (molecules) in the medium after the propagation of the first pump pulse. Such a partial emission is called optical induction (e.g., see [64]). Optical induction also arises after the propagation of the second pump pulse through a medium.

To minimize the energy emitted after the action of the first pump pulse, we should ensure conditions when the irreversible transverse relaxation time T_2 is much more than the reversible transverse relaxation time $T_2{}^*$. In other words, to keep a nonzero vector $\boldsymbol{P}(\tau)$ by the moment of time when the medium is irradiated with the second pump pulse, we should produce a photon echo using a resonant transition with an inhomogeneously broadened spectral line:

$$1/T_2{}^* >> 1/T_2. \qquad (2.3.29)$$

Note that the inhomogeneous broadening of the spectral line of the resonant transition is the necessary condition of the formation of the photon echo and all its modifications. We emphasize also that the polarization vector of the medium $\boldsymbol{P}(\tau)$ acquires the phase $kv_y\tau$ by the end of the first field-free time interval.

Within the time interval $0 \le \xi - \tau - \widetilde{T}_1 \le \widetilde{T}_2$, the second pump pulse (2.3.2) and (2.3.4) passes through the point y of a gas medium. Linearized equations (2.3.8)–(2.3.11) in such a situation are solved in the same way as in the case when the first pump pulse propagates through the medium. Solutions to the set (2.3.8)–(2.3.11) taken at the moment of time $t = \tau + \widetilde{T}_1 + y/c$ are used as the initial conditions in this case. One of these solutions is derived from (2.3.28) with $\xi = \tau + \widetilde{T}_1$. The expression thus obtained should be multiplied by $\exp[i(\Phi_1 - \Phi_2)]$, since, in accordance with (2.3.1) and (2.3.2), the pump pulses have different constant phase shifts. Note that the initial condition derived from (2.3.28) with $\xi = \tau + \widetilde{T}_1$ contributes to the electric field strength in the photon echo signal.

At the moment of time $\xi = \tau + \widetilde{T}_1 + \widetilde{T}_2$, when the second pulse (2.3.2) and (2.3.4) emerges from the gas medium at the point y, the quantity $\widetilde{\psi}_q^{(\kappa)}(\widetilde{T}_2)$, which contributes to the echo signal, is written as

$$\widetilde{\psi}_q^{(1)}(\widetilde{T}_2) = (-1)^{q-1} d^{-1} p_q(\widetilde{T}_2), \tag{2.3.30}$$

where

$$p_q(\widetilde{T}_2) = (2/9)\cos\psi\exp\{-i[(kv_y - \Delta\omega(v))\tau + \Phi_1 - \Phi_2]\}p^{(2)}\,\delta_{q,0},$$

$$p^{(2)} = -i(2\hbar^3)^{-1}|d|^4 e^{(1)}\,\widetilde{T}_1\, e^{(2)\,2}\,\widetilde{T}_2^{\,2}\, N_0 f(\mathbf{v}) p_1 |q_2|^2 \exp[-\gamma_{ab}^{(1)}(v)\,\tau].$$

Here, p_1 can be determined from (2.3.19) with $n = 1$, q_2 can be found from

$$q_n = 2(\Omega_n \widetilde{T}_n)^{-1}\exp[-\tfrac{1}{2}i(kv_y - \omega + \omega_0)\,\widetilde{T}_n]\sin(1/2\Omega_n \widetilde{T}_n) \tag{2.3.31}$$

with $n = 2$, and the quantity Ω_n is given by formula (2.3.20).

Note that, in writing (2.3.30), we keep only the term involving the phase $-kv_y\tau$ and omit the terms involving the phase $kv_y\tau$, as well as the terms that do not involve the phase dependent on $kv_y\tau$. This is due to the fact that only the phase $-kv_y\tau$ can be compensated in the subsequent evolution of the system in time. Thus, for the part of the polarization vector of the medium $\boldsymbol{P}(\widetilde{T}_2)$ involving the phase $-kv_y\tau$, the action of the second pump pulse is reduced to effective time reversal. As can be seen from (2.3.30), this part of $\boldsymbol{P}(\widetilde{T}_2)$ is written as

$$\boldsymbol{P}(\widetilde{T}_2) = \boldsymbol{p}(\widetilde{T}_2)\exp\{i[\omega(\tau + \widetilde{T}_1 + \widetilde{T}_2) + \Phi_2]\} + \text{c.c.},$$

where $\boldsymbol{p}(\widetilde{T}_2)$ is the vector with the circular component $p_q(\widetilde{T}_2)$.

The quantity defined by (2.3.30) is used as the initial condition in the solution of equation (2.3.9) within the interval $\xi \geq \tau + \widetilde{T}_1 + \widetilde{T}_2$ after the propagation of the second pump pulse. This approach yields

$$\widetilde{\psi}_q^{(1)}(t') = \widetilde{\psi}_q^{(1)}(\widetilde{T}_2)\exp[-\gamma_{ab}^{(1)}(v)\,t']\exp[i(kv_y - \Delta\omega(v))t'], \tag{2.3.32}$$

where

$$t' = \xi - \tau - \widetilde{T}_1 - \widetilde{T}_2. \tag{2.3.33}$$

Consequently, the vectors

$$\boldsymbol{P}(t') = (2/9)\cos\psi\, p^{(2)}\, \boldsymbol{l}_z \exp[-\gamma_{ab}^{(1)}(v)\, t']\exp\{i[(kv_y - \Delta\omega(v))(t' - \tau) + \omega\xi + 2\Phi_2 - \Phi_1]\} + \text{c.c.} \tag{2.3.34}$$

produced by different groups of atoms (molecules) with different velocity projections on the Y axis have different phases after the propagation of the second pump pulse. Here, $\boldsymbol{l}_z$ is the unit vector corresponding to the Cartesian axis. Therefore, electromagnetic waves emitted by these groups of atoms (molecules) are incoherent. However, as can be seen from (2.3.34), different $\boldsymbol{P}(t')$ vectors are matched at the moment of time $t \approx t_e$, where

$$t_e = 2\tau + 1.5\widetilde{T}_1 + \widetilde{T}_2 + y/c, \tag{2.3.35}$$

and the system of excited atoms (molecules) at each point y is transferred into a superradiant state, resulting in a spontaneous coherent emission – a photon echo (Fig. 2.2). Independent photon echo pulses emitted by different areas of the medium add up together with their delay times to produce the resulting electromagnetic pulse.

The final expression for the electric field strength in the photon echo signal can be derived from equation (2.3.8) with allowance for (2.3.30) and (2.3.32):

$$\begin{aligned}\boldsymbol{E}^e = -\pi\omega L c^{-1}\hbar^{-3}|d|^4 e^{(1)}\,\widetilde{T}_1\, e^{(2)2}\,\widetilde{T}_2^{\,2} N_0 S \boldsymbol{e}^e \exp[i(\omega t - ky + \\ + 2\Phi_2 - \Phi_1)] + \text{c.c.}\end{aligned} \tag{2.3.36}$$

Here, the nonvanishing component of the vector $\boldsymbol{e}^e$, characterizing polarization properties of the photon echo signal, is given by

$$e_z^e = (2/9)\cos\psi. \tag{2.3.37}$$

The quantity S, which characterizes the shape and the decay of the photon echo signal, is written as

$$\begin{aligned}S = \int d\mathbf{v}\, p_1|q_2|^2 f(\mathbf{v})\exp[-\gamma_{ab}^{(1)}(v)\,(t - y/c)]\exp[i(kv_y - \\ - \Delta\omega(v))(t - t_e + \tfrac{1}{2}\widetilde{T}_1)],\end{aligned} \tag{2.3.38}$$

where the parameters p_1 and q_2 can be found from formulas (2.3.19) and (2.3.31).

In the next section, we will provide a detailed discussion of expressions (2.3.36)–(2.3.38). Here, we should mention only that the photon echo signal

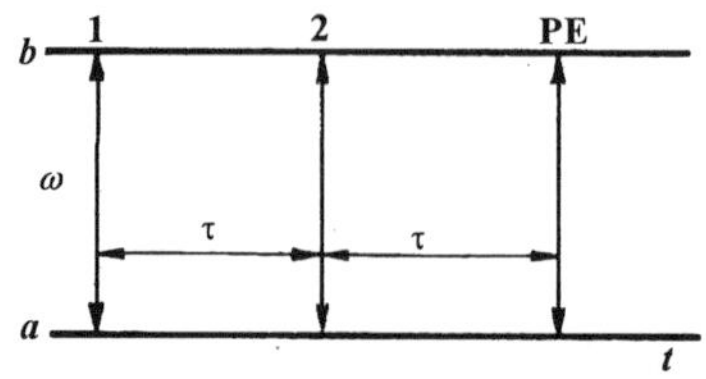

Fig. 2.2. Diagram of the formation of a photon echo signal: 1, 2, pump pulses; PE, photon echo signal.

(2.3.36)–(2.3.38) is linearly polarized along the polarization vector of the second pump pulse and propagates with the carrier frequency ω in the direction of incidence of both pump pulses.

2.4 Characteristic Parameters of the Theory of the Photon Echo

As mentioned above, the area of the pump pulse and the ratio of the inhomogeneous width $1/\tilde{T}_2{}^*$ of the spectral line of the resonant transition to the spectral bandwidth $1/\tilde{T}_n$ of the pump pulse are two out of three characteristic parameters in the theory of the photon echo. The area of the pump pulse was defined by formula (2.3.21) only for a linearly polarized pump pulse and for optically allowed resonant transition with angular momenta equal to $J_a = 0$ and $J_b = 1$. For optically allowed transitions with arbitrary angular momenta of resonant levels, the area θ_n of the n-th linearly polarized pump pulse can be conveniently defined as [14]

$$\theta_n = 2J^{1/2} e^{(n)} \hbar^{-1} |d| \tilde{T}_n [(2J+1)(J+1)]^{-1/2} \qquad (2.4.1)$$

for $J \rightarrow J$ transitions,

$$\theta_n = 2(J+1)^{1/2} e^{(n)} \hbar^{-1} |d| \tilde{T}_n [(2J+1)(2J+3)]^{-1/2} \qquad (2.4.2)$$

for $J \rightleftarrows J+1$ transitions with integer J, and

$$\theta_n = e^{(n)}\hbar^{-1}|d|\widetilde{T}_n(J+1)^{-1/2} \tag{2.4.3}$$

for $J \rightleftarrows J+1$ transitions with half-integer J. Here, as before, $e^{(n)}$ and $\widetilde{T}_n$ are the amplitude and the duration of the n-th pump pulse, and d is the reduced matrix element of the dipole moment operator for the resonant transition. Thus, the area of a pump pulse is given by some numerical factor multiplied by the Rabi frequency $e^{(n)}|d|\hbar^{-1}$ and the pulse duration.

The ratio of the Rabi frequency to the inhomogeneous width $1/\widetilde{T}_2{}^*$ of the spectral line of the resonant transition is the third important parameter in the theory of the photon echo. The limiting case

$$e^{(n)}|d|\hbar^{-1} >> 1/\widetilde{T}_2{}^*, \tag{2.4.4}$$

is usually referred to [72] as the strong-field limit. The opposite case,

$$e^{(n)}|d|\hbar^{-1} << 1/\widetilde{T}_2{}^*, \tag{2.4.5}$$

is usually referred to [72] as the weak-field limit.

Let us analyze the quantity S, which is defined by formula (2.3.38) and which characterizes the shape and the decay of the photon echo signal (2.3.36)–(2.3.38) produced by rectangular pump pulses (2.3.1)–(2.3.4).

The integral in formula (2.3.38) involves the dependence on the velocity $\mathbf{v}$ of resonant atoms (molecules) through the functions $\gamma_{ab}^{(1)}(v)$, $\Delta_{ab}^{(1)}(v)$, $f(\mathbf{v})$, $\Omega_n - \Omega_n(v_y)$, $\sin(\frac{1}{2}\Omega_n\widetilde{T}_n)$, and $\cos(\frac{1}{2}\Omega_n\widetilde{T}_n)$. This formula also involves an explicit dependence on v_y through the oscillating exponential and through a factor. Therefore, the integral in (2.3.38) cannot be calculated analytically in the general case. Obviously, numerical simulations can be performed, if necessary, with some specific model determining the functions $\gamma_{ab}^{(1)}(v)$ and $\Delta_{ab}^{(1)}(v)$. In what follows, we will consider some approximations that allow several analytical expressions to be derived for the quantity S.

Note that the function $f(\mathbf{v})$ noticeably decays for the values of v_y on the order of the mean quadratic velocity u of resonant atoms (molecules). Due to the fact that the integral in (2.3.38) involves the functions $\Omega_n(v_y)$, $\sin(\frac{1}{2}\Omega_n\widetilde{T}_n)$, and $\cos(\frac{1}{2}\Omega_n\widetilde{T}_n)$, the values of v_y around the point $(\omega - \omega_0)/k$ with a distribution width of v_y on the order of $\theta_n/(k\widetilde{T}_n)$ provide a considerable contribution to the integral in v_y. Here, θ_n is the area of the n-th pump pulse and $k = \omega/c$. Let w be the characteristic scale of variation of the functions $\gamma_{ab}^{(1)}(v)$ and $\Delta_{ab}^{(1)}(v)$. Under these conditions, if the widths u and $\theta_n/(k\widetilde{T}_n)$ are comparable, then the quantity w can be compared with each of these quantities. However, if one of these widths is much greater than the other one, then w should be compared with the smallest width. The comparison of the widths u and $\theta_n/(k\widetilde{T}_n)$ allows us to distinguish between strong (2.4.4) and weak (2.4.5) fields of pump pulses.

If min(ku, $\theta_n/\widetilde{T}_n$) is much less than kw, then the functions $\gamma_{ab}^{(1)}(v)$ and $\Delta_{ab}^{(1)}(v)$ taken at the point corresponding to the maximum of the steepest function, i.e., either at $v_y = 0$ or at $v_y = (\omega - \omega_0)/k$, can be placed outside the integral sign in (2.3.38).

Let us first consider the case when the field of pump pulses is strong. Then, the Maxwell distribution is the steepest function under the integral sign in (2.3.38). Since only the functions $\gamma_{ab}^{(1)}(v)$, $\Delta_{ab}^{(1)}(v)$, and $f(\mathbf{v})$ depend on v_x and v_z in the integral in (2.3.38), integration in these variables is reduced to the replacement of v_x and v_z in the arguments of $\gamma_{ab}^{(1)}(v)$ and $\Delta_{ab}^{(1)}(v)$ by zeroes. Variable v_y in the arguments of $\gamma_{ab}^{(1)}(v)$ and $\Delta_{ab}^{(1)}(v)$ can be also replaced by zeroes. Then, expression (2.3.38) can be rewritten as

$$S = \Lambda \exp[-\gamma_{ab}^{(1)}(0)\,(t - y/c) - i(\omega - \omega_0 - \Delta_{ab}^{(1)}(0))(t - t_e + \tfrac{1}{2}\widetilde{T}_1)], \qquad (2.4.6)$$

where

$$\Lambda = 8\int dv_y\,(\Omega_1\widetilde{T}_1)^{-1}(\Omega_2\widetilde{T}_2)^{-2}\sin(\tfrac{1}{2}\Omega_1\widetilde{T}_1)\sin^2(\tfrac{1}{2}\Omega_2\widetilde{T}_2)[\cos(\tfrac{1}{2}\Omega_1\widetilde{T}_1) -$$
$$- i(kv_y - \omega + \omega_0)\Omega_1^{-1}\sin(\tfrac{1}{2}\Omega_1\widetilde{T}_1)]f(v_y)\exp[ikv_y(t - t_e + \tfrac{1}{2}\widetilde{T}_1)], \qquad (2.4.7)$$

the quantity Ω_n can be found from (2.3.20), and t_e is defined by formula (2.3.35).

The integral in (2.4.7) describes the profile of the photon echo signal in the absence of depolarizing collisions. Depending on the relation between the parameters ku and $1/\widetilde{T}_n$ ($n = 1, 2$), this integral may correspond to either a narrow (2.3.24) or a broad (2.3.25) spectral line of the resonant transition. Specifically, when the spectral line of the resonant transition involved in the formation of the photon echo signal is narrow (2.3.24) for both pump pulses and we deal with the case of an exact resonance ($\omega = \omega_0$), then formula (2.4.7) yields

$$\Lambda = 4\theta_1^{-1}\theta_2^{-2}\sin(\theta_1)\sin^2(\tfrac{1}{2}\theta_2)\exp[-\tfrac{1}{4}(T_2^*)^{-2}(t - 2\tau - y/c)^2]. \quad (2.4.8)$$

Thus, the maximum of the photon echo signal in the case under study reaches the observation point y at the moment of time $t = 2\tau + y/c$. This signal has a Gaussian shape, and its width is on the order of T_2^*. The intensity of the photon echo signal, which is determined by (2.3.36), (2.3.37), (2.4.6), and (2.4.8) and which is proportional to $\sin^2(\theta_1)\sin^4(\tfrac{1}{2}\theta_2)$, reaches its maximum when the areas of the first and second pump pulses are equal, for example, to $\theta_1 = \pi/2$ and $\theta_2 = \pi$. Such pump pulses are usually called 90° and 180° pulses, respectively. This terminology stems from the theory of the spin echo [1–5]. Recall that, as mentioned in Section 2.1, the case when one or both pump pulses have small areas is best suited for spectroscopic purposes.

Since the photon echo signal is always produced in the case of an inhomogeneously broadened spectral line (2.3.29), the intensity I_e of this signal, as it follows from (2.3.36), (2.3.37), (2.4.6), and (2.4.8), decays as

$$I_e(\tau) \sim \exp(-4\gamma_{ab}^{(1)}\tau), \quad (2.4.9)$$

where $\gamma_{ab}^{(1)} = \gamma_{ab}^{(1)}(0)$, with the growth in the time interval τ between the pump pulses for $J_b = 1 \to J_a = 0$ transition. Experimental applications of formula (2.4.9) will be discussed in detail in Chapter 4.

Now, let us consider the case when the field of the pump pulses is weak, and inequality (2.4.5) is satisfied. Then, the quantities $\gamma_{ab}^{(1)}(v)$ and $\Delta_{ab}^{(1)}(v)$ taken at the point $v_y = (\omega - \omega_0)/k$ can be placed outside the integral in v_y in (2.3.38). Then, expression (2.3.38) can be rewritten as

$$S = \Lambda \int dv_x dv_z f(v_x) f(v_z)\exp[-\gamma_{ab}^{(1)}(\xi)\,(t - y/c)]\exp[-i(\omega - \omega_0 -$$
$$- \Delta_{ab}^{(1)}(\xi))(t - t_e + \tfrac{1}{2}\tilde{T}_1)], \quad (2.4.10)$$

where

$$\xi = v_x^2 + v_z^2 + (\omega - \omega_0)^2/k^2,$$

while the quantity Λ, which can be determined, similar to the case considered above, from formula (2.4.7), describes the profile of the photon echo signal in the absence of elastic depolarizing collisions. As can be seen from formulas (2.3.36), (2.3.37), and (2.4.10), the intensity of the photon echo signal depends on the detuning $(\omega - \omega_0)$ of the frequency from the exact resonance.

Formula (2.4.10) can be simplified if the inequality $kw >> ku$ is satisfied. Then, the integration in v_x and v_z in (2.4.10) is reduced to the replacement $v_x = v_z = 0$ in the arguments of $\gamma_{ab}^{(1)}(\xi)$ and $\Delta_{ab}^{(1)}(\xi)$, and formula (2.4.10) yields

$$S = \Lambda \exp[-\gamma_{ab}^{(1)}(\eta)(t - y/c) - i(\omega - \omega_0 - \Delta_{ab}^{(1)}(\eta))(t - t_e + \tfrac{1}{2}\widetilde{T}_1)], \qquad (2.4.11)$$

where

$$\eta = (\omega - \omega_0)^2/k^2. \qquad (2.4.12)$$

These expressions make it possible to experimentally investigate the dependence $\gamma_{ab}^{(1)}(v)$ by measuring the intensity of the photon echo signal as a function of the frequency detuning $(\omega - \omega_0)$ from the exact resonance and processing the results of such measurements with the formula

$$I_e(\omega - \omega_0) \sim \exp[-4\gamma_{ab}^{(1)}(\eta)\,\tau], \qquad (2.4.13)$$

where the quantity η is defined by (2.4.12). Note that formula (2.4.13) has been already applied for the processing of the experimental data by the authors of [66].

As mentioned above, the quantity Λ, which is involved in (2.4.10) and which is defined by formula (2.4.7), describes the profile of an echo signal in the absence of elastic depolarizing collisions. Similar to the case of a strong field, we can consider the limits of narrow (2.3.24) and broad

(2.3.25) spectral lines of an inhomogeneously broadened resonant transition when we perform calculations using (2.4.7). Specifically, when the photon echo signal is produced due to a spectral line that is broad for both pump pulses, integration in (2.4.7) can be usually performed only with the use of numerical methods. The only exceptions are associated with the case of small-area pulses ($\theta_1 << 1$ and $\theta_2 << 1$) and the case of an exact resonance ($\omega = \omega_0$). These circumstances were pointed out for the first time by the authors of [53]. In the case under study, formula (2.4.7) with $T_2 \geq T_1$ yields [56]

$$\Lambda = \pi^{1/2} T_2^* \tilde{T}_1^{-1} b^{-2}\{(a+b)^2[\theta(a+b) - \theta(a+b-1)] +$$
$$+ (2a+2b-1)[\theta(a+b-1) - \theta(a)] - (2a^2 - 2a - 2b + 1)[\theta(a) - \theta(a-1)] +$$
$$+ (-2a+2b+1)[\theta(a-1) - \theta(a-b)] + [(a-b)^2 - 2a +$$
$$+ 2b + 1][\theta(a-b) - \theta(a-b-1)]\}, \tag{2.4.14}$$

where

$$a = (t - t_e + \tfrac{1}{2}\tilde{T}_1)\tilde{T}_1^{-1}, \quad b = \tilde{T}_2 \tilde{T}_1^{-1},$$

and the function $\theta(x)$ is defined by formula (2.3.5).

As can be seen from (2.4.14), in the case when $\omega = \omega_0$ and the spectral line involved in the formation of the photon echo signal is broad for both rectangular small-area pump pulses, the signal reaches its maximum at the moment of time $t = t_e$, where t_e is given by formula (2.3.35). The duration of the photon echo signal in this case is on the order of the maximum of the pump pulse durations.

Note that the delay of the maximum of the photon echo signal relative to the moment of time $t = 2\tau + \tilde{T}_1 + \tilde{T}_2 + y/c$ was observed for the first time by Abella, Kurnit, and Hartmann [10] in experiments with ruby. This delay was also observed in experiments [73], which were also performed with ruby. Vasilenko and Rubtsova [74] were the first to observe the delay of the maximum of the photon echo signal relative to the moment of time $t = 2\tau +$ $+ \tilde{T}_1 + \tilde{T}_2 + y/c$ in a gas medium. We emphasize that formula (2.4.14) gives an analytical expression for the delay time $t_d = 0.5\,\tilde{T}_1$.

Since, deriving (2.3.36), (2.3.37), and (2.4.10), we assumed that the durations of the pump pulses are small as compared with the times of irreversible relaxation (2.3.16), the intensity of the photon echo signal produced through $J_b = 1 \rightarrow J_a = 0$ transition with $\omega = \omega_0$ decays in accordance with formula (2.4.9) with the growth in the time interval τ in the case when the spectral line involved in the formation of the photon echo is broad for both pump pulses.

If the functions $\gamma_{ab}^{(1)}(v)$ and $\Delta_{ab}^{(1)}(v)$ are steeper than the functions $f(\mathbf{v})$ and $\Omega_n(v_y)$, i.e., kw is less than $min(ku,\ \theta_2/\widetilde{T}_2)$, then, as it follows from (2.3.38), the influence of depolarizing collisions on the photon echo signal becomes weaker.

As can be seen from the analysis performed above, the dependence of relaxation characteristics on the modulus of the velocity of resonant atoms (molecules) may have a considerable influence on the photon echo signal. Analysis of the photon echo is often performed within the framework of approximation where this dependence is ignored, i.e., the collision integral involved in the right-hand side of equations (1.5.5)–(1.5.7) is replaced by an integral averaged over the modulus of the velocity of resonant atoms (molecules). Comparing the results obtained with such an approach with the results of the analysis performed above, we find that this simplified approach is applicable with certain restrictions on the dependence of relaxation characteristics on the modulus of the velocity of resonant atoms (molecules). Hereinafter, we will assume, except for Sections 4.5 and 4.6, that these requirements are met, and we can employ the collision integral avereged not only in the direction, but also in the modulus of the velocity of resonant atoms (molecules).

Closing this section, we emphasize that polarization properties of the photon echo signal (2.3.36)–(2.3.38) produced through $1 \rightarrow 0$ transition are independent of the areas of the pump pulses. We deal with a similar situation when we consider $0 \rightarrow 1$, $1 \rightarrow 1$, $1/2 \rightarrow 1/2$, and $1/2 \rightleftarrows 3/2$ transitions [13, 14, 20]. For other transitions, polarization properties of the photon echo signal substantially depend on the areas of the pump pulses (e.g., see [54]). Only in the limiting case of small area pump pulses, polarization properties of the photon echo signal produced through an optically allowed transition with arbitrary angular momenta of resonant levels are independent of θ_1 and θ_2. As shown in [67], polarization properties of the photon echo signal in this case are also independent of the shape of the pump pulses. The former of these properties makes it possible to introduce an experimental criterion of the smallness of pump-pulse areas (see Section 4.1 for details). Both of these properties make small-area pump pulses rather promising for polarization echo-spectroscopy measurements (see Chapter 4).

2.5 Specific Features of the Formation of Stimulated Photon Echo Signals in Gases

In this section, we will discuss the formation of stimulated photon echo signals in a gas medium. Similar to Section 2.3, we will consider the case when the angular momenta of the levels involved in an optically allowed resonant transition are equal to $J_a = 0$ and $J_b = 1$. In our analysis, we will employ the collision integral calculated in the model of elastic depolarizing collisions and averaged not only in the direction, but also in the modulus of the velocity of resonant atoms (molecules).

Suppose that the signal of stimulated photon echo is produced under the action of three rectangular linearly polarized pump pulses with a carrier frequency ω, which is resonant to the frequency ω_0 of an atomic (molecular) transition $b \to a$. We assume that these pulses propagate through the gas medium in the same direction (Fig. 2.3). Similar to Section 2.3, we characterize the quantum-mechanical states of resonant atoms (molecules) by their energies, total angular momenta J, and projections m of these momenta on a quantization axis. The strengths of the electric field in pump pulses can be written as

$$\boldsymbol{E}_1 = e^{(1)}\boldsymbol{\ell}_1 g_1(\xi)\exp[i(\omega t - ky + \Phi_1)] + \text{c.c.}, \tag{2.5.1}$$

$$\boldsymbol{E}_2 = e^{(2)}\boldsymbol{\ell}_2 g_2(\xi - \tau_1 - \widetilde{T}_1)\exp[i(\omega t - ky + \Phi_2)] + \text{c.c.}, \tag{2.5.2}$$

$$\boldsymbol{E}_2 = e^{(3)}\boldsymbol{\ell}_3 g_3(\xi - \tau_2 - \tau_1 - \widetilde{T}_2 - \widetilde{T}_1)\exp[i(\omega t - ky + \Phi_3)] + \text{c.c.} \tag{2.5.3}$$

Here, $e^{(n)}$, Φ_n, $\widetilde{T}_n$, and $\boldsymbol{\ell}_n$ ($n = 1, 2, 3$) are the constant amplitude, phase, duration, and polarization vector of the n-th pump pulse, respectively; $k = \omega/c$ and the function g_n describes the waveform of the n-th pump pulse. For the considered case of rectangular pump pulses, the functions g_1, g_2, and g_3 are written as

$$g_1(\xi) = \theta(\xi) - \theta(\xi - \widetilde{T}_1), \tag{2.5.4}$$

$$g_2(\xi - \tau_1 - \widetilde{T}_1) = \theta(\xi - \tau_1 - \widetilde{T}_1) - \theta(\xi - \tau_1 - \widetilde{T}_1 - \widetilde{T}_2), \tag{2.5.5}$$

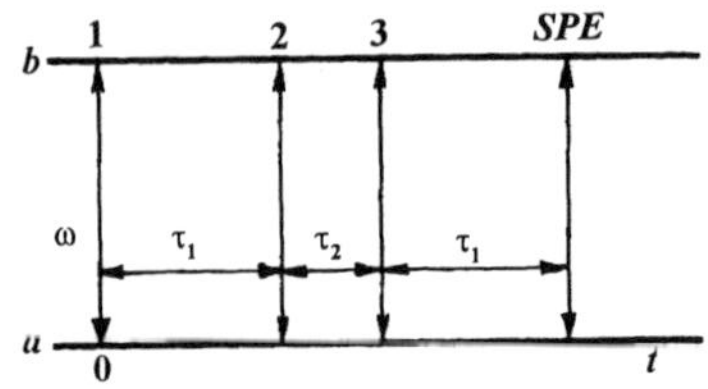

Fig. 2.3. Diagram of the formation of a stimulated photon echo signal: 1–3, pump pulses; SPE, stimulated photon echo signal.

$$g_3(\xi - \tau_2 - \tau_1 - \tilde{T}_2 - \tilde{T}_1) = \theta(\xi - \tau_2 - \tau_1 - \tilde{T}_2 - \tilde{T}_1) - \\ - \theta(\xi - \tau_2 - \tau_1 - \tilde{T}_2 - \tilde{T}_1 - \tilde{T}_3), \tag{2.5.6}$$

where the function $\theta(x)$ is defined by formula (2.3.5), and τ_1 (τ_2) is the time interval between the first and second (second and third) pump pulses.

In what follows, we will assume for definiteness that the Z axis is directed along the polarization vector of the third pump pulse ($\boldsymbol{\ell}_3 = (0, 0, 1)$), while the polarization vectors of the first and second pump pulses make angles ψ_1 and ψ_2 with the polarization vector of the third pulse. We measure these angles clockwise when looking along the Y axis. Therefore, we have $\boldsymbol{\ell}_1 = (\sin\psi_1, 0, \cos\psi_1)$ and $\boldsymbol{\ell}_2 = (\sin\psi_2, 0, \cos\psi_2)$.

The electric field strength in the stimulated photon echo signal produced through the $J_b = 1 \rightarrow J_a = 0$ transition with pump pulses described by (2.5.1)–(2.5.6) is calculated in the same way as it was done in the case of the primary photon echo (see Section 2.3).

We should emphasize, however, that, in contrast to the primary photon echo, the coherence induced by the first two pump pulses in the multipole moments of resonant levels b and a, rather than the coherence induced in the dipole moment of the resonant transition $b \rightarrow a$, plays an important role in the formation of the stimulated photon echo. In the case under study, these multipole moments within the time interval between the second and third pump pulses decay with characteristic times equal to $1/\gamma_a^{(0)}$, $1/\gamma_b^{(0)}$, $1/\gamma_b^{(1)}$, and $1/\gamma_b^{(2)}$, where $1/\gamma_{a,b}^{(0)}$ is the relaxation time of the population in the level a (b), $1/\gamma_b^{(1)}$ is the orientation relaxation time of the level b, and $1/\gamma_b^{(2)}$ is the alignment relaxation time of the level b. Recall that the decay time of the magnetic moment of a level is determined by the orientation relaxation time, while the decay time of the quadrupole moment of a level is determined by the alignment relaxation time.

In the regime of stimulated photon echo, the third pump pulse is employed to transfer the coherence from multipole moments of resonant levels to the moments of the resonant transition. Therefore, some of the terms in the multipole moments related to the resonant transition have a phase factor $\exp(-ikv_y\tau_1)$ by the moment of time when the third pump pulse emerges the gas medium at the point y. Note that, with the increase in the time interval τ_2 between the second and third pump pulses, these terms decay with characteristic times equal to $1/\gamma_a^{(0)}$, $1/\gamma_b^{(0)}$, $1/\gamma_b^{(1)}$, and $1/\gamma_b^{(2)}$.

As the third pump pulse leaves the gas medium at the point y, the dipole moment of the polarization of the medium, which contributes to the signal of stimulated photon echo, oscillates because of Doppler dephasing. Therefore, some part of the dipole moment of the polarization of the medium contributing to the stimulated photon echo signal is proportional to

$$\exp[ikv_y(t - 2\tau_1 - \tau_2 - \widetilde{T}_2 - \widetilde{T}_1 - \widetilde{T}_3 - y/c)].$$

Thus, when a stimulated photon echo signal is produced in a gas medium, emitters at the point y are phase-matched at the moment of time $t \approx t_{se}$, where

$$t_{se} = 2\tau_1 + \tau_2 + \widetilde{T}_2 + \widetilde{T}_1 + \widetilde{T}_3 + y/c. \tag{2.5.7}$$

These phase-matched emitters generate the stimulated photon echo signal, which propagates in the same direction as the pump pulses, has the carrier frequency ω, and contains the information concerning the relaxation times $1/\gamma_a^{(0)}$, $1/\gamma_b^{(0)}$, $1/\gamma_b^{(1)}$, and $1/\gamma_b^{(2)}$ of the resonant levels with $J_a - 0$ and $J_b = 1$. The latter circumstance was highlighted for the first time by the authors of [75].

To find the electric field strength in the stimulated photon echo signal produced by pump pulses described by (2.5.1)–(2.5.6), we should solve the set of equations (2.3.8)–(2.3.11). Performing this procedure with a collision integral averaged not only in the direction, but also in the modulus of the velocity of resonant atoms (molecules), we derive the following expression for the stimulated photon echo signal produced through the $J_b = 1 \to \to J_a = 0$ transition [75, 76]:

$$\boldsymbol{E}^{se} = -\pi \omega L c^{-1} \hbar^{-3} |d|^4 e^{(1)} \widetilde{T}_1 e^{(2)} \widetilde{T}_2 e^{(3)} \widetilde{T}_3 N_0 \boldsymbol{e}^{se} \exp[i(\omega t - ky + \\ + \Phi_3 + \Phi_2 - \Phi_1) - \gamma_{ab}^{(1)} (t - \tau_2 - y/c)] + \text{c.c.} \tag{2.5.8}$$

Here, $\gamma_{ab}^{(1)} = T_2^{-1}$ is the homogeneous half-width of the spectral line of an optically allowed transition $b \to a$; d is the reduced matrix element of the dipole moment operator for this transition; N_0 is determined by formula (2.3.18); and the nonvanishing components of the vector $\boldsymbol{e}^{se}$, which characterizes polarization properties and the waveform of the stimulated photon echo signal, are given by

$$e_x^{se} = (1/9)\{\sin(2\psi_2)\cos(\psi_1 - \psi_2)\exp(-\gamma_b^{(2)} \tau_2) M(p_2, q_3^*) + \\ + \sin(\psi_1 - \psi_2)[\cos(2\psi_2)\exp(-\gamma_b^{(2)} \tau_2) - \exp(-\gamma_b^{(1)} \tau_2)] M(q_2, q_3^*)\}, \tag{2.5.9}$$

$$e_z^{se} = (2/9)\{\cos(\psi_1 - \psi_2)[(\gamma_b^{(0)} - \gamma_a^{(0)} - \gamma_{ab})(\gamma_b^{(0)} - \\ - \gamma_a^{(0)})^{-1} \exp(-\gamma_a^{(0)} \tau_2) + (1/3)(\gamma_b^{(0)} - \gamma_a^{(0)} + 3\gamma_{ab})(\gamma_b^{(0)} - \\ - \gamma_a^{(0)})^{-1} \exp(-\gamma_b^{(0)} \tau_2) + (1/6)(1 + 3\cos(2\psi_2))\exp(-\gamma_b^{(2)} \tau_2)] M(p_2, p_3^*) - \\ - \tfrac{1}{2}\sin(\psi_1 - \psi_2)\sin(2\psi_2)\exp(-\gamma_b^{(2)} \tau_2) M(q_2, p_3^*)\}. \tag{2.5.10}$$

Here, $1/\gamma_{ab}$ is the lifetime of the state b with respect to the spontaneous decay to the state a; the quantities p_n and q_n ($n = 1, 2, 3$) are determined by formulas (2.3.19) and (2.3.31), respectively; and the quantity $M(\chi, \varphi)$ is given by

$$M(\chi, \varphi) = \int dv_y \, \chi \varphi \, p_1 f(v_y) \exp[i(kv_y - \omega + \omega_0 + \Delta_{ab}^{(1)})(t - t_{se})]. \tag{2.5.11}$$

Generally, the stimulated photon echo signal (2.5.8)–(2.5.11) has an elliptical polarization and propagates with a carrier frequency ω in the direction of incidence of the pump pulses. We emphasize that, even for

$J_b = 1 \rightarrow J_a = 0$ transition, polarization properties of the stimulated photon echo signal depend on the areas θ_1, θ_2, and θ_3 of the pump pulses. Such a dependence is observed almost always for various optically allowed resonant transitions. Only in the limiting case of small-area pump pulses, polarization properties of the stimulated photon echo signal become independent of the areas of the pump pulses (see Section 4.8 for the details).

In the elementary case when all the pump pulses are polarized in the same plane ($\psi_1 = \psi_2 = 0$), the stimulated photon echo signal (2.5.8)–(2.5.11) becomes linearly polarized. As it follows from (2.5.9) and (2.5.10), the direction of the polarization vector of this signal in such a situation coincides with the direction of the polarization vectors of the pump pulses. Indeed, using (2.5.9) and (2.5.10) with $\psi_1 = \psi_2 = 0$, we have $e_x^{se} = 0$ and

$$e_z^{se} = (2/9[(\gamma_b^{(0)} - \gamma_a^{(0)} - \gamma_{ab})(\gamma_b^{(0)} - \gamma_a^{(0)})^{-1}\exp(-\gamma_a^{(0)}\tau_2) +$$
$$+ (1/3)(\gamma_b^{(0)} - \gamma_a^{(0)} + 3\gamma_{ab})(\gamma_b^{(0)} - \gamma_a^{(0)})^{-1}\exp(-\gamma_b^{(0)}\tau_2) +$$
$$+ (2/3)\exp(-\gamma_b^{(2)}\tau_2)]M(p_2, p_3^*). \tag{2.5.12}$$

Below, we will analyze for the sake of simplicity expressions (2.5.8) and (2.5.12) describing the electric field strength in the stimulated photon echo signal produced by pump pulses that are linearly polarized in the same plane. First, we will investigate the expression for the quantity $M(p_2, p_3^*)$, which is involved in (2.5.12) and which characterizes in the case under study the waveform of the stimulated photon echo signal. If the spectral line involved in the formation of the photon echo signal (2.5.8) and (2.5.12) is narrow (2.3.24) for all the pump pulses, then expression (2.5.11) yields in the strong-field limit (2.4.4) in the case of an exact resonance ($\omega = \omega_0 + \Delta_{ab}^{(1)}$)

$$M(p_2, p_3^*) = \theta_1^{-1}\theta_2^{-1}\theta_3^{-1}\sin(\theta_1)\sin(\theta_2)\sin(\theta_3)\exp[-\tfrac{1}{4}(T_2^*)^{-2}(t -$$
$$- 2\tau_1 - \tau_2 - y/c)^2]. \tag{2.5.13}$$

Thus, the maximum of the stimulated photon echo signal in the case under consideration reaches the observation point y at the moment of time $t = 2\tau_1 + \tau_2 + y/c$. This signal has a Gaussian shape, and its duration is on the order of T_2^*. The intensity of the stimulated photon echo signal described by (2.5.8), (2.5.12), and (2.5.13), which is proportional to

$\sin^2(\theta_1)\sin^2(\theta_2)\sin^2(\theta_3)$, reaches its maximum, for example, when the areas of the pump pulses are equal to $\theta_1 = \theta_2 = \theta_3 = \pi/2$.

Since spectral lines involved in the formation of the stimulated photon echo signal, similar to the primary photon echo, are always inhomogeneously broadened (2.3.29), we can make the following conclusions from the analysis of (2.5.8), (2.5.12), and (2.5.13). The intensity I_{se} of the stimulated photon echo signal related to a narrow (2.3.24) spectral line corresponding to the $J_b = 1 \rightarrow J_a = 0$ transition decays as

$$I_{se}(\tau_1) \sim \exp(-4\gamma_{ab}^{(1)}\tau_1) \qquad (2.5.14)$$

with the growth in the time interval τ_1 between the first and second pump pulses. Experimental applications of formula (2.5.14) will be discussed in Chapter 4.

Under the same conditions, the intensity of the stimulated photon echo signal as a function of the time interval τ_2 between the second and third pump pulses, which is described by (2.5.8), (2.5.12), and (2.5.13), decays as

$$I_{se}(\tau_2) \sim [(\gamma_b^{(0)} - \gamma_a^{(0)} - \gamma_{ab})(\gamma_b^{(0)} - \gamma_a^{(0)})^{-1}\exp(-\gamma_a^{(0)}\tau_2) +$$
$$+ (1/3)(\gamma_b^{(0)} - \gamma_a^{(0)} + 3\gamma_{ab})(\gamma_b^{(0)} - \gamma_a^{(0)})^{-1}\exp(-\gamma_b^{(0)}\tau_2) +$$
$$+ (2/3)\exp(-\gamma_b^{(2)}\tau_2)]^2. \qquad (2.5.15)$$

Such a behavior of the intensity of the stimulated photon echo signal in a gas even in the elementary case of $J_b = 1 \rightarrow J_a = 0$ transition considerably differs from the behavior of the intensities of stimulated spin echoes and stimulated photon echoes in solids. In the latter cases, we have [1–5, 64]

$$I_{se}(\tau_2) \sim \exp(-2T_1^{-1}\tau_2), \qquad (2.5.16)$$

where T_1 is the time of irreversible longitudinal relaxation.

Obviously, as the pressure of the resonant gas lowers, the times $1/\gamma_b^{(2)}$ and $1/\gamma_b^{(0)}$ may become approximately equal to each other. Under these conditions, formula (2.5.15) yields

$$I_{se}(\tau_2) \sim (\gamma_b^{(0)} - \gamma_a^{(0)})^{-2}[(\gamma_b^{(0)} - \gamma_a^{(0)} - \gamma_{ab})\exp(-\gamma_a^{(0)}\tau_2) + \\ + (\gamma_b^{(0)} - \gamma_a^{(0)} + \gamma_{ab})\exp(-\gamma_b^{(0)}\tau_2)]^2. \qquad (2.5.17)$$

Recall that, as mentioned above, relaxation characteristics $\gamma_b^{(0)}$ and $\gamma_a^{(0)}$ are related to radiative decay and inelastic gas-kinetic collisions. In particular, the quantity $\gamma_b^{(0)}$ involves the relaxation characteristic γ_{ab}. Therefore, even when the multipole moments of the upper resonant level with $\kappa = 2$ and $\kappa = 0$ relax with the same rate, i.e., the quantity $\Gamma_b^{(2)}$ related to elastic depolarizing collisions is small ($\Gamma_b^{(2)}\tau_2 << 1$), the intensity of the stimulated photon echo signal corresponding to $J_b = 1 \rightarrow J_a = 0$ transition is not described by formula (2.5.16), and we cannot introduce the time of irreversible longitudinal relaxation. The time of irreversible longitudinal relaxation ($T_1 = 1/\gamma_a^{(0)} = 1/\gamma_b^{(0)}$) can be introduced only in the case when $\gamma_b^{(0)}$ is close to $\gamma_a^{(0)}$ ($|\gamma_b^{(0)} - \gamma_a^{(0)}|\tau_2 << 1$).

In the case when the spectral line involved in the formation of the stimulated photon echo signal (2.5.8) and (2.5.12) is broad (2.3.25) for all three pump pulses, integration in $M(p_2, p_3{*})$ can be usually performed only numerically. The only exception is the case when all three pump pulses have small areas ($\theta_n << 1$, $n = 1, 2, 3$), the field is weak (2.4.5), and we deal with an exact resonance ($\omega = \omega_0 + \Delta_{ab}^{(1)}$). This circumstance was pointed out for the first time by the authors of [75]. In this case, formula (2.5.11) with $\widetilde{T}_1 = \widetilde{T}_2 = \widetilde{T}_3 = \widetilde{T}$ yields

$$M(p_2, p_3{*}) = \pi^{1/2} T_2{*}\, \widetilde{T}^{-1}\{(a+1)^2[\theta(a+1) - \theta(a)] - \\ - (2a^2 - 2a - 1)[\theta(a) - \theta(a-1)] + (a-2)^2[\theta(a-1) - \theta(a-2)]\}, \qquad (2.5.18)$$

where $a = (t - t_{se})\widetilde{T}^{-1}$ and t_{se} can be determined from (2.5.7) with $\widetilde{T}_1 = \widetilde{T}_2 = \widetilde{T}_3 = \widetilde{T}$. Thus, the duration of the stimulated photon echo signal produced in the case when the relevant spectral line is broad for all the pump pulses is $\sim \widetilde{T}$, and the signal reaches its maximum at the moment of time $t = 2\tau_1 + \tau_2 + 3.5\widetilde{T} + y/c$. Formula (2.5.18) was derived for the first time by Yevseyev *et al.* [75].

Since we derived formulas (2.5.8)–(2.5.11) assuming that the durations of the pump pulses are small as compared with irreversible relaxation times, the decay of the intensity of the photon echo signal with the growth in the time interval τ_2 in the case when the spectral line corresponding to the

$J_b = 1 \rightarrow J_a = 0$ transition involved in the formation of the stimulated photon echo is broad for all three pump pulses is described by general relationship (2.5.15) rather than by formula (2.5.16) even when $\psi_1 = \psi_2 = 0$.

Thus, a simple example of the generation of the stimulated photon echo signal through the $J_b = 1 \rightarrow J_a = 0$ transition shows that the decay of the intensity of the photon echo signal with the growth in the time interval τ_2 between the second and third pump pulses usually deviates from that predicted by formula (2.5.16), which describes the characteristic behavior of stimulated spin echo [1–4]. To extract the spectroscopic information with the use of the stimulated photon echo, one has to apply a theory that would differ from the theory of echo spectroscopy based on the spin echo. Such a theory is called the theory of polarization echo spectroscopy and is presented in Chapter 4.

REFERENCES

1. N.M. Pomerantsev, *Usp. Fiz. Nauk* **65**: 87–110 (1958).
2. A. Löshe, *Kerninduktion* (Berlin: Wissenschaften, 1957) (in German).
3. A. Abraham, *The Principles of Nuclear Magnetism* (London: Oxford Univ. Press, 1961).
4. K.M. Salikhov, A.G. Semenov, Yu.D. Tsvetkov, *Elektronnoe spinovoe ekho i ego primenenie* (*Electron Spin Echo and Its Applications*) (Novosibirsk: Nauka, 1976) (in Russian).
5. S. Fernbach, W.G. Proctor, *J. Appl. Phys.*: **26**: 170–181 (1955).
6. U.Kh. Kopvillem, V.R. Nagibarov, *Fiz. Metallov Metallovedenie* **15**: 313–315 (1963).
7. N.A. Kurnit, I.D. Abella, S.R. Hartmann, *Phys. Rev. Lett.* **13**: 567–568 (1964).
8. N.A. Kurnit, S.R. Hartmann, *Bull. Am. Phys. Soc.* **11**: 112 (1966).
9. C.K.N. Patel, R.E. Slusher, *Phys. Rev. Lett.* **20**: 1087–1089 (1968).
10. I.D. Abella, N.A. Kurnit, S.R. Hartmann, *Phys. Rev.* **141**: 391–406 (1966).
11. M. Scully, M.J. Stephen, D.C. Burnham, *Phys. Rev.* **171**: 213–214 (1968).
12. V.V. Samartsev, *Ukr. Fiz. Zh.* **14**: 1045–1046 (1969).
13. A.I. Alekseyev, I.V. Yevseyev, *Zh. Eksp. Teor. Fiz.* **56**: 2118–2128 (1969).
14. J.P. Gordon, C.H. Wang, C.K.N. Patel, *et al.*, *Phys. Rev.* **179**: 294–309 (1969).
15. V.R. Nagibarov, V.V. Samartsev, *Opt. Spektrosk.* **27**: 467–473 (1969).
16. A. Dienes, *IEEE J. Quantum Electron.* **5**: 246–249 (1969).
17. A.I. Alekseyev, *Pis'ma Zh. Eksp. Teor. Fiz.* **9**: 472–475 (1969).
18. V.V. Samartsev, *Opt. Spektrosk.* **28**: 178–181 (1970).
19. V.V. Samartsev, *Ukr. Fiz. Zh.* **15**: 160–161 (1970).
20. A.I. Alekseyev, I.V. Yevseyev, *Zh. Eksp. Teor. Fiz.* **57**: 1735–1744 (1969).
21. A.I. Alekseyev, I.V. Yevseyev, *Pis'ma Zh. Eksp. Teor. Fiz.* **10**: 105–108 (1969).
22. C.V. Heer, R.H. Kohl, *Phys. Rev.* **A 1**: 693–709 (1970); *Phys. Rev.* **A 2**: 549 (1970).
23. V.R. Nagibarov, V.V. Samartsev, *Prib. Tekh. Eksp.* No. 3: 189–190 (1970).
24. A.I. Alekseyev, I.V. Yevseyev, *Phys. Lett.* **A 30**: 242–243 (1969).
25. C.H. Wang, *Phys. Rev.* **B 1**: 156–163 (1970).

26. I.V. Yevseyev, *Polyarizatsionnye effekty foton-ekha v gazovoi srede* (*Polarization Effects in the Photon Echo in a Gas Medium*) (Moscow: Candidate of Phys. and Math. Science Dissertation, 1970) (in Russian).
27. V.V. Samartsev, *Voprosy teorii svetovykh i fononnykh induktsii i ekho* (*Problems of Theory of Optical and Photon Induction and Echo*) (Kazan, Candidate of Phys. and Math. Science Dissertation, 1970) (in Russian).
28. A.I. Alekseyev, *Phys. Lett.* **A 31**: 495–496 (1970).
29. V.V. Samartsev, A.G. Shagidullin, *Opt. Spektrosk.* **30**: 171–174 (1971).
30. A.I. Alekseyev, E.A. Manykin, *Phys. Lett.* **A 35**: 87–88 (1971).
31. I.A. Nagibarova, V.V. Samartsev, *Phys. Lett.* **A 35**: 231–232 (1971).
32. A.I. Alekseyev, I.V. Yevseyev, *Opt. Spektrosk.* **36**: 1187–1192 (1974).
33. A.I. Alekseyev, I.V. Yevseyev, *Phys. Lett.* **A 43**: 465–466 (1973).
34. I.V. Yevseyev, *Phys. Lett.* **A 44**: 163–164 (1973).
35. A.I. Alekseyev, I.V. Yevseyev, *Izv. Akad. Nauk SSSR, Ser. Fiz.* **37**: 2046–2055 (1973).
36. A.I. Alekseyev, I.V. Yevseyev, *Izv. Akad. Nauk SSSR, Ser. Fiz.* **37**: 2134–2137 (1973).
37. L.A. Nefediev, V.V. Samartsev, *Opt. Spektrosk.* **37**: 1171–1174 (1974).
38. V.R. Nagibarov, V.V. Samartsev, L.A. Nefediev, *Phys. Lett.* **A 43**: 195–196 (1973).
39. A.I. Sirasiev, V.V. Samartsev, *Opt. Spektrosk.* **39**: 730–734 (1975).
40. L.A. Nefediev, V.V. Samartsev, A.I. Sirasiev, *Spectrosc. Lett.* **7**: 285–292 (1974).
41. A.I. Alekseyev, I.V. Yevseyev, *Zh. Eksp. Teor. Fiz.* **68**: 456–464 (1975).
42. C.V. Heer, R.J. Nordstrom, *Phys. Rev.* **A 11**: 536–548 (1975).
43. L.A. Nefediev, V.V. Samartsev, *Ukr. Fiz. Zh.* **20**: 1567–1568 (1975).
44. I.V. Yevseyev, V.M. Yermachenko, *Phys. Lett.* **A 60**: 187–189 (1977).
45. A.I. Alekseyev, I.V. Yevseyev, V.M. Yermachenko, *Zh. Eksp. Teor. Fiz.* **73**: 470–480 (1977).
46. T. Baer, I.D. Abella, *Phys. Rev.* **A 16**: 2093–2100 (1977).
47. V.K. Matskevich, I.V. Yevseyev, V.M. Yermachenko, *Opt. Spektrosk.* **45**: 17–22 (1978).
48. L.A. Nefediev, V.V. Samartsev, *Opt. Spektrosk.* **46**: 537–542 (1979).
49. T. Baer, *Phys. Rev.* **A 18**: 2570–2579 (1978).
50. D.S. Bakaev, I.V. Yevseyev, V.M. Yermachenko, *Zh. Eksp. Teor. Fiz.* **76**: 1212–1225 (1979).
51. I.V. Yevseyev, V.M. Yermachenko, *Pis'ma Zh. Eksp. Teor. Fiz.* **28**: 689–692 (1978).
52. L.A. Nefediev, V.V. Samartsev, *Opt. Spektrosk.* **47**: 571–574 (1979).
53. I.V. Yevseyev, V.M. Yermachenko, *Zh. Eksp. Teor. Fiz.* **76**: 1538–1546 (1979).
54. I.V. Yevseyev, V.M. Yermachenko, *Polarization Echo Spectroscopy*, in Interaction of Electromagnetic Field with Condensed Matter (Singapore: World Sci., 1990), Vol. 7, pp. 162–209.
55. S.S. Alimpiev, N.V. Karlov, *Zh. Eksp. Teor. Fiz.* **63**: 482–490 (1972).
56. I.V. Yevseyev, V.M. Yermachenko, *Opt. Spektrosk.* **47**: 1139–1144 (1979).
57. I.V. Yevseyev, V.A. Reshetov, *Opt. Spektrosk.* **57**: 869–874 (1984).
58. I.V. Yevseyev, V.A. Reshetov, *Opt. Acta* **29**: 119–130 (1982).
59. A.I. Alekseyev, A.M. Basharov, *Opt. Spektrosk.* **53**: 631–636 (1982).
60. A.V. Yevseyev, I.V. Yevseyev, V.M. Yermachenko, *Opt. Spektrosk.* **50**: 77–84 (1981).

61. J.R. Meckley, C.V. Heer, *Phys. Lett.* **A 46**: 41–42 (1973).
62. R.J. Nordstrom, W.M. Gutman, C.V. Heer, *Phys. Lett.* **A 50**: 25–26 (1974).
63. W.M. Gutman, C.V. Heer, *Phys. Lett.* **A 51**: 437–438 (1975).
64. E.A. Manykin, V.V. Samartsev, *Opticheskaya ekho-spektroskopiya* (*Optical Echo Spectroscopy*) (Moscow: Nauka, 1984) (in Russian).
65. I.V. Yevseyev, S.V. Ivliev, *Proc. VII All-Union Symp. High- and Ultrahigh-Resolution Molec. Spectrosc.* (Tomsk: Sib. Div. USSR Acad. Sci., 1986) (in Russian), Part 2, pp. 222–226.
66. L.S. Vasilenko, N.N. Rubtsova, V.P. Chebotayev, *Pis'ma Zh. Eksp. Teor. Fiz.* **38**: 391–393 (1983).
67. I.V. Yevseyev, V.A. Reshetov, *Opt. Spektrosk.* **53**: 796–799 (1982).
68. T. Kohmoto, M. Nido, Y. Fukida, M. Matsuoka, *Rev. Laser Eng.* **10**: 378–384 (1982).
69. M.I. D'yakonov, *Zh. Eksp. Teor. Fiz.* **47**: 2213–2221 (1964).
70. M.I. D'yakonov, V.I. Perel', *Opt. Spektrosk.* **20**: 472–480 (1966).
71. C.H. Wang, W.J. Tomlinson, R.T. George, *Phys. Rev.* **181**: 125–136 (1969).
72. S.O. Elyutin, S.M. Zakharov, E.A. Manykin, *Zh. Eksp. Teor. Fiz.* **76**: 835–845 (1979).
73. V.V. Samartsev, R.G. Usmanov, G.M. Ershov, B.Sh. Khamidullin, *Zh. Eksp. Teor. Fiz.* **74**: 1979–1987 (1978).
74. L.S. Vasilenko, N.N. Rubtsova, *Investigation of Relaxation Processes in a Gas with the Use of Coherent Transient Processes*, in Lazernye sistemy (Laser Systems) (Novosibirsk: Sib. Div. USSR Acad. Sci., 1982) (in Russian), pp. 143–154.
75. I.V. Yevseyev, V.M. Yermachenko, V.A. Reshetov, *Zh. Eksp. Teor. Fiz.* **78**: 2213–2221 (1980).
76. I.V. Yevseyev, V.N. Tsikunov, *Opt. Spektrosk.* **59**: 1372–1373 (1985).

Chapter 3

EXPERIMENTAL APPARATUS AND TECHNIQUE FOR OPTICAL COHERENT SPECTROSCOPY OF GASES

This chapter is devoted to the description of the main methods employed in the investigation of optical transient processes in gas media and the apparatus used in these studies. Recently, a new direction of laser spectroscopy has appeared. This direction is based on several methods of extracting the information concerning a resonant medium, including: *a*) analysis of decay curves for the intensities of optical coherent responses within certain time intervals; *b*) analysis of temperature lowering, i.e., the dependencies of the intensities of coherent responses on the temperature of a gas medium under study; *c*) investigation of the modulated structures of optical coherent responses related to the hyperfine coupling of unpaired electrons with nuclei of molecules or atoms in the gas (modulation coherent spectroscopy), including the regime of double resonances; *d*) polarization coherent spectroscopy associated with the identification of the type of the energy transition being studied and the determination of characteristics of this transition from the polarization properties of optical coherent responses; *e*) the method of optical "coherent averaging," which is closely related to a multipulse excitation of gas particles and the regime of double resonances; and *f*) the method of angular spectroscopy, which is based on the spatial features of the generation of optical coherent responses and superradiance.

By now, these methods have been used in a large number of experiments [1–203] and have proved to be reliable. These methods can be modified, depending on the character of coherent excitation of a gas medium, which, in its turn, can be implemented in several ways.

3.1 The Methods of Excitation of Optical Coherent Responses in Gas Media

3.1.1 The Pulsed Method

The pulsed method (Fig. 3.1*a*) was employed for the first time by Kurnit *et al.* [204]. This method implies that a short laser pulse whose carrier frequency is equal to the frequency of transition between a pair of energy levels in a medium under study is divided by a beam splitter (BS) into two

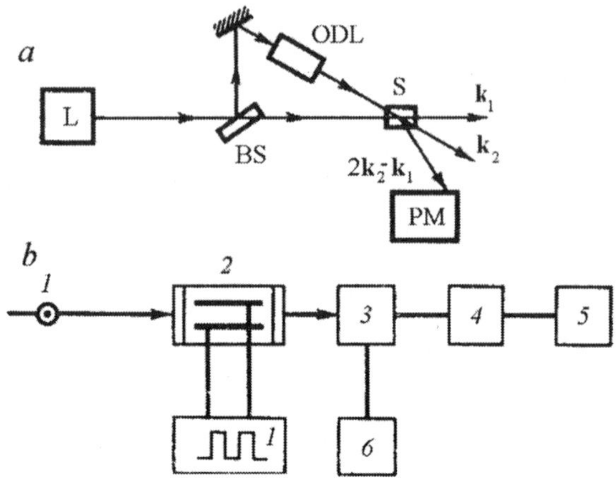

Fig. 3.1. Methods of excitation of photon echo signals. *a*) Pulsed excitation: L – laser; ODL – optical delay line; BS – beam splitter; PM – photomultiplier. *b*) Stark switching: 1– continuous-wave laser; 2 – Stark gas cell; 3 – photodetector; 4 – integrator; 5 – recorder; 6 – oscilloscope; 7 – pulse generator.

beams. One of these beams irradiates a resonant medium, propagating in the direction $\boldsymbol{k}_1$, while the other beam is preliminarily delayed in an optical delay line (ODL) and irradiates the same medium in the direction $\boldsymbol{k}_2$ within the time interval τ. An echo signal is emitted by the resonant medium in the direction $\boldsymbol{k}_e = 2\boldsymbol{k}_2 - \boldsymbol{k}_1$ and is detected by an ELU-FT fast-response photodetector.

To illustrate the pulsed method of photon-echo excitation in gas media, we will briefly describe the photon-echo setup employed in [14]. An SF_6 gas in a cell with a length of 3 m at a pressure of 10^{-3}–10^{-4} mm Hg was used as a resonant medium. The diagram of the setup is shown in Fig. 3.2. This setup includes two pulsed CO_2-lasers with a common rotating mirror *15* (a plane-parallel glass plate covered with gold from both sides). This mirror was used to synchronize the emission pulses of the lasers. The time interval τ between the pump pulses was smoothly varied within the range of 0–100 μs by the rotation of diffraction gratings. This time interval depends on the aperture of the diffraction grating and the rotation speed of mirror *15*. Rotation of diffraction gratings *8* and *16* around the direction of their grooves made it possible to spectrally tune lasers *5* and *23*. The beam

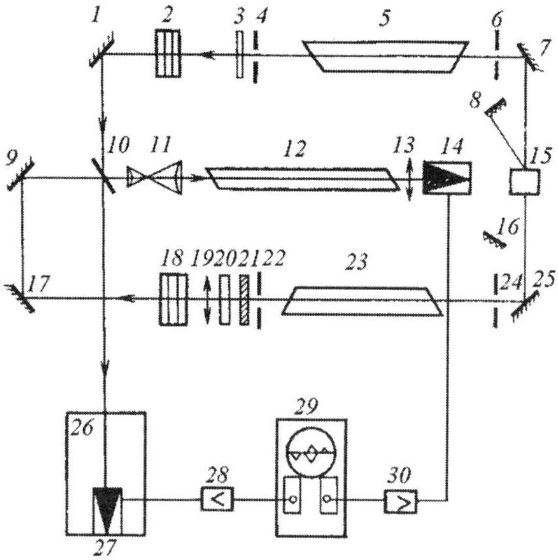

Fig. 3.2. Diagram of an infrared photon-echo setup [14]: (*1*, *7*, *9*, *17*, *25*) – rotating mirrors; (*2*, *18*) – attenuators; (*3*, *21*) – output couplers of laser cavities; (*4*, *6*, *22*, *24*) – intracavity diaphragms; (*5*, *23*) – active media of pump lasers; (*8*, *16*) – diffraction gratings of laser cavities; (*10*) – beam splitter; (*11*) – telescope consisting of two NaCl lenses; (*12*) – cell with a gas under study; (*13*, *19*) – polarizers; (*14*, *27*) – photodetectors; (*15*) – spinning mirror; (*20*) – quarter-wave phase plate; (*26*) – monochromator; (*28*, *30*) – matching units; (*29*) – S1-42 oscilloscope.

splitter *10* directs attenuated output laser pulses into the cell with a gas under study (*12*). Echo signals and attenuated pump pulses were detected with a photodetector (*14*). The frequency and the intensity of these signals were measured with a detector (*27*) placed at the output of a monochromator (*26*). Polarization properties of the photon echo were investigated with the use of a quarter-wave plate (*20*), which was placed at the output of one of the lasers and which transformed linear polarization of laser radiation into circular polarization. Rotation of the polarizer *19* made it possible to rotate the polarization in one of the laser beams with respect to the polarization in the other beam. This setup was employed for the investigation of the photon (optical) echo in SF_6 and BCl_3 gases [14].

3.1.2 The Method of Stark Switching

An important disadvantage of the pulsed technique described above is associated with the fact that the shortening of pump pulses leads to the increase in the spectral width of these pulses, which complicates the solution of different problems in high- and ultrahigh-resolution optical echo spectroscopy. Therefore, the idea of performing echo experiments with cw-laser radiation with a small spectral width (on the order of tens of megahertz or even a few megahertz) seems rather natural. To ensure the excitation of optical echo signals within such a narrow spectral band, all the operations associated with the implementation of pulsed excitation should be performed with an object under study (atoms or molecules in a gas) rather than with laser radiation. This regime of excitation was implemented for the first time in 1971 by Brewer and Shoemaker [4], who pumped an NH_2D gas using radiation of a cw CO_2-laser with a wavelength of 944.1948 cm^{-1}. The wavelength of laser radiation remained unchanged during the entire echo experiment, but the applied electric field gave rise to a shift and splitting of resonant levels due to the Stark effect. For this purpose, a gas-filled Stark cell (see Fig. 3.1*b*) was used in these experiments. A dc electric field with a magnitude up to 3660 V/cm was applied to the plates of the Stark cell within short time intervals. Recall that the inhomogeneous width of spectral lines in gases is determined by the Doppler effect and may reach 10^9 s^{-1}. In principle, there is no need in increasing the inhomogeneous width. One subsystem of gas species is excited during Stark pulses, and another subsystem is excited in the absence of these pulses. Each of these subsystems emits optical coherent responses at its own frequencies. The selection of responses related to Stark pulses does not encounter any technical difficulties (this problem can be solved either by means of optical heterodyning, since the waveforms of optical responses are modulated due to the interference of pump radiation and the response on a photodetector, or by using a frequency filter).

We should note that this technique also has some drawbacks. One of these drawbacks is associated with the fact that this method can be applied only to molecules possessing a permanent electric moment. The question is whether it is possible to modify the method of Stark switching in such a way as to allow this technique to be used for the investigation of particles without a permanent electric moment. Such a modification was done by Brewer *et al.* [38], who switched the laser frequency instead of switching the resonant frequency. The rate of switching in this scheme remained unchanged. For this purpose, an electrooptical crystal was placed inside the

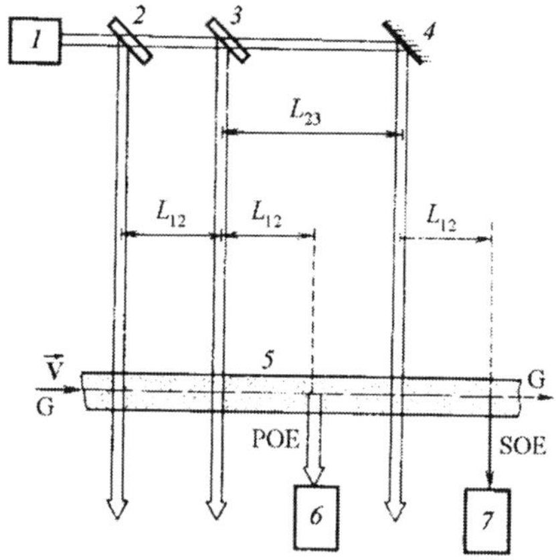

Fig. 3.3. The kinetic method of photon-echo excitation [205]: (*1*) – continuous-wave laser; (*2*, *3*) – beam splitters; (*4*) – rotating totally reflecting mirror; (*5*) – cell with a gas flow G moving with a velocity **V**; (*6*, *7*) – photomultipliers.

laser cavity, and an electric field was applied to this crystal during controlled time intervals. This field changed the refractive index of the crystal, which resulted in a variation in the optical path of light in the cavity and, as a consequence, in the change in the laser frequency. We should emphasize that the technique of intracavity frequency modulation has received broad applications in optical communication systems, where the possibility of changing the laser frequency with a high rate was experimentally demonstrated. Note also that the frequency change in this regime is on the order of 0.03–0.05 mHz per 1 V of applied voltage.

3.1.3 The Kinetic Method

The kinetic method was proposed by Nagibarov and Samartsev [205] for the investigation of gas flows (Fig. 3.3). Similar to the method of Stark switching discussed above, excitation was performed with the use of a cw laser, while a pulsed character of excitation was achieved due to the fact that a gas flow propagated in a direction close to the normal to the laser beams. Obviously, the carrier frequency of laser radiation should be resonant to the frequency of a certain energy transition of atoms or molecules in a gas. Such a double or triple laser irradiation of a gas flow within the distances L_{12} and L_{23} gives rise to the generation of primary (POE) and stimulated (SOE) optical echo signals at the points corresponding to $2L_{12}$ and $L_{23} + 2L_{12}$, respectively, provided that the following inequality is satisfied:

$$D/V \leq T_1, T_2, \quad (3.1.1)$$

where D is the diameter of the laser beam, V is the velocity of the gas flow, and T_1 and T_2 are the times of longitudinal and transverse irreversible relaxation, respectively. It can be easily shown that, with $D = 10^{-2} \div 10^{-1}$ cm and $V = 10^5 \div 10^6$ cm/s, this inequality can be readily satisfied.

We should also mention here yet another method of excitation, which was proposed by Nagibarova and Samartsev [206] and which was called the collisional method. As compared with the kinetic method, laser beams in this case are replaced by flows of excited particles. When nonexcited and excited flows (with equal splitting values of the working levels) collide, a coherent migration of energy occurs, e.g., due to the nonsecular part of the dipole-dipole interaction operator. Double or triple collisions of such gas flows give rise to the emission of primary and stimulated optical echo signals by particles in the initially nonexcited flow. A similar method of excitation was experimentally implemented by Zewail *et al.* [207], who detected changes in the incoherent spontaneous background rather than the coherent response itself.

3.1.4 The Method of Studying Coherent Radiation in Time-Separated Fields

The method of studying coherent radiation in time-separated fields was developed and experimentally implemented in [40–42]. The diagram of the experimental setup described in [208] is presented in Fig. 3.4.

Radiation of a cw CO_2-laser propagated through an ML-8 electrooptical modulator located in the waist of a collimator consisting of mirrors M_1 and M_2 with equal curvature radii (2 m). Next, laser radiation passed through a nonlinear filter implemented as a cell filled with an SF_6 gas at a pressure of about 10^{-1} Torr. A collimator expanding the laser beam by a factor of three (up to a diameter of 2 cm) was placed behind the filtering cell. Finally, collimated laser radiation passed through a cell with the studied SF_6 gas (the diameter of the cell was 4 cm, and the length of the cell was 2 m). An important feature of the method of studying coherent radiation in time-separated fields is associated with the fact that this approach employs standing waves. Advantages and specific features of using standing waves for the excitation of photon (optical) echo signals were analyzed by Nagibarov and Samartsev [209]. One of the main advantages of this technique is associated with low powers of laser beams. In the setup described above, the standing wave was produced in the working cell through the retroreflection of laser radiation from mirrors M_5 at a small angle $\theta \sim 10^{-3}$ rad. Note that the angular separation of the forward and backward waves in these experiments made it possible to select the backward wave and to direct this wave to a short-focal-length lens L placed in front of the photodetector D with the use of mirrors M_6–M_8. A Hg–Cd–Te photoresistor was employed as a photodetector (the time constant of the detector with a preamplifier was equal to 0.1 μs). The signals produced by the photodetector were fed to the input of an oscilloscope (*OSC*), which was triggered by an oscillator *1*, controlling the generation of the first pump pulse. The second pulse is controlled by an oscillator *2*, which is triggered by the oscillator *1* with a delay time τ. The signals produced by both master oscillators are fed to a high-voltage amplifier *4*, which is connected to the ML-8 modulator. The detection of coherent responses by means of a plotter was performed with the use of an "electronic gate" scheme, which made it possible to select some section of a signal within any time interval determined by a gate pulse produced by the oscillator. This scheme operated with a repetition rate from 40 Hz up to 5 kHz. The section of the signal selected by this scheme was integrated and was fed to the plotter. This scheme provided an opportunity, for example, to separate the n-th response of a medium at the moment of time $t = \tau + n\tau$. The reference signal from the oscillator *3* delayed in time by $\tau + n\tau$ with respect to the pulse produced by the oscillator *1* was applied to the electronic gate scheme *5*. The output signal from the scheme *5* was inputted into a two-coordinate automatic plotter, whose X input was loaded with a signal proportional to the detuning of the laser carrier frequency from the center of the resonant line corresponding to the transition in the gas under study (SF_6). The laser frequency was scanned due to changes in the current through magnetostriction coils *MS* wound around invar rods fixing the laser fittings. The current in the coils was controlled by a saw-tooth voltage

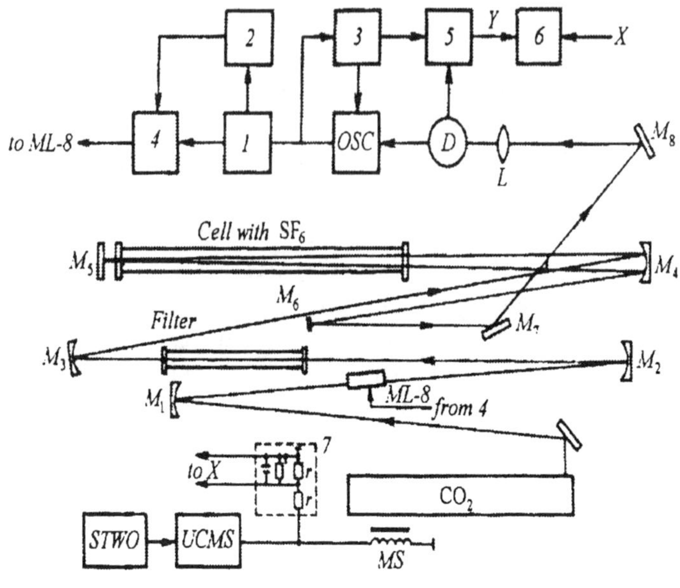

Fig. 3.4. Diagram of the setup for the investigation of coherent radiation signals in time-separated fields [208] (see text for details).

oscillator (*STWO*), which was connected to the unit controlling magnetostriction (*UCMS*). The voltage applied to the *X* input of the plotter was produced by a scheme *7*, which consisted of a divider R_1, R_2 (connected in parallel with magnetostriction coils *MS*) and an offset scheme, which made it possible to vary the potential at *X* within a broad range. Before recording the time-resolved signal of coherent radiation as a function of the frequency detuning parameter, the laser was tuned to a frequency around the center of the relevant spectral line with the use of the UCMS. The tuning procedure also employed the maximum of the first response observed on the screen of the oscilloscope. Next, the pen of the plotter was switched, and the oscillator of saw-tooth voltage was turned on. The amplitude of the saw-tooth voltage was chosen in such a way that the linear section of frequency tuning coincided in its length with the section where the frequency dependence of the response was observed. The typical time required to record a single signal was equal to 10 s. The laser under these conditions was tuned to a frequency around 10 MHz.

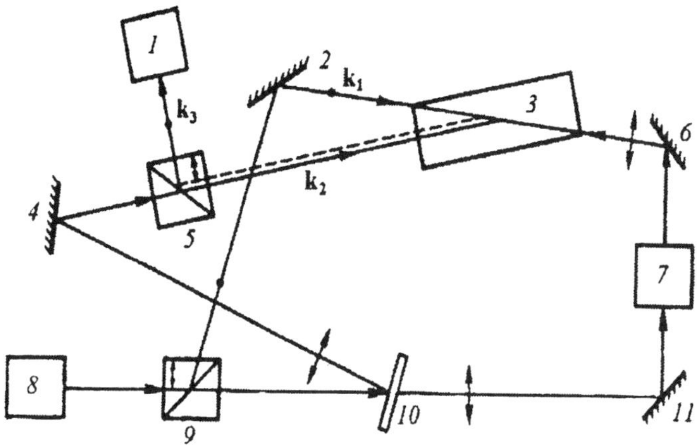

Fig. 3.5. Diagram of the setup for the generation of backward photon echo signals [62]: (*1*) – photomultiplier; (*2*, *4*, *6*, *11*) – mirrors; (*3*) – cell with sodium vapor; (*5*, *9*) – polarization Glan prisms where radiation is coupled out through the side surface; (*7*) – optical delay line; (*8*) – dye laser; (*10*) – beam splitter.

3.1.5 Excitation of Backward Optical Coherent Responses

The method of excitation of backward optical coherent responses was implemented for the first time in experiments [62, 80, 95]. The diagram of the setup employed in these studies is presented in Fig. 3.5.

This method is based on the specific features of the spatial phase matching of the photon echo in gas media. As demonstrated in [62], the spatial phase-matching condition for the stimulated photon echo is written as

$$\boldsymbol{k}_s = -\boldsymbol{k}_2 + \boldsymbol{k}_1 + \boldsymbol{k}_3, \qquad (3.1.2)$$

where $\boldsymbol{k}_n$ is the wave vector of the n-th pump pulse, (n = 1, 2, 3) and $\boldsymbol{k}_s$ is the wave vector of the stimulated photon echo. Note that with $\boldsymbol{k}_1 = -\boldsymbol{k}_3$, we have

$$\boldsymbol{k}_s = -\boldsymbol{k}_2. \qquad (3.1.3)$$

Backward optical responses are excited when condition (3.1.3) is satisfied.

Let us consider the experimental setup in greater detail. The frequency-tunable laser *8* generated radiation pulses with a duration of 2 ns, a power of 1 kW, and the spectral bandwidth of 0.5 cm^{-1} at the wavelength of 5896 Å. A Glan prism *9* split the laser beam into two linearly polarized beams whose polarization vectors were orthogonal to each other. These beams irradiated the cell *3* filled with a gas under study (sodium vapor) at an angle of 20 mrad. The length of the cell was equal to 30 cm, and the area where the gas was heated in the cell had a length of 3 cm. A buffer gas (argon) was placed between the cell windows and sodium vapor to ensure the transparency of cell windows. A part of the laser pulse coming out through the side face of a prism *9* irradiated the cell *3* in the direction of the wave vector $\boldsymbol{k}_1$. The second part of the laser pulse, transmitted through the prism *9* in the forward direction, is directed to a beam splitter *10*. The fraction of this beam reflected from the beam splitter passes through the Glan prism *5* and irradiates the cell *3* in the direction of $\boldsymbol{k}_2$.

The fraction of the laser beam transmitted through the beam splitter in the forward direction passes an optical delay line *7* and enters the cell with the gas under study in the direction of $\boldsymbol{k}_3$. The directions of polarization vectors of the pump pulses and the echo signal are shown with an arrow or a dot in Fig. 3.5. The stimulated photon echo signal is emitted by a sodium vapor in the direction opposite of the direction of the second pulse. One of the advantages of this scheme is associated with a fact that the weak (as compared with pump pulses) echo signal is separated in space from the pump pulses, which prevents a photomultiplier from being saturated with high-power pulses. In addition, this method is very convenient for the investigation of the wave front of coherent responses, including experiments on echo holography.

An original modification of this scheme was implemented in experiments [108] with a vapor of molecular iodine at the wavelength of 18790 cm^{-1} at a temperature of 20°C. The specific feature of these experiments is that this scheme employs the spike structure of the first two pump pulses. The third, single pulse propagated in the direction opposite of the first pulse. Under these conditions, iodine molecules emitted a decaying series of stimulated photon echo signals in the direction opposite of the second pulse. Thus, such an approach makes it possible to promptly measure the decay curve of the stimulated echo. We should note that the use of the spike structure of laser pulses in echo experiments offers several advantages, which were pointed out for the first time by Nagibarov and Samartsev [210]. One of these advantages is a small duration of each spike, which

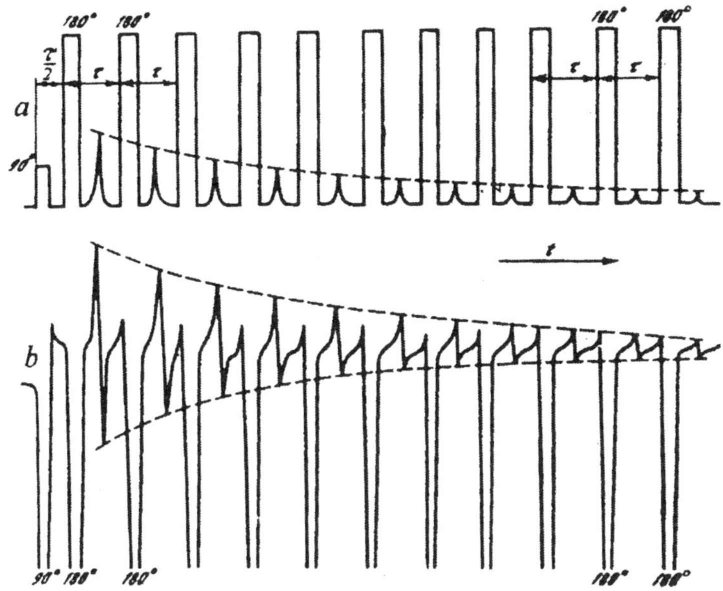

Fig. 3.6. The Carr–Parcell method of excitation of a sequence of photon echo signals [19]: (*a*) the sequence of pump pulses; (*b*) oscilloscope trace of the implementation of the Carr–Parcell technique in a gas.

makes it possible to investigate gases with a high density. In the above-described experiments [108], the spike structure of YAG-laser radiation was used. The duration of each spike was 140 ps, and the spike power was 100 W. The homogeneous line width determined from the measured decay curve was equal to 21 ns^{-1}.

3.1.6 The Carr–Parcell Method

In optics, the Carr–Parcell method was experimentally implemented for the first time by Schmidt *et al.* [19]. The sequence of pump pulses is shown in Fig. 3.6*a*. First, the gas under study (CH_3F) is irradiated with a 90-degree pulse. Then a series of 180-degree pulses comes, where the time interval between the first and second pulses is two times less than the time interval between the 180-degree pulses. With such an excitation, a resonant gas medium emits a sequence of photon echo signals within time intervals between

180-degree pulses. As can be seen from the oscilloscope trace presented in Fig. 3.6*b*, this sequence of signals decays with a characteristic time T_2. Thus, this method of excitation provides an opportunity to simultaneously measure the decay curve and to find the relaxation time T_2.

After the general overview of the methods employed for coherent resonant excitation of gas media performed in this section, it seems appropriate to consider some specific example of a photon-echo relaxometer. Such a consideration will provide a deeper insight into the experience of overcoming technical difficulties in photon-echo experiments. Therefore, the next section will be devoted to the description of an optical echo relaxometer, allowing the investigation of spectral and relaxation parameters of molecular iodine.

3.2 Optical Echo Relaxometer of Gas Media with Remote-Controlled Tuning

In this section, we will describe an optical echo relaxometer with smooth remote-controlled tuning of the frequency of pumping laser radiation created by Popov and Bikbov [188]. It seems appropriate to start with specifying the main parameters of this device: (1) the duration of pump laser pulses is 10 to 15 ns, (2) the time interval between the pump laser pulses ranges from 0 to 100 ns, (3) the spectral bandwidth of pumping laser radiation is 0.1 Å, (4) the frequency range covered by the echo relaxometer stretches from 400 to 800 nm, and (5) the wavelength can be smoothly tuned within the range from 500 to 600 nm.

Such an echo relaxometer was intensively employed for the investigation of spectral and relaxation parameters of a molecular iodine vapor [192, 194, 196, 199–201, 203]. This device is designed in a modern way and currently includes software for the processing of the experimental data. The frequency of pumping radiation produced by this device can be smoothly tuned, and the system under study allows a remote control of this process. The main element of the considered echo relaxometer is an optical oscillator generating double laser pulses. This oscillator consists of a frequency-tunable dye laser and a modified LTIPCh-5 laser, which is employed for the pumping of the dye laser. Coaxial lines are employed to introduce delay times between the control electric pulses. Two laser pulses separated in time are produced by a dye laser pumped by two time-separated pump pulses generated by the LTIPCh-5 laser. The time interval between the pump pulses produced by LTIPCh-5 is determined by the length of the coaxial delay line. Spectral characteristics of both pump pulses can be changed by tuning the dye-laser cavity, which can be done from a control panel through an electric motor and a reducer, rotating the diffraction grating in the dye-laser cavity. This approach ensures the identity of polarization and spectral characteristics, as well as the directions of wave vectors of pump pulses within the entire range of frequency tuning.

The considered echo relaxometer has the following advantages over other experimental systems employed for the observation of the (optical) photon echo:

tunable control of the time interval between nanosecond laser pulses without changes in the optical scheme due to the use of coaxial and biaxial delay lines;
an original method of the formation of photon echo signals in parallel laser beams, which makes it possible to investigate extended gas media due to the coincidence of the wave vectors in the pump laser pulses and the echo signal under conditions when both pump pulses are generated by the same cavity;
remote-controlled smooth frequency tuning of pumping laser radiation, which provides an opportunity of studying the hyperfine structure in the spectra of samples in a semiautomatic mode;
the control of the intensities of pump laser pulses, which allows the energy dependences of responses of a gas medium to be investigated;
the use of an electrooptical switch (EOS) to prevent the saturation of a photoelectric multiplier by high-power nanosecond pump pulses, coming with small time intervals, due to the formation of an echo signal in parallel laser beams.

The diagram of the echo relaxometer is presented in Fig. 3.7. This system includes an oscillator generating doubles laser pulses. Radiation produced by this oscillator is coupled out of the cavity through the mirror M_1 and is focused by a spherical lens L_1 into a sample under study, where a photon echo signal is generated. The pump pulse transmitted through the sample and the echo signal are focused by a spherical lens L_2 on an electro-optical switch, which consists of three crossed Glan prisms with two electrooptical elements placed between these prisms. The switch remains closed when high-power pump pulses pass through this switch, which leads to the attenuation of these pulses. The switch is open when an echo signal comes. For the synchronization of the electrooptical switch, a high-voltage electric control pulse from the switch of the second emitting unit in the modified

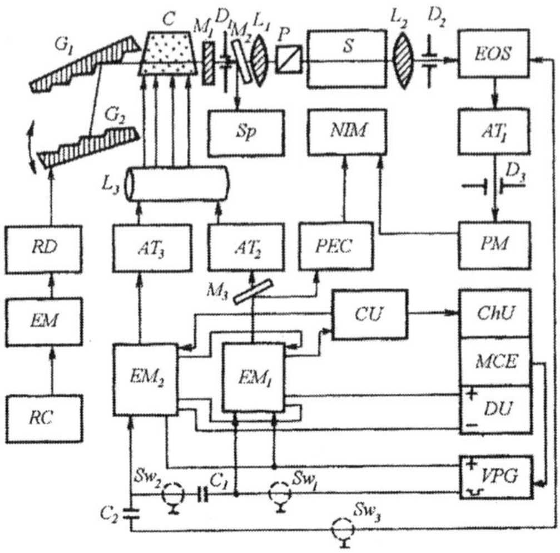

Fig. 3.7. Diagram of an optical echo relaxometer [188]: G_1, grazing-incidence grating; G_2, Littrow grating; C, cell with dyes; M_1, output coupler of the cavity; M_2, M_3, semitransparent mirrors; D_1–D_3, diaphragms; P, polarizer; S, sample; L_1, L_2, focusing lenses; L_3, cylindrical lens; EOS, electrooptical switch; AT_1–AT_3, attenuators; Sp, spectrograph (DFS-452); NIM, instrument for measuring nanosecond time intervals (I2-7); RD, reducer; PEC, photoelectric calorimeter (FEK-22); PM, photoelectric multiplier (ELU-FTS); EM, ac electric motor; CU, cooling unit (BO-1); ChU, charging unit; EM_1, EM_2, emitters of the YAG laser; MCE, modulator-control unit; RC, remote control; DU, discharge unit; VPG, voltage pulse generator (MGIN-5); C_1, C_2, coupling capacitors; Sw_1–Sw_3, switches.

LTIPCh-5 laser, which is employed to produce the second pump pulse for the frequency-tunable dye laser, is applied to the commutating switch through the coaxial delay line. Attenuated pump pulses and the echo signal pass through an attenuator AT_1, which consists of a set of neutral-density filters, and reach the photomultiplier (ELU–FTS). Electric signals produced by the photomultiplier enter a unit for measuring nanosecond time intervals, which is synchronized by a pulse coming from the photoelectric calorimeter (*PEC*). To produce the input signal for this calorimeter, a semitransparent mirror M_3 splits the first pump beam of the dye laser generated by the emitting unit EU_1. A semitransparent mirror M_2 directs some fraction of pump radiation to a DFS-452 spectrograph, which allows the spectrum of pump radiation to be measured. Diaphragms D_1–D_3 block the stray light due to the scattering from various elements of the echo relaxometer.

The oscillator generating doubled laser pulses consists of a remote-controlled dye laser and a modified LTIPCh-5 laser, which is employed to pump the dye laser. Rhodamine 6G is used as a dye in the dye laser (other dyes can be also employed if necessary). The dye laser includes an output coupler M_1 and 1200-grooves/mm diffraction gratings G_1 and G_2. The grating G_1 is employed in the regime of grazing incidence, which improves the resolution of this grating, narrows the width of the laser line, and prevents the surface of the grating from damage. The grating G_2 is employed in the Littrow-grating regime. The radiation frequency is tuned by rotating the grating G_2 around its axis parallel to its grooves. The grating G_2 is rotated by a dc electric motor (*EM*) through a reducer (*R*). This process is controlled from the control panel (*CP*). Such a dye laser generates radiation with a bandwidth of 0.1 Å. The dye laser is pumped with a modified scheme of a commercially available LTIPCh-5 laser. This laser includes an emitting unit EU_1 (IZ-9), a cooling unit (*CU*), a charging unit (*ChU*), a discharge unit (*DU*), a switch-control unit (*SCU*), and a voltage-pulse oscillator (*VPO*). The modification of the LTIPCh-5 laser implied the introduction of the second emitting unit EU_2 (IZ-9) into the scheme of this device. The flashlamps of the emitting units EU_1 and EU_2 were connected in series to the discharge unit *DU*. The water-cooling circuits were connected in series to the cooling unit *CU*. A dc high voltage on the order of +4.5 kV produced by the voltage-pulse oscillator *VPO* is applied in parallel to the electrooptical switches of both emitting units. The high-voltage pulse triggering the switch in the first emitting unit comes directly from the voltage-pulse oscillator, while the pulse triggering the second emitting unit passes through a decoupling capacitor C_1 and the coaxial line, which plays the role of a delay line. The delay time was varied within the range from 0 to 100 ns. The high-voltage pulse from the second emitting unit is applied to the electrooptical switch through a decoupling capacitor C_2 and the second coaxial delay line. Time-separated second-harmonic pulses of the emitting units EU_1 and EU_2 with λ = 532 nm (the third harmonic with λ = 354 nm can be also employed if necessary) pass through attenuators AT_2 and AT_3 and are focused with a cylindrical lens L_3 into the cell of the dye laser. Attenuators AT_2 and AT_3 consist of sets of neutral-density filters intended to adjust the intensity of pump laser pulses through the change in the intensity of laser pulses pumping the dye laser. The intensity of the latter pulses can be also controlled by changing the accumulator voltage in the switch-control unit.

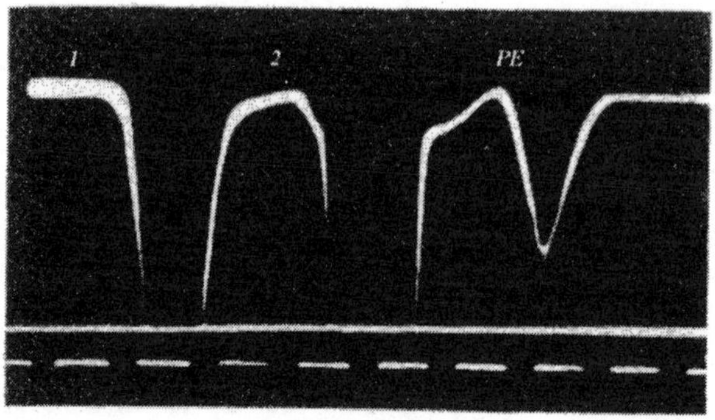

Fig. 3.8. Oscilloscope trace of the primary photon (optical) echo in molecular iodine vapor generated with pump radiation with tunable frequency [188]: 1, 2, the first and second pump pulses; PE, photon echo signal. Time markers correspond to 10 ns.

The above-described optical echo relaxometer was employed to perform a series of experiments on the investigation of specific features of the photon echo in a molecular iodine (I_2) vapor. The frequency of pump radiation in these experiments was varied within the range of 560–585 nm. These studies have also allowed the homogeneous widths of spectral lines to be measured.

A characteristic oscilloscope trace, illustrating the behavior of the primary photon echo signal, is shown in Fig. 3.8, where the signal on the right corresponds to the echo signal.

Some of the results of the experiments devoted to the investigation of specific features of the photon echo in I_2 vapors will be discussed in the next sections of this chapter.

3.3 Non-Faraday Polarization Rotation in Photon Echo

This section presents the results of experimental investigations devoted to the specific (non-Faraday) rotation of the polarization vector of the photon echo in a molecular I_2 vapor [201, 203, 211]. This effect was observed for $X^1\Sigma_g^+ - B^3\Pi_{0u}^+$ energy transitions at the wavelength of 571 nm in the presence of a longitudinal magnetic field. Before proceeding with the discussion of the experimental data, we should note that the well-known vector model of the photon echo [204], operating with rotations, precession, and nutation of pseudoelectric dipoles in a pseudoelectric field under the action of pump pulses, is applicable only in the case when the electric-dipole mechanism plays a dominant role in the interaction of particles with the fields of these pulses, while the role of other interaction mechanisms is not significant. The vector model was illustrated by experiments [204], where the photon echo was observed for ${}^4A_2(\pm1/2) \to {}^2E(\bar{E};\pm1/2)$ transitions of ruby in a longitudinal magnetic field **H**. This field was directed along the optical axis of the crystal, lifting the degeneracy in spin. The relevant four-level system involved in the formation of the photon echo can be divided under these conditions into two two-level subsystems, ${}^4A_2(M=-1/2)-{}^2E(M=-1/2)$ and ${}^4A_2(M=+1/2)-{}^2E(M=+1/2)$, with equal energy splitting values. Thus, the echo signal in experiments [204] was produced through 1/2 → 1/2 transitions. Due to this fact, the polarization of the photon echo was the same as the polarization of pump pulses and no specific polarization rotation was observed in the presence of the magnetic field if the pump pulses were linearly polarized in the same direction. As demonstrated by Alekseev [212], the polarization of the photon echo signal for 0 ⇔ 1 transitions in gases is subject to a specific rotation of the polarization vector different from the Faraday rotation. Consequently, a similar effect should be observed in solids when 0 ⇔ 1 transitions are involved in the formation of the photon echo.

Yevseyev and Yermachenko [213] considered an elementary case when the g-factors (g_a and g_b) of resonant levels are close to each other and demonstrated that the polarization of the photon echo in gases is subject to specific rotation for all optically allowed resonant transitions, except for 1/2 → 1/2 transitions. The same conclusion was made in [214] for different g_a and g_b. Consequently, the necessary condition for the specific rotation of the polarization vector of the photon echo is that the resonant transition involved in the formation of the photon echo signal should differ from a 1/2 → 1/2 transition regardless of the ratio of the g-factors related to the resonant levels and the type of the resonant medium.

The non-Faraday rotation of the polarization vector of the photon echo was observed for the first time by the authors of [36] in an atomic cesium vapor at the wavelength of 459.3 nm. The electron angular momenta of the resonant levels in [36] were equal to 1/2. The nuclear spin of ${}^{133}Cs$ atoms is equal to 7/2. Therefore, the total angular momenta of the levels in the hyperfine structure of resonant transitions in experiments [36] were equal to 3 and 4.

Thus, the angular momenta of resonant levels in experiments [36] and in all the subsequent experiments were low. In addition, the angle of specific

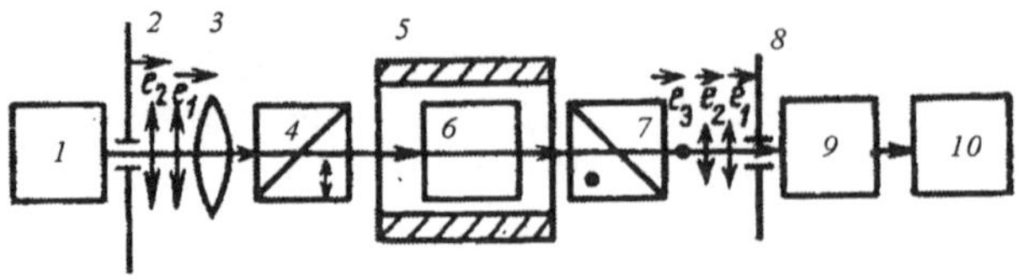

Fig. 3.9. Diagram of the experimental setup for the demonstration and investigation of non-Faraday rotation of the polarization vector of photon echo in molecular iodine vapor [203, 211]: (1) generator of doubled laser pulses; (2, 8) diaphragms; (3) converging lens; (4, 7) polarization Glan prisms; (5) solenoid; (6) cell with iodine vapor; (9) photodetector (ELU-FTS); (10) fast-response oscilloscope (I2-7).

rotation of the polarization vector of the photon echo in atomic gases, which is proportional to $\mu_0 gH\hbar^{-1}\tau$ [212–214] (μ_0 is the Bohr magneton, and τ is the time interval between the pump pulses), was observed in such gases when the strengths H of the external longitudinal magnetic field were on the order of several oersteds.

A different situation arises in the case of molecular gases. First, rotational quantum numbers J_a and J_b of the levels involved in resonant transitions in molecular gases are usually high. Second, substantially higher magnetic fields are required for the observation of the specific rotation of the polarization vector of the photon echo in molecular gases, because μ_0 in the formula for the specific rotation angle should be understood in this case as the nuclear magneton.

The diagram of the setup employed for these measurements is presented in Fig. 3.9. An iodine vapor filled a quartz cell in the dynamic equilibrium state, i.e., the rate of vapor evacuation at the output of the cell was equal to the evaporation rate of iodine at the input of the cell. The cell was placed inside a solenoid, which induced a longitudinal dc magnetic field with a strength up to 5000 Oe. The iodine vapor was irradiated with two time-separated laser pulses with a duration of 12 ns and with a variable time interval τ between these pulses, which could reach 40 ns. The pulses were linearly polarized in the same direction and propagated in the same direction. The peak power of each of the pulses was equal to 10 kW. These pulses were produced by an oscillator of doubled laser pulses *1*, which was described in the previous section. Passing through a diaphragm *2*, these pulses irradiated the cell with an I_2 vapor. The cell was placed between two crossed Glan prisms. The polarization plane of the prism *4*, located in front of the cell, coincided with the polarization of the pump pulses. The pump pulses and the echo signal induced in the cell with an iodine vapor and propagating in the same direction as the pump pulses passed through the prism *7* and diaphragm *8* and were registered with a photodetector connected to the input of a fast-response oscilloscope. We should note that the saturation of the photodetector with high-power pump laser pulses was prevented due to the attenuation of these pulses by crossed polarization prisms *4* and *7*. Since the polarization direction of the photon echo signal in the absence of the longitudinal magnetic field H_0 coincides with the direction of polarization vectors of the pump pulses and the intensity is two orders of magnitude lower, the polarization prism *7* attenuates the echo signal down to a zero level (while the pump pulses are attenuated by the prism *7* by two to three orders of magnitude).

The oscilloscope traces of the signals observed in this experiment are presented in Fig. 3.10. The upper trace illustrates the generation of a primary photon echo signal with rotation of the polarization vector (with respect to the polarization vector of the pump pulses) in a longitudinal magnetic field. The lower oscilloscope trace indicates the absence of the rotation of the polarization vector of an echo signal in a zero magnetic field. Since the experiments were performed with resonant levels with high rotational quantum numbers and close g -factors, the results of [213] were employed to process the experimental data. In accordance with [213], when $J \gg 1$ and the echo is produced within the Q branch, the specific rotation angle φ of the polarization vector of the echo signal with respect to the polarization of the pump pulses can be found from the equation

$$\tan\varphi = 2\sin(2\varepsilon\tau)/[1 + 2\cos(2\varepsilon\tau)], \quad (3.3.1)$$

where $\varepsilon = (\varepsilon_a + \varepsilon_b)/2$ and $\varepsilon_{a,b} = \mu_0 g_{a,b} H\hbar^{-1}$.

Thus, in the case under consideration, the rotation angle φ may reach 90°, i.e., the polarization vector of the photon echo may become perpendicular to the polarization vectors of the pump pulses. When the photon echo is produced within the P or R branch with $J \gg 1$, the angle φ can be determined from the equation [213]

$$\tan\varphi = \sin(2\varepsilon\tau)/[3 + \cos(2\varepsilon\tau)]. \quad (3.3.2)$$

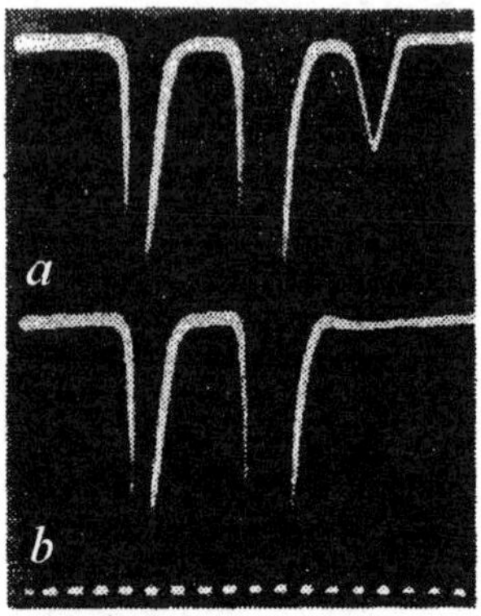

Fig. 3.10. Oscilloscope traces illustrating the non-Faraday rotation of the polarization vector of photon echo in molecular iodine vapor in a longitudinal magnetic field [203, 211]: (*a*) the presence of the photon echo signal (on the right) with a tilted polarization vector in a longitudinal magnetic field; (*b*) the absence of a photon echo signal with a tilted polarization vector in a zero magnetic field. Time markers correspond to 10 ns.

Thus, in this case, the angle φ cannot be made equal to 90° by increasing the parameter $\varepsilon\tau$. The angle φ in this case is limited by a much smaller value of $\arctan(2)^{-3/2}$.

Since the maximum projection of the electric field strength in the photon echo signal on the direction orthogonal to the polarization vectors of the pump pulses as a function of H was less than the strength of the total echo signal, we can infer that the observed echo signal was produced within the P or R branch.

We should emphasize that, as it follows from [213], for low angular momenta of the levels involved in the resonant transition, we can always adjust H in such a way as to make the polarization vector of the photon echo signal orthogonal to the polarization vectors of the pump pulses. The maximum angle φ is comparatively small for the studied transition. This is the third important distinctive feature of experiments with molecular gases as compared with studies performed in atomic gases, where a maximum specific rotation angle less than 90° was never observed. Measurements performed with a variable longitudinal magnetic field H have shown that the function $\tan\varphi$ reached its maximum within the range $H = 4000 \pm 200$ Oe. Thus, the sum of gyromagnetic ratios for a given value of $\tau = 40$ ns can be estimated as

$$g_a + g_b = [\pi/2 + 2\pi k + \arctan(2)^{-3/2}]\hbar(\mu_0 H\tau)^{-1}, \quad (3.3.3)$$

where $k = 0, 1, 2, \ldots$ and $\mu_0 = 5\cdot 10^{-24}$ erg/Oe. This formula gives an estimate close to 2.5. Since the spectroscopic studies give grounds to assume that the g-factors in the ground and excited states are equal to each other, we can find the absolute values of the g-factors: $g_a = g_b = 1.25$. We should note also that the Faraday rotation of the polarization of the echo signal can be estimated in accordance with the formula [212]

$$\varphi_{\mathrm{Far}} = N\lambda^2\Omega\gamma T_2{}^* l, \quad (3.3.4)$$

where N and $1/\gamma$ are the density and the radiative lifetime of iodine molecules, $1/T_2{}^*$ is the Doppler width of the level, λ is the wavelength of the working transition, l is the length of the cell, and $\Omega = g\mu_0 H_0\hbar^{-1}$. Such an estimate gives a very small value of the rotation angle.

The effect of the non-Faraday rotation of the polarization vector is inherent to the stimulated photon echo also. Firstly this effect is theoretically investigated by Yevseyev and Reshetov in paper [214]. They show, that angle ψ_s of the non-Faraday rotation of the polarization vector of stimulated photon echo may be calculated by means of the formulas:

$$\tan\psi_s = [\sin(2\varepsilon\tau) + \sin(2\varepsilon\tau + 2\varepsilon T)][1 + \cos(2\varepsilon\tau) + \cos(2\varepsilon\tau + 2\varepsilon T)]^{-1} \quad (3.3.5)$$

(for $J \to J$ transitions, when $J >> 1$) and

$$\tan\psi_s = [6\sin(2\varepsilon\tau) + \sin(2\varepsilon\tau + 2\varepsilon T)][1 + 6\cos(2\varepsilon\tau) + \cos(2\varepsilon\tau + 2\varepsilon T)]^{-1} \quad (3.3.6)$$

(for $J \to J + 1$ transitions, when $J >> 1$), where τ and T are the time intervals between the first and second pulses and between the second and third pulses correspondingly; $\varepsilon = \mu g H\hbar^{-1}$, μ is the Bohr magneton; g is the factor of the spectroscopy splitting; H is the value of the longitudinal permanent magnetic field.

The first experimental observation of this effect is carried out in paper [215] on the molecular iodine vapor (I_2) at room temperature. The excitation of the stimulated photon echo is carried out on the energetical transitions $B^3\Pi^+_{0u} - X^1\Sigma^+_g$ ($\lambda = 571.5$ nm) of the vibrational-rotational band of molecular iodine and in the range of molecular iodine pressures 10–70 mTorr. The value of the longitudinal magnetic field is equal to 2700 Oe. The estimations of the ε parameter under the conditions of this experiment ($H = 2.7 \times 10^3$ Oe, $\mu = 5 \times 10^{-24}$ erg·Oe^{-1}; $g = 1.25$) show that $\varepsilon = 1.7 \times 10^7$ s^{-1}. In this case the angle ψ_s may be smaller than 90°, when $\tau = 25$ ns and $T = 35$ ns.

Consider the experimental setup and results of investigations. The scheme of the setup is shown on Fig. 3.11. The main node of this scheme is the dye laser *1*, the pumping of which is implemented by means of the second harmonic of the two pulsed YAG lasers (*2* and *3*). The definite time interval τ between the first and second pulses is secured by the coaxial delay line *19*, but the moment of the influence of the third pulse is regulated by means of the optical delay line *14*. We used rhodamine 6*G* as the dye. The dye laser is constructed in the scheme that guaranteed the transverse pumping of the dye. Since the lifetime of the electron–vibrational levels is very small, the pulses of the dye laser *1* fully repeat the pulses of the pumping lasers (*2* and *3*). As a result, we received a sequence of three excited pulses with vertical polarization. These pulses propagate either in the same optical way or in a few parallel ways. It is secured by the generation of all three pulses in the same cavity with the dye. The construction of the laser cavity can be different, but in this experiment it consists of a diffraction grating, a mirror, and a cell with dye. The diffraction grating operates in the regime of grazing incidence. Since the laser cavity is the same for all pulses, the parameters of these pulses virtually coincide. This pulsed sequence is directed (by means of the mirror *4* and the Glan polarizer *5*) to the cell with molecular iodine vapor 6, situated in the solenoid *7* (35 cm long), which endures a longitudinal magnetic field of more than 7000 Oe. In this experiment the field was near 2700 Oe. The non-Faraday rotation of the polarization vector of the stimulated photon echo takes place in this magnetic field. In the zero magnetic field the effect is absent. The Glan polarizer *8* is adjusted for the echo-signal propagation with the turned polarization, but it locks the optical way for the radiation with the other polarization angles. After the Glan polarizer *8*, the coherent radiation is registered by the photodetector *9*. The signals are observed by the high-speed oscilloscope *10*. The work of this oscilloscope is synchronized by the part of radiation of the first pulse by means of the mirror *12* and the photoregister *11*. The control of all process of the echo excitation is carried out by means of the pulley *20*. The

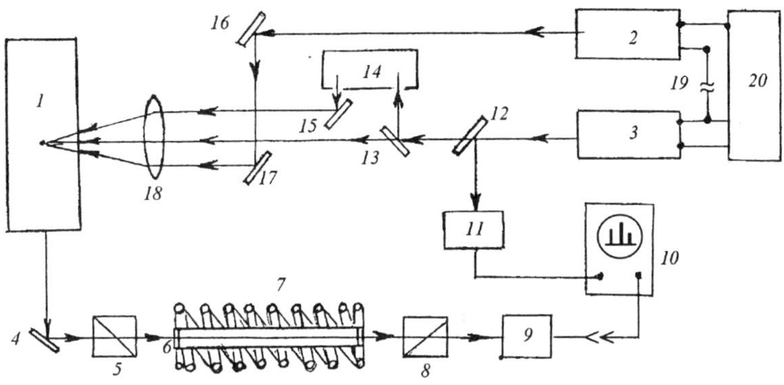

Fig 3.11. The block-scheme of arrangement for the discovering of the non-Faraday polarization turning of the stimulated photon echo in the molecular iodine vapor: (13) beam splitter; (14) delay line; (15, 16, 17) mirrors; (18) converging lens.

duration of the excited pulses is equal to 12 ns, but the time intervals between pulses must be smaller than the relaxation times: $T_1 = 0.2 \div 0.95$ μs, $T_2 = 41 \div 69$ ns.

Oscillograms of the observed signals are shown on Fig. 3.12. The upper oscillogram shows the stimulated photon echo with turned polarization vector is absent in zero magnetic field. The lower oscillogram illustrates the generation of the stimulated photon echo (first signal to the right) with the turned polarization vector in a longitudinal magnetic field. The angle of the rotation of the stimulated photon echo's polarization ψ_s is equal to 7°. Thus, in paper [215] the effect of the non-Faraday rotation of the stimulated photon echo's polarization is firstly observed. The angle of this rotation depends on the time intervals τ and T. One excited pulse may be chosen as the "echelon" of the coding signals. In this case the rotation angle ψ_s may be considered as a key in the retrieval of information in the gas optical echo-processor.

Closing this section, we should emphasize that the considered non-Faraday rotation of the polarization vector of the photon echo is of fundamental importance for optical echo spectroscopy, because this effect allows the modulated dependence of the echo signal intensity on the longitudinal magnetic field H to be used to extract the information concerning the hyperfine structure of lines masked with Doppler broadening.

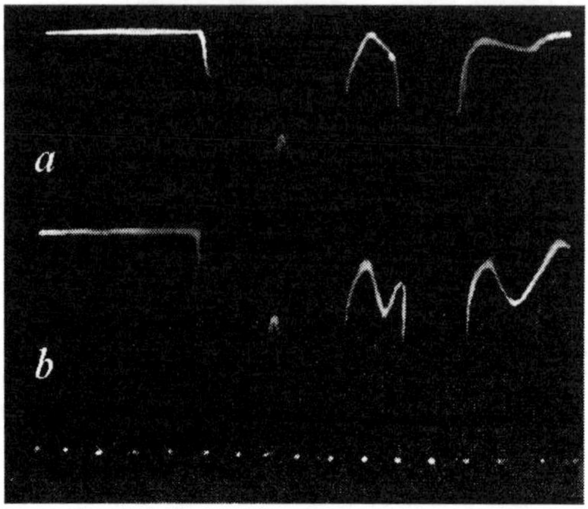

Fig 3.12. Oscillograms, illustrating the observation of the non-Faraday turning of the SPE polarization vector in the iodine vapor, situated in the longitudinal magnetic field: (*a*) $H_0 = 0$ Oe; (*b*) $H_0 = 2700$ Oe. Time markers correspond to 10 ns.

3.4 The Method of Measurement of Homogeneous Spectral Line Widths by Means of Photon Echo Signals

In this section, we will consider an example of molecular iodine vapor to demonstrate the method of extracting the information concerning homogeneous spectral line widths and relaxation times with the use of an optical photon echo [194, 216]. Investigations were performed for $B^3\Pi^+_{0u} - X^1\Sigma^+_g$ energy transitions in the vibrational-rotational band of molecular iodine at a temperature of 25°C within the range of wavelengths from 560 to 600 nm, where the absorption spectrum was close to a quasicontinuum. These experiments have revealed a reliable formation of intense photon echo signals in a pure saturated I_2 vapor at 570.8, 571.5, and 590 nm and in a rarefied I_2 vapor at 571 and 590 nm. Photon echo signals were excited with small-area pump pulses through spectroscopic transitions with high angular momenta J. The time interval τ between the pump pulses was varied in a discrete way within the range from 40 to 120 ns. The power of the pump pulses was on the order of 10 kW (i.e., it was much lower than the bleaching threshold for I_2, which is equal to 1 MW).

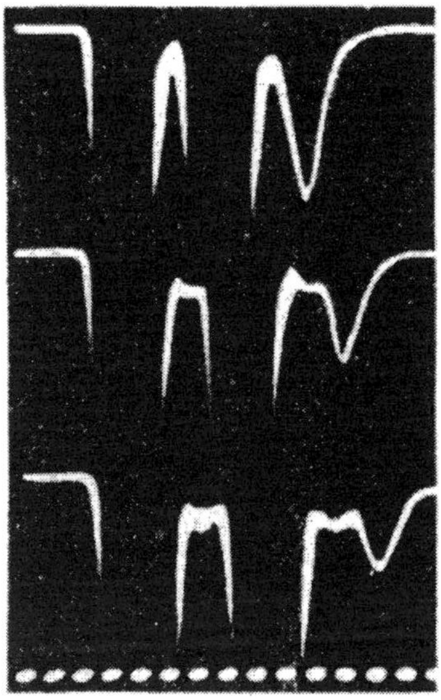

Fig. 3.13. Oscilloscope traces illustrating the intensity decay of the photon echo in molecular iodine vapor with the growth in the time interval τ between the pump pulses [216]. Time markers correspond to 10 ns.

In accordance with the theory of photon echo in gas media (e.g., see [217]), the intensity $I_p(\tau)$ of the primary photon echo is proportional to the following dissipation factor:

$$I_p(\tau) \sim \exp(-4\gamma^{(1)}\tau) = \exp[-4\tau(A + BP)], \tag{3.4.1}$$

where $\gamma^{(1)}$ is the homogeneous line width ($\gamma^{(1)} = 1/T_2$), A is the homogeneous width due to the radiative (spontaneous) decay of the energy state, BP is the homogeneous width due to elastic depolarizing and inelastic gas-kinetic collisions of gas particles, P is the pressure, and τ is the time interval between the pulses. Thus, the dependence of the intensity of the primary (two-pulse) photon echo on the time interval τ is measured in experiments under conditions when other parameters are fixed (Fig. 3.13). Figure 3.14 displays the dependence $I_p(\tau)$ measured at 571 nm. Similar dependences were obtained for other wavelengths specified above. The measured dependences

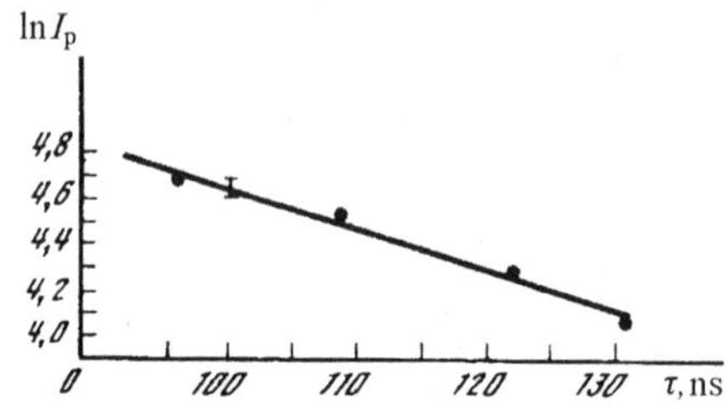

Fig. 3.14. Intensity (in arbitrary units) of the primary photon echo as a function of the time interval τ [194, 216]. The wavelength is $\lambda = 571$ nm, and the pressure is 27 mTorr.

were processed with the use of a computer. This procedure yielded the following homogeneous widths and transverse irreversible relaxation times T_2 for spectral lines in the vibrational–rotational band of the $B^3\Pi^+_{0u} - X^1\Sigma^+_g$ electron transition in a saturated I_2 vapor:

λ, nm	$\gamma^{(1)}$, MHz	T_2, ns
570.8	2.30	69
571.5	2.18	73
590.0	3.84	41

The following parameters were obtained for a rarefied I_2 vapor:

λ, nm	P, mTorr	$\gamma^{(1)}$, MHz	T_2, ns
571	27	0.97	163
590	52	0.68	235

The increase in the homogeneous width for a rarefied I_2 vapor relative to the homogeneous widths characteristic of a saturated vapor is apparently due to the presence of buffer gases in a rarefied vapor, which penetrate into of the above-specified parameters did not exceed 10%. The error of experimental measurements is consistent with the experimental data presented

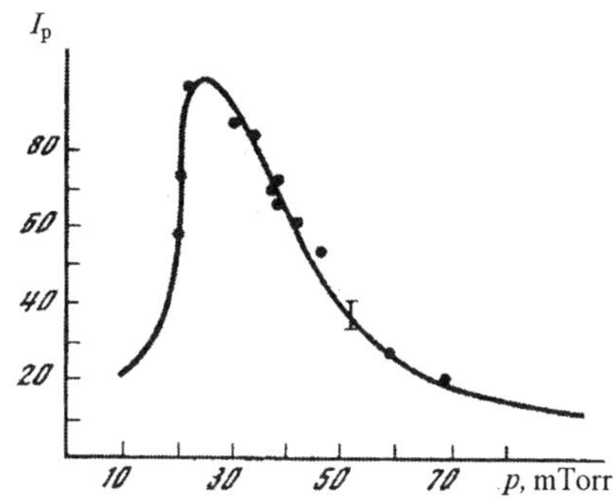

Fig. 3.15. Intensity (in arbitrary units) of the primary photon echo as a function of the pressure of iodine vapor at the wavelength of 571 nm for the time interval $\tau = 86$ ns [194, 216].

in [26, 108] (for a saturated vapor at a pressure of 0.31 Torr, the relaxation time at the wavelength of 532.5 nm was equal to 62 ns, while for the wavelength of 590 nm, $T_2 = 44$ ns).

The intensity of the primary photon echo was measured as a function of the iodine vapor pressure (Fig. 3.15) (in fact, here we deal with the concentration dependence of the echo signal intensity) to find the range of pressures where the homogeneous line width is determined to a considerable extent by collisions between iodine molecules and between iodine molecules and iodine atoms. The behavior of this dependence is determined by the competition of the quadratic concentration dependence of I_p and the exponential dissipation decay of this quantity due to collisions. These two factors compensate each other around the maximum. Then, the role of collisions increases with the growth in the pressure. Therefore, all the measurements of the collisional width were carried out for pressures exceeding the pressure corresponding to the maximum of the dependence $I_p(P)$. We can find the parameter B from the ratio of the echo signal intensities $I_p(\tau)$ and $I'_p(\tau)$ corresponding to two experimental values of the pressure P_1 and P_2 [see expression (3.4.1)] using the following formula:

$$B = \ln[I'_p(\tau)/I_p(\tau)]/4\,\tau\,(P_1 - P_2), \tag{3.4.2}$$

where τ is a fixed value of the time interval between the pulses.

On the other hand, using the value of $\gamma^{(1)}$ determined from the analysis of the decay curve, we can find the parameter $A = \gamma^{(1)} - BP$. Finally, this procedure gives the following results:

$$\lambda = 570.8 \text{ nm}, \qquad \gamma^{(1)} = 2.70 + 38P,$$

$$\lambda = 571.5 \text{ nm}, \qquad \gamma^{(1)} = 4.70 + 29P,$$

$$\lambda = 590.0 \text{ nm}, \qquad \gamma^{(1)} = 1.19 + 74P,$$

where P is the pressure of the I_2 vapor in Torrs. These values of $\gamma^{(1)}$ are consistent with the results of [108]: $\lambda = 590$ nm, $\gamma^{(1)} = 0.79 + 71P$.

Closing this section, we will consider the technique that employs the signals of stimulated optical photon echo. As is well known [217], the intensity I_s of these signals is proportional to

$$I_s \sim \exp(-2\gamma^{(0)}T), \tag{3.4.3}$$

where $\gamma^{(0)} = 1/T_1$, T_1 is the lifetime of the excited state, and T is the time interval between the second and third pulses.

Thus, we experimentally measure the dependence $I_s(T)$ (Fig. 3.16). Analysis of this dependence (for a pressure of the I_2 vapor equal to $P = 310$ mTorr) gives the following values of parameters:

$$\lambda = 564.9 \text{ nm}, \qquad T_1 = 0.95\ \mu\text{s},$$

$$\lambda = 590{,}0 \text{ nm}, \qquad T_1 = 0.20\ \mu\text{s}.$$

As is well known, the technique of stimulated photon echo makes it possible to distinguish between the dissipation contributions of elastic and inelastic collisions to the homogeneous width [218]. Specifically, the use of this approach allowed the dissipation contribution of inelastic gas-kinetic collisions to be determined. This quantity was equal to 19.3 and 3.64 μs^{-1} at the wavelengths of 564.9 and 590 nm, respectively.

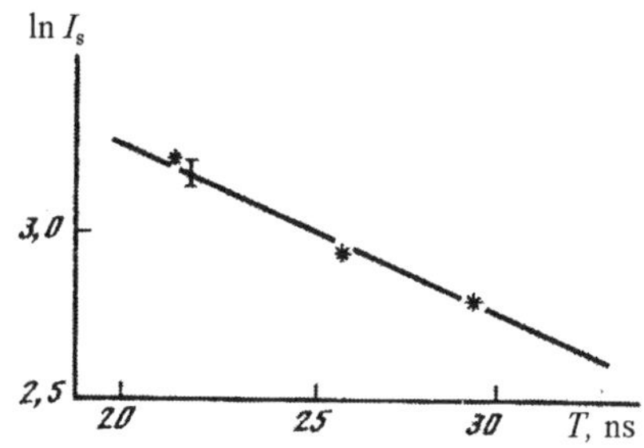

Fig. 3.16. Intensity (in arbitrary units) of the stimulated photon (optical) echo as a function of the time interval T between the second and third pulses [216].

3.5 Self-Induced Transparency and Self-Compression of a Pulse in a Resonant Gas Medium

Self-induced transparency is a physical phenomenon associated with the optical bleaching of a resonant medium under the action of a laser pulse with an area $\theta \geq \pi$ and a duration less than the characteristic irreversible relaxation times. This phenomenon was experimentally observed and theoretically explained by McCall and Hahn [219]. The physics of this effect is closely related to the response of a resonant medium to a short laser pulse propagating through this medium discussed in the previous section. The energy absorbed in a resonant medium from the leading edge of the pulse is then reemitted by the medium into other parts of the pulse, distorting the pulse waveform until a "soliton" (a 2π-pulse) arises in the medium. The resulting soliton propagates in the medium in the absence of resonant absorption. Since absorption and reemission processes are characterized by some finite time, the group velocity of soliton propagation in a resonant medium is less than the phase velocity of light.

Let us point to two processes that impede the identification of self-induced transparency in a gas medium. The first process is the saturation that bleaches a resonant medium under the action of a pulse of any duration due to the equalization of populations in resonant levels. The second process is the dissociation of molecules into atoms under the action of high-power laser radiation, when the optical bleaching of a medium is due to the fact that the resonance vanishes. Each of these processes is characterized by its own kinetics of the distortion of the pulse waveform in a medium different from the kinetics of self-induced transparency (SIT). The specific features of the distortion of the pulse waveform in the case of SIT can be summarized in the following way. First, the increase in the pulse power at the input of a medium (and, correspondingly, the growth in the area of the pulse within the range of $\pi \le \theta < 2\pi$) leads to the broadening of the pulse at the output of the medium for the compensation of the increase in the area up to the soliton value of 2π. A further increase in the power (and the area within the range of $2\pi \le \theta < 3\pi$) results in the self-partitioning of the pulse into several solitons. Consequently, as the pulse power increases, its waveform should first display self-broadening, then self-compression, and then self-partitioning.

Theoretical description of SIT is based on the solution of the self-consistent problem, which is usually performed with the use of the set of coupled differential Maxwell–Bloch equations (see Appendix 1). We will use the following approximations: (1) the condition of exact resonance is satisfied for all the particles; (2) characteristic irreversible relaxation times are sufficiently large as compared with the pulse duration, i.e., T_1, $T_2 = \infty$; and (3) the phase velocity is assumed to be constant, and the change in the phase Φ accompanying the propagation of a pulse in a resonant medium is neglected.

When these conditions are satisfied, the set of coupled Maxwell–Bloch equations is reduced to (ε_0 is the amplitude of the pulse electrical field)

$$\partial U/\partial t = 0, \tag{3.5.1}$$

$$\partial V/\partial t = -(k^2/\omega)\varepsilon_0 W, \tag{3.5.2}$$

$$\partial W/\partial t = V\varepsilon_0\omega, \tag{3.5.3}$$

$$\partial\varepsilon_0/\partial z = (\eta/c)\,\partial\varepsilon_0/\partial t - (2\pi\omega/\eta c)V. \tag{3.5.4}$$

The solution to equation (3.5.1) subject to the initial condition $U_0 = 0$ is obvious: $U = 0$. The solutions to equations (3.5.2) and (3.5.3) under the above-specified conditions can be found in any handbook on differential equations (e.g., see [220]):

$$V = Np\sin\psi(z, t), \tag{3.5.5}$$

$$W = W_0 \cos \psi(z, t), \tag{3.5.6}$$

where

$$\psi(z, t) = \frac{k}{2} \int_{-\infty}^{t} dt' \varepsilon_0(z, t'). \tag{3.5.7}$$

Recall that the total rotation angle of the pseudoelectric dipole can be written as

$$\theta = \frac{k}{2} \int_{-\infty}^{\infty} dt' \varepsilon_0(z, t'). \tag{3.5.8}$$

Since $\partial\varepsilon_0/\partial z = -v^{-1}\partial\varepsilon_0/\partial t$, the solution to differential equation (3.5.4) can be written as

$$\varepsilon_0(z, t) = (2/k\tau)\mathrm{sech}[(t - z/v)/\tau)], \tag{3.5.9}$$

where $\tau = [(1/Q')(1/v - 1/c)]^{1/2}$, $Q' = 2\pi k\omega Np/(\eta c)$, v is the group velocity of the pulse, c is the phase velocity of light in the resonant medium, and N is the number of active particles in the gas. Using formula (3.5.7), we find that

$$\partial\psi/\partial t = k\varepsilon_0. \tag{3.5.10}$$

Invoking formula (3.5.9), we can obtain the following expression for the derivative of the amplitude of the electric field:

$$\partial\varepsilon_0/\partial t = [2\pi\omega Np/\eta c(1/v - 1/c)]\sin\psi. \tag{3.5.11}$$

Then, taking yet another time derivative in expression (3.5.10), we find that

$$\partial^2 \psi/\partial t^2 = \tau^{-2}\sin\psi. \tag{3.5.12}$$

Thus, we arrived at the equation of a physical pendulum that starts its motion from the upper unstable equilibrium state. It is well known that all the solutions to this equation have a periodic character, except for the one solution corresponding to a pulse with a finite duration and energy:

$$\tan(\psi/4) = \exp[(t - z/v)/\tau)]. \tag{3.5.13}$$

Using this solution, we can derive formula (3.5.9) for the amplitude of the electric field in the laser pulse.

The fact that the theoretical analysis of SIT has brought us to the equation of a physical pendulum (3.5.12) characterizes the physics behind the absorption and reemission of light. The "particles + field" system is reminiscent of a swing. First, the field of a pulse, which plays the role of a driving force, acts on a resonant medium (a "swing"). Then, the absorbed energy is transferred into the field energy again (i.e., the source of the driving force is swung by the swing). The reemission (response) of a resonant medium distorts the waveform of the pulse until a soliton appears. The shape of this soliton, as it follows from (3.5.9), is described by a hyperbolic secant (which is close to a Gaussian function, but has more extended wings).

Using expression (3.5.3), we can derive a differential equation that is called the theorem of areas:

$$d\theta(z)/dz = (\alpha/2)\sin\theta(z), \tag{3.5.14}$$

where $\alpha = 8\pi^2 N p^2 \omega g(0)/(\hbar\pi c)$. Note that, for a small-area pulse (i.e., in the case when $\sin\theta \approx \theta$), we arrive at the well-known Bouguer–Beer law:

$$\varepsilon_0(t)^2 = \varepsilon_0(0)^2\exp(-\alpha z), \tag{3.5.15}$$

where α is understood as the coefficient of resonant absorption.

For strong pulses, the solution to differential equation (3.5.14) is written as

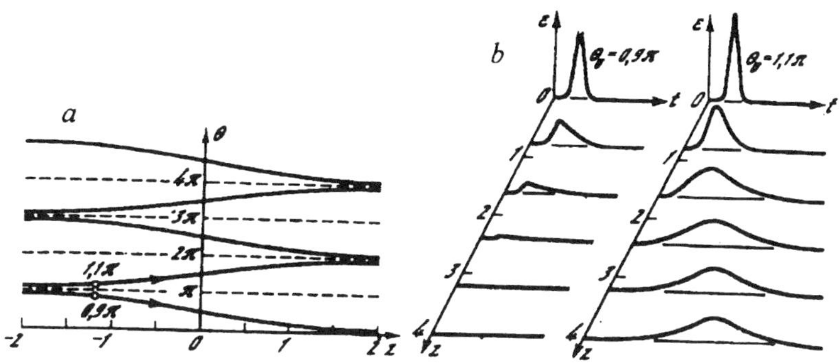

Fig. 3.17. Numerical analysis of the distortion kinetics of a laser pulse in a resonant medium under conditions of nonlinear coherent interaction [219]: (*a*) dependence $\theta(z)$; (*b*) evolution of the waveform of a laser pulse in a medium for two different values of the initial area θ_0.

$$\theta(z) = 2\arctan[\tan(\theta_0/2)\exp(-\alpha z/2)], \tag{3.5.16}$$

where θ_0 is the area of the pulse at the input of the resonant medium.

The numerical analysis of solution (3.6.16) was performed by McCall and Hahn [219]. Some results of this analysis are presented in Fig. 3.17, which shows that, with $\theta_0 < \pi$, the pulse decays as it propagates through a resonant medium, while with $\theta_0 \geq \pi$, a stationary 2π-pulse (soliton) arises in the medium.

As can be seen from the formula $\tau = [1/Q'(1/v - 1/c)]^{1/2}$, the group velocity of a pulse in a resonant medium is given by

$$1/v = \eta/c + \alpha\tau/2, \tag{3.5.17}$$

while the delay time of a pulse in a medium with a length L is

$$t_d = \alpha\tau L/2. \tag{3.5.18}$$

The numerical analysis of the kinetics of pulse distortion in a medium,

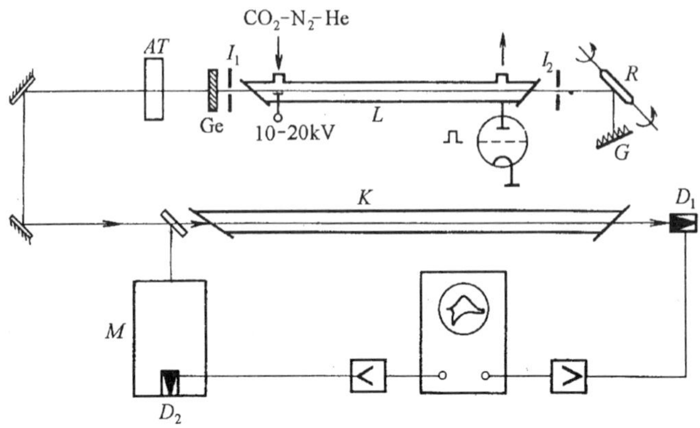

Fig. 3.18. Diagram of the setup for studying self-induced optical transparency in a BCl_3 gas [223] (see text for details).

which was carried out in [219] (see also the review [221]), has shown that pulses with $\theta > 3\pi$ are divided into two and more solitons as they propagate through a medium. Analysis performed in this paper has also demonstrated that relaxation processes have the following influence on the characteristics of a pulse in a medium:

$$\theta \approx 2\pi(1 - \tau/T_2), \tag{3.5.19}$$

$$t_d = -T_{eff}\ln(1 - \alpha\Delta t_{in}L/2T_{eff}), \tag{3.5.20}$$

where $T^{-1}_{eff} = 8/3(2/T_2 + 1/T_1)$, Δt_{in} is the duration of the pulse at the input of the medium, and L is the length of the resonant medium.

The first experiment on the investigation of SIT in an SF_6 gas medium was performed by Patel and Slusher [222]. Later, this phenomenon was studied by Alimpiev and Karlov [223] for a BCl_3 gas. Let us discuss experiments [223] in greater detail in order to demonstrate the apparatus and the technique of SIT investigations in gases. Experiments were performed with a setup shown in Fig. 3.18. A CO_2 laser operating in the double Q-switching regime was employed as a source of light pulses. Modulation was implemented with the use of a mirror R, which was rotated with a frequency of 200–400 Hz, and power-supply pulses with a duration of 1 μs and a repetition rate of 25–50 Hz, which were synchronized with the rotation of the mirror. The laser cavity consisted of a plane-parallel germanium plate with a dielectric coating, rotating mirror *R*, and a diffraction multilayer grating *G* (100 grooves/mm). One of the faces of the mirror *R* was covered with a multilayer dielectric coating. The second face of the mirror was antireflection-coated in such a way as to prevent additional selection in the cavity due to the output mirror. The laser is tuned in oscillation lines with the use of the diffraction grating. Two iris diaphragms I_1 and I_2 placed inside the cavity ensured lasing in the TEM_{00} fundamental transverse mode and made it possible to vary the duration of laser pulses within the range from 100 to 150 ns for many oscillation lines of the CO_2 laser. Rotating mirrors directed laser pulses to a 3 m-long cell filled with the gas under study. The cell was evacuated down to the required pressure (from 1 to 100 mTorr) with the use of a diffusion pump, and the gas pipelines were blocked then. The laser radiation intensity was varied in a discrete way with the use of an attenuator *AT* (or two crossed polarizers). Some fraction of the laser beam was directed to a monochromator *M* in order to measure the frequency and the intensity of laser radiation. Radiation was detected with two Cd–Hg–Te infrared detectors D_1 and D_2. The SIT phenomenon was thoroughly studied by the authors of [223] for a BCl_3 gas at the frequency of 951.16 cm^{-1}, which corresponded to the *P*12 line of CO_2-laser radiation. The BCl_3 gas is characterized by the maximum absorption coefficient $\alpha = 0.13\ cm^{-1}\cdot Torr^{-1}$ at this frequency. The intensity I_{out} of the output pulse was measured in SIT experiments as a function of the intensity I_{in} of the pulse at the input of the gas cell. The kinetics of the pulse-waveform distortion was measured under conditions when the pulse intensity was increased at the input of the cell. The delay time of the pulse in the gas cell relative to a pulse passing the same path outside the gas cell was also determined. The dependence $I_{out}(I_{in})$ is presented in Fig. 3.19. The curves shown in this figure display bending corresponding to a pulse with an area $\theta = \pi$. Therefore, using these plots to determine the intensity I_b at the bending point, one can estimate the modulus of the electric dipole moment of the resonant transition in accordance with the formula

$$p = \pi\hbar/[8\pi I_b(\Delta t)^2/c]. \tag{3.5.21}$$

For BCl_3, such an estimate yields $p = 5\cdot 10^{-20}$ CGSE.

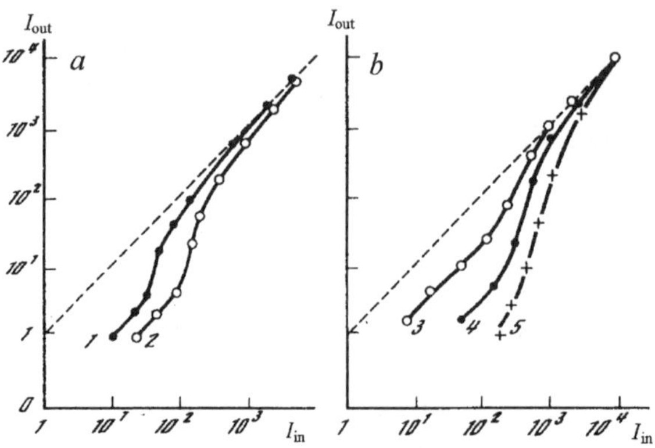

Fig. 3.19. Dependence of the pulse intensity at the output of a resonant medium (BCl_3 gas) on the input intensity (in arbitrary units) for (*a*) different pulse durations and (*b*) for different gas pressures in the cell [223]: (*a*) (*1*) Δt = 360 ns, (*2*) 80 ns; (*b*) (*3*) P = 40 mTorr, (*4*) 75 mTorr, (*5*) 120 mTorr.

The kinetics of the distortion of the pulse waveform accompanying the growth in the pulse intensity at the output of the medium is shown in Fig. 3.20 (the input pulses are shown by lower traces). The pulse waveform considerably changes for pulse intensities close to the bending point (Fig. 3.20*c*), where the delay time reaches its maximum. The relaxation considerably changes for pulse intensities close to the bending point (Fig. 3.20*c*), where the delay time reaches its maximum. The relaxation time estimated from these data is T_2 = 30 ns·Torr.

Let us consider in a greater detail yet another experiment [224] on the investigation of SIT in a molecular rubidium Rb_2 vapor. This experiment is of interest because it revealed a noticeable self-compression of a pulse accompanying SIT. Note that the authors of [225–227] have demonstrated that ruby-laser radiation (λ = 6943 Å) leads to energy transitions from vibrational levels $v'' = 5 \div 10$ of the ${}^1\Sigma_g^+$ ground state to vibrational levels $v' = 2 \div 7$ of the ${}^1\Pi_u^+$ excited state of rubidium molecules. For intensities of pump pulses lower than 1 kW/cm^3, the number of excited molecules N is proportional to the intensity, while for intensities ranging from 10^5 to 10^6 W/cm^2, the number of active molecules is proportional to the square root of the intensity. The radiative relaxation time in the case under consideration is 19 ns. The time of relaxation due to collisions of Rb_2 molecules with each other and with Rb atoms depends on the temperature (and, naturally, on the pressure, which is related to the temperature). Under

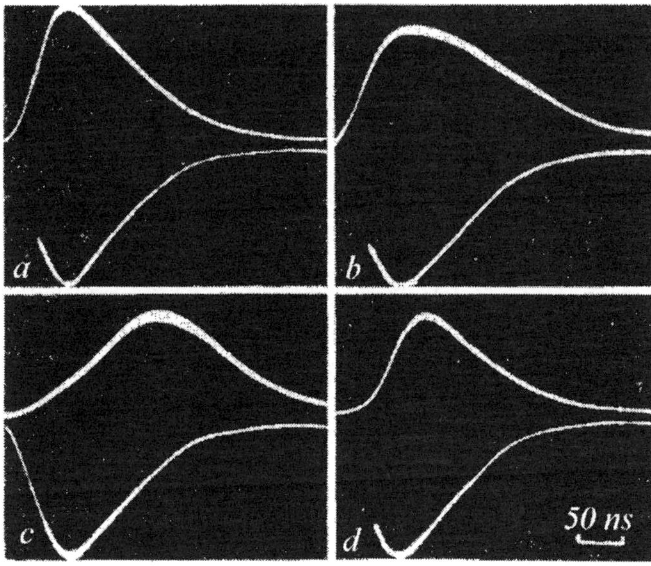

Fig. 3.20. Oscilloscope traces illustrating the distortion kinetics of the waveform of a pulse with the growth in the pulse area [223]. The oscilloscope trace in (*c*) corresponds to a π-pulse. The lower traces display the input pulses.

experimental conditions of [224], this relaxation time was no less than 40 ns. Kostin and Khodovoi [225, 226] have demonstrated that the absorption spectrum of rubidium display a clearly pronounced dip at the wavelength of 6943 Å under the action of a ruby-laser pulse (Δt = 20 ns). The amplitude of this dip depends on the intensity of laser radiation. For laser radiation intensities on the order of several milliwatts per square centimeter, a uniform (≈100 Å) bleaching of the Rb_2 vapor due to the dissociation of molecules is observed along with a narrow dip (≈1 Å). For $I \leq 0.18$ mW/cm^2, the width of the dip was less than 0.15 Å, and the detection of this dip was limited by the resolution of the experimental equipment. The existence of a narrow dip is associated with the saturation effect. The authors of [226] employed the ratio of fluorescence intensities corresponding to weak and strong excitation to estimate the modulus of the electric dipole moment of resonant transitions. Such estimates yielded a value of the dipole moment equal to 10^{-19} CGSE.

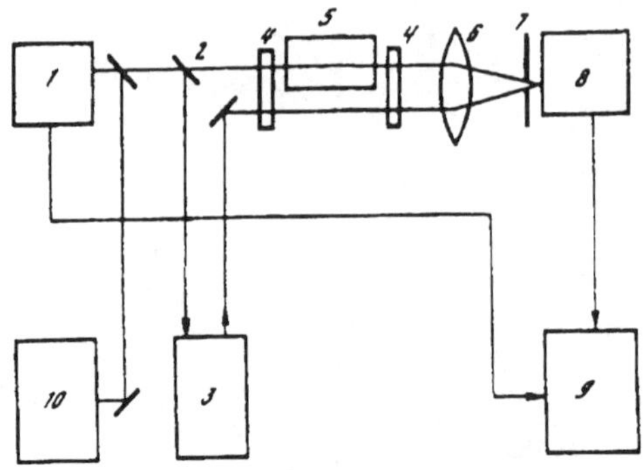

Fig. 3.21. Diagram of the setup for studying the self-compression of a pulse in molecular rubidium vapor [218]: (*1*) ruby laser; (*2*) beam splitter; (*3*) delay line; (*4*) filters; (*5*) cell with rubidium vapor; (*6*) focusing lens; (*7*) diaphragm; (*8*) fast-response photomultiplier ELU-FT; (*9*) instrument for measuring time intervals I2-7; (*10*) helium-neon laser.

Nagibarov *et al.* [224] experimentally studied SIT in a molecular rubidium vapor in a cell (with a length of 110 cm) with sapphire windows at a temperature of 490–520 K. The diagram of the setup employed for these experiments is shown in Fig. 3.21. The experimental setup includes a *Q*-switched ruby laser *1*, a beam splitter *2*, a set of accessory filters *3*, a cell with a rubidium vapor *4*, an attenuating filter *5*, an ELU-FT fast-response photomultiplier *6* (with a resolution no lower than 2.7 ns), and an I2-7 unit for measuring time intervals. A helium-neon laser *10* was employed to adjust the experimental setup. The cell with sapphire windows was placed inside a housing and was heated with the use of an electric spiral. The temperature at the ends and in the middle of the cell was measured by means of a thermocouple. The temperature near the windows of the cell was 15–20°C higher than the temperature in the middle of the cell, which prevented the deposition of Rb_2 vapor on the windows of the cell. The duration of the laser pulse at the input of the cell with rubidium vapor was 17–20 ns, and the laser radiation power was varied within the range from 10^5 to 10^6 W/cm^2 with the use of optical filters. These experiments revealed the bleaching of a molecular rubidium vapor under the action of laser pulses with the above-specified parameters. The bleaching curves for different vapor temperatures can be found in [224]. Here, we restrict ourselves to the illustration of this dependence measured at a temperature of

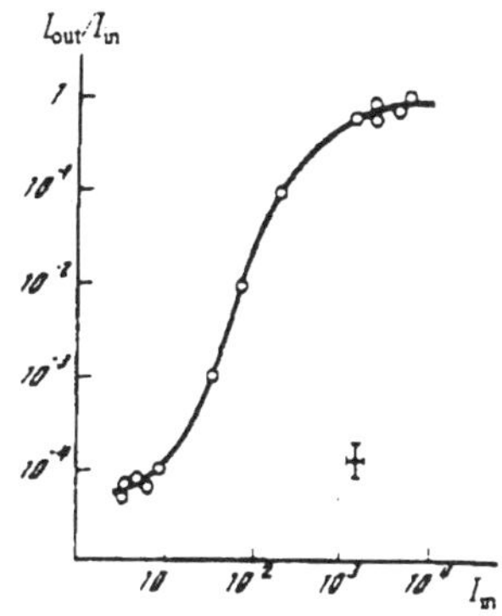

Fig. 3.22. The ratio I_{out}/I_{in} as a function of I_{in} (in arbitrary units) for molecular rubidium vapor at 495 K [232] (I_{in} and I_{out} are the intensities of light pulses at the input and at the output of the cell, respectively).

495 K (see Fig. 3.22). In addition, these experiments also demonstrated that a light pulse is slightly (by approximately 3 ns) delayed in a resonant medium. In principle, the observed effect of optical bleaching may be due to three factors: SIT, dissociation of molecules into atoms, and the saturation effect. As mentioned above, the dissociation of molecules occurs at intensities higher than several megawatts per square centimeter, while the intensities achieved in the considered experiments did not exceed 10^5 W/cm^2. The influence of the saturation effect on SIT was investigated in [227], where it was demonstrated that the ratio of the threshold power (W) of bleaching due to the saturation effect (s) to the SIT threshold power (SIT) can be calculated with the use of the following formula:

$$W_s/W_{SIT} = (10^{-7}\nu/5)(T/M)^{1/2}\Delta t^2/T_1, \qquad (3.5.22)$$

where ν is the frequency of the resonant transition, M is the molecular weight of active particles, T is the absolute temperature of such particles, Δt is the pulse duration, and T_1 is the longitudinal relaxation time (radiative relaxation time).

The estimates show that, under conditions of experiments [224], this ratio is on the order of one even with $T = 500$ K, which points to a considerable

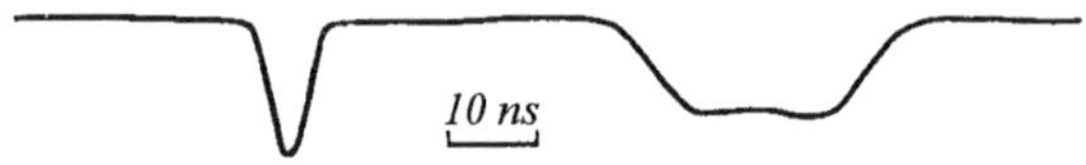

Fig. 3.23. Self-compression of a ruby-laser pulse in molecular rubidium vapor [224]. The input pulse is shown on the right, and the pulse at the output of the gas cell is shown on the left.

role of the saturation effect, which increases, in accordance with (3.5.22) with the growth in the temperature. Therefore, to reduce the role of saturation in the bleaching of a medium, it would be appropriate to consider the bleaching curves corresponding to temperatures lower than 500 K.

Let us consider the curve shown in Fig. 3.22. As mentioned above, the bending point in the bleaching curve $I_{out}(I_{in})$ approximately corresponds to the intensity of a π-pulse in a resonant medium. Using the intensity corresponding to the bending point in the considered plot, we can find the modulus of the electric dipole moment of the relevant transition with the use of formula (3.5.21): $p = (0.8 \div 1) \cdot 10^{-19}$ CGSE. This estimate is consistent with the value of p obtained in [225] from the ratio of fluorescence signals in the case of weak and strong excitation.

Note that the ultimate goal of experiments [224] was to implement the self-compression of a pulse in the SIT regime. According to SIT theory, the self-compression of a pulse should be observed when the condition $2\pi < \theta_0 < 3\pi$ is satisfied. Such a self-compression was detected in experiments [224]. The pulse at the input of the cell filled with rubidium vapor had a duration of 17–20 ns, while the duration of the pulse emerging from the cell was less than the duration of the input pulse by a factor of 2.5–3. The oscilloscope trace illustrating this self-compression effect is presented in Fig. 3.23. Due to the simplicity of experimental implementation and the large magnitude, this phenomenon holds much promise for technical applications.

In conclusion, we should note that, in this chapter, we considered only a small part of a broad variety of optical coherent phenomena. Some more examples are discussed in the appendices.

REFERENCES

1. C.K.N. Patel, R.E. Slusher, *Phys. Rev. Lett.* **20**: 1087–1089 (1968).
2. B. Bölger, J.C. Diels, *Phys. Lett. A* **28**: 401–402 (1968).

3. J.P. Gordon, C.H. Wang, C.K.N. Patel, *et al.*, *Phys. Rev.* **179**: 294–309 (1969).
4. R.G. Brewer, R.L. Shoemaker, *Phys. Rev. Lett.* **27**: 631–634 (1971).
5. R.L. Shoemaker, R.G. Brewer, *Bull. Am. Phys. Soc.* **17**: 66 (1972).
6. S.S. Alimpiev, N.V. Karlov, *Zh. Eksp. Teor. Fiz.* **63**: 482–490 (1972).
7. N.V. Karlov, S.S. Alimpiev, *Proc. VI All-Union Conf. Nonlinear Optics* (Minsk: BSSR Acad. Sci., 1972) (in Russian), pp. 115–116.
8. R.G. Brewer, *Bull. Am. Phys. Soc.* **17**: 1129 (1972).
9. S.S. Alimpiev, N.V. Karlov, *Kratk. Soobshch. Fiz.* No. **1**: 17–21 (1973).
10. R.G. Brewer, *Bull. Am. Phys. Soc.* **18**: 400 (1973).
11. S.S. Alimpiev, *Proc. I All-Union Symp. Opt. Echo* (Kazan: KGPI, 1973) (in Russian), pp. 4–5.
12. J. Schmidt, P.R. Berman, R.G. Brewer, *Phys. Rev. Lett.* **31**: 1103–1106 (1973).
13. J.R. Meckley, C.V. Heer, *Phys. Lett. A* **46**: 41–42 (1973).
14. S.S. Alimpiev, N.V. Karlov, *Izv. Akad. Nauk SSSR, Ser. Fiz.* **37**: 2022–2031 (1973).
15. S.S. Alimpiev, *Investigation of the Effects of Coherent Interaction of Infrared Radiation Pulses with Molecular Gases* (Moscow: Candidate of Phys. Math. Science Dissertation, 1973) (in Russian).
16. P.R. Berman, J. Schmidt, R.G. Brewer, *Bull. Am. Phys. Soc.* **18**: 1518 (1973).
17. R.L. Shoemaker, F.A. Hopf, *Phys. Rev. Lett.* **33**: 1527–1530 (1974).
18. R.J. Nordstrom, W.M. Gutman, C.V. Heer, *Phys. Lett. A* **50**: 25–26 (1974).
19. J. Schmidt, P.R. Berman, R.G. Brewer, *Coherent Transient Study of Velocity-Changing Collisions, in Laser Spectroscopy* (New York: Plenum, 1974), pp. 283–291.
20. C.V. Heer, R.J. Nordstrom, *Phys. Rev. A* **11**: 536–548 (1975).
21. P.R. Berman, J.M. Levy, R.G. Brewer, *Phys. Rev. A* **11**: 1668–1688 (1975).
22. W.M. Gutman, C.V. Heer, *Phys. Lett. A* **51**: 437–438 (1975).
23. W.M. Gutman, C.V. Heer, *Bull. Am. Phys. Soc.* **20**: 671 (1975).
24. W.M. Gutman, C.V. Heer, *Bull. Am. Phys. Soc.* **21**: 151 (1976).
25. S.S. Alimpiev, *Trudy FIAN* **87**: 92–133 (1976).
26. R.G. Brewer, A.Z. Genack, Phys. *Rev. Lett.* **36**: 959–962 (1976).
27. B. Bölger, L. Baede, H.M. Gibbs, *Opt. Commun.* **18**: 67–68 (1976).
28. T. Baer, I.D. Abella, *Phys. Lett. A* **59**: 371–372 (1976).
29. T. Baer, I.D. Abella, *Bull. Am. Phys. Soc.* **21**: 1246 (1976).
30. A.Z. Genack, R.G. Brewer, *Bull. Am. Phys. Soc.* **21**: 1283 (1976).
31. R.G. Brewer, *Coherent Optical Spectroscopy, in Very High Resolution Spectroscopy* (London: Academic, 1976), pp. 127–142.
32. S.B. Grossman, A. Schenzle, R.G. Brewer, *Phys. Rev. Lett.* **38**: 275–278 (1977).
33. R.G. Brewer, *Bull. Am. Phys. Soc.* **22**: 32 (1977).
34. A.H. Zewail, T.E. Orlowski, K.B. Jones, *et al.*, *Chem Phys. Lett.* **48**: 256–261 (1977).
35. W.M. Gutman, C.V. Heer, *Phys. Rev. A* **16**: 659–665 (1977).
36. T. Baer, I.D. Abella, *Phys. Rev. A* **16**: 2093–2100 (1977).
37. T. Mossberg, A. Flusberg, R. Kachru, S.R. Hartmann, *Phys. Rev. Lett.* **39**: 1523–1526 (1977).
38. R.G. Brewer, A.Z. Genack, S.B. Grossman, *Coherent Transients and Pulse Fourier Transform Spectroscopy, in Laser Spectroscopy III* (Berlin: Springer, 1977), Vol. 7, pp. 220–230.

39. A.H. Zewail, *Coherent Optical Spectroscopy of Molecules Undergoing Resonance Scattering and Radiationless Transitions: The Right-Angle Photon Echo, in Laser Spectroscopy III* (Berlin: Springer, 1977), Vol. 7, pp. 268–278.
40. V.P. Chebotayev, N.M. Dyuba, M.N. Skvortsov, L.S. Vasilenko, *Appl. Phys.* **15**: 319–322 (1978).
41. L.S. Vasilenko, N.M. Dyuba, M.N. Skvortsov, *Kvantovaya Elektron.* **5**: 1725–1730 (1978).
42. L.S. Vasilenko, N.M. Dyuba, M.N. Skvortsov, V.P. Chebotayev, *Pis'ma Zh. Tekh. Fiz.* **4**: 278–282 (1978).
43. T.E. Orlowski, K.E. Jones, A.H. Zewail, *Chem. Phys. Lett.* **54**: 197–202 (1978).
44. A. Flusberg, T. Mossberg, S.R. Hartmann, *Opt. Commun.* **24**: 207–210 (1978).
45. A.Z. Genack, R.G. Brewer, *Phys. Rev. A* **17**: 1463–1473 (1978).
46. T. Mossberg, A. Flusberg, R. Kachru, S.R. Hartmann, *Bull. Am. Phys. Soc.* **23**: 33 (1978).
47. A. Flusberg, T. Mossberg, R. Kachru, S.R. Hartmann, *Phys. Rev. Lett.* **41**: 305–308 (1978).
48. L.S. Vasilenko, N.M. Dyuba, M.N. Skvortsov, V.P. Chebotayev, *Proc. IX All-Union Conf. Coherent Nonlinear Optics Dedicated to the Memory of Acad. R.V. Khokhlov* (Moscow: Moscow St. Univ., 1978) (in Russian), Part **2**, p. 162.
49. A. Flusberg, T. Mossberg, R. Kachru, S.R. Hartmann, *J. Opt. Soc. Am.* **68**: 696 (1978).
50. P. Hu, H.M. Gibbs, *J. Opt. Soc. Am.* **68**: 1630–1631 (1978).
51. T. Baer, I.D. Abella, *Opt. Lett.* **3**: 170–171 (1978).
52. N.C. Griffen, C.V. Heer, *Appl. Phys. Lett.* **33**: 865–866 (1978).
53. R.L. Shoemaker, R.E. Scotti, B. Comaskey, *J. Opt. Soc. Am.* **68**: 1388 (1978).
54. R.L. Shoemaker, *Coherent Transient Infrared Spectroscopy, in Laser and Coherence Spectroscopy* (New York: Plenum, 1978), pp. 197–371.
55. H.M. Gibbs, P. Hu, *Picosecond Self-Induced Transparency and Photon Echoes in Sodium Vapor, in Picosecond Phenomena* (Berlin: Springer, 1978), pp. 336–338.
56. A. Flusberg, R. Kachru, T. Mossberg, S.R. Hartmann, *Phys. Rev. A* **19**: 1607–1621 (1979).
57. N.C. Griffen, C.V. Heer, *Bull. Am. Phys. Soc.* **24**: 529 (1979).
58. S. Aoki, *Phys. Rev. A* **20**: 2013–2021 (1979).
59. T. Mossberg, A. Flusberg, R. Kachru, S.R. Hartmann, *Phys. Rev. Lett.* **42**: 1665–1669 (1979).
60. T. Mossberg, A. Flusberg, R. Kachru, S.R. Hartmann, *Bull. Am. Phys. Soc.* **24**: 25 (1979).
61. R. Kachru, A. Flusberg, T.W. Mossberg, S.R. Hartmann, *Bull. Am. Phys. Soc.* **24**: 25 (1979).
62. M. Fujita, H. Nakatsuka, H. Nakanishi, M. Matsuoka, *Phys. Rev. Lett.* **42**: 974–977 (1979).
63. T. Baer, *Phys. Rev. A* **20**: 2610–2618 (1979).
64. T. Mossberg, R. Kachru, A. Flusberg, S.R. Hartmann, *Bull. Am. Phys. Soc.* **24**: 560 (1979).
65. R. Kachru, T.W. Mossberg, S.R. Hartmann, *Opt. Commun.* **30**: 57–62 (1979).
66. T.W. Mossberg, R. Kachru, E. Whittaker, S.R. Hartmann, *Phys. Rev. Lett.* **43**: 851–855 (1979).
67. R. Kachru, T.W. Mossberg, E. Whittaker, S.R. Hartmann, *Opt. Commun.* **31**: 223–227 (1979).

68. R. Kachru, T.W. Mossberg, S.R. Hartmann, *J. Opt. Soc. Am.* **69**: 1407 (1979).
69. T. Shimizu, N. Morita, T. Kasuga, et al., *Appl. Phys.* **21**: 29–34 (1980).
70. N.C. Wong, S.S. Kano, R.G. Brewer, *Phys. Rev. A* **21**: 260–267 (1980).
71. R. Kachru, T.W. Mossberg, S.R. Hartmann, *Phys. Rev. A* **21**: 1124–1133 (1980).
72. Y. Fukuda, K. Yamada, T. Hachi, *J. Phys. Soc. Jpn.* **48**: 1403–1404 (1980).
73. T.W. Mossberg, R. Kachru, S.R. Hartmann, *Phys. Rev. Lett.* **44**: 73–77 (1980).
74. R. Kachru, T.W. Mossberg, S.R. Hartmann, *J. Phys. B* **13**: L363–L368 (1980).
75. R. Kachru, T.W. Mossberg, S.R. Hartmann, *Phys. Rev. A* **22**: 1953–1957 (1980).
76. T.W. Mossberg, E. Whittaker, R. Kachru, S.R. Hartmann, *Phys. Rev. A* **22**: 1962–1969 (1980).
77. L.S. Vasilenko, N.M. Dyuba, N.N. Rubtsova, *Proc. X All-Union Conf. Coherent Nonlinear Optics* (Moscow: Moscow State Univ., 1980) (in Russian), p. 234.
78. M. Matsuoka, H. Nakatsuka, M. Fujita, *Picosecond Backward Echo in Sodium Vapor – Relaxation and Quantum Beat Modulation, in Picosecond Phenomena II* (Berlin: Springer, 1980), Vol. **14**, pp. 357–361.
79. B. Comaskey, R.E. Scotti, R.L. Shoemaker, *Opt. Lett.* **6**: 45–47 (1981).
80. H. Nakatsuka, M. Fujita, M. Matsuoka, *Opt. Commun.* **36**: 234–236 (1981).
81. C. Tai, H.J. Simon, *Bull. Am. Phys. Soc.* **26**: 711 (1981).
82. C. Tai, H.J. Simon, *Opt. Lett.* **6**: 133–135 (1981).
83. Y. Fukuda, K. Yamada, T. Hashi, *Opt. Commun.* **37**: 299–302 (1981).
84. K.P. Leung, T.W. Mossberg, S.R. Hartmann, *Bull. Am. Phys. Soc.* **26**: 278 (1981).
85. K.P. Leung, T.W. Mossberg, S.R. Hartmann, *Bull. Am. Phys. Soc.* **26**: 280 (1981).
86. J.A. Kash, E.L. Hahn, *Phys. Rev. Lett.* **47**: 167–170 (1981).
87. L.S. Vasilenko, N.N. Rubtsova, *Pis'ma Zh. Tekh. Fiz.* **7**: 1079–1082 (1981).
88. R. Kachru, T.J. Chen, S.R. Hartmann, *et al.*, *Phys. Rev. Lett.* **47**: 902–905 (1981).
89. L.S. Vasilenko, *Proc. II All-Union Symp. on Optical Echo* (Kazan: USSR Acad. Sci., 1981) (in Russian), p. 43.
90. M.N. Skvortsov, Nonstationary *Coherent Resonant Phenomena in a Gas in the Interaction with Optical Fields* (Novosibirsk: Candidate of Phys. Math. Science Dissertation, 1981) (in Russian).
91. T.W. Mossberg, R. Kachru, K.P. Leung, *et al.*, *The Measurement of Collisional Lineshapes Using Optical Echo Techniques, in Spectral Line Shapes* (Berlin: De Gruyter, 1981), pp. 1093–1111.
92. R. Kachru, T.J. Chen, T.W. Mossberg, S.R. Hartmann, *Phys. Rev. A* **25**: 1546–1549 (1982).
93. K.P. Leung, T.W. Mossberg, S.R. Hartmann, *Phys. Rev. A* **25**: 3097–3101 (1982).
94. L.S. Vasilenko, N.N. Rubtsova, *Kvantovaya Elektron.* **9**: 2243–2248 (1982).
95. M. Fujita, S. Asaka, H. Nadatsuka, M. Matsuoka, *J. Phys. Soc. Jpn.* **51**: 2582–2590 (1982).
96. K.P. Leung, T.W. Mossberg, S.R. Hartmann, *Opt. Commun.* **43**: 145–150 (1982).
97. L.S. Vasilenko, N.N. Rubtsova, *Investigation of Relaxation Processes in a Gas by Means of Coherent Transient Processes, in Lazernye sistemy* (Laser Systems) (Novosibirsk: Sib. Div. USSR Acad. Sci., 1982) (in Russian), pp. 143–154.

98. J. Langelaar, D. Bebelaar, J.D.W. Van Voorst, *Appl. Phys. B* **28**: 274–275 (1982).
99. L.S. Vasilenko, N.N. Rubtsova, *Izv. Akad. Nauk SSSR, Ser. Fiz.* **46**: 2021–2025 (1982).
100. W.S. Warren, A.H. Zewail, *J. Chem. Phys.* **78**: 2279–2297 (1983).
101. Y. Fukuda, K. Yamada, T. Hashi, *Opt. Commun.* **44**: 297–300 (1983).
102. R.A. Forber, L. Spinelli, J.E. Thomas, M.S. Feld, *Phys. Rev. Lett.* **50**: 331–335 (1983).
103. R. Beach, B. Brody, S.R. Hartmann, *Phys. Rev. A* **27**: 2925–2929 (1983).
104. M.R. Woodworth, *Opt. Lett.* **8**: 307–309 (1983).
105. N.W. Carlson, A.G. Yodh, T.W. Mossberg, *Phys. Rev. Lett.* **51**: 35–38 (1983).
106. C. Tai, *Phys. Rev. A* **28**: 3459–3466 (1983).
107. S. Asaka, H. Yamada, H. Nakatsuka, M. Matsuoka, *J. Phys. Soc. Jpn.* **52**: 3029–3038 (1983).
108. H. Nakatsuka, S. Asaka, M. Tomita, M. Matsuoka, *Opt. Commun.* **47**: 65–69 (1983).
109. N.W. Carlson, L.J. Rothberg, A.G. Yodh, *et al.*, *Opt. Lett.* **8**: 483–485 (1983).
110. N.N. Rubtsova, *Investigation of Population and Dipole-Moment Relaxation through Collisions of Molecules by Means of Stationary and Nonstationary Spectroscopy* (Novosibirsk, Candidate of Phys. Math. Science, 1983) (in Russian).
111. N.W. Carlson, W.R. Babbitt, T.W. Mossberg, *Opt. Lett.* **8**: 623–625 (1983).
112. J.E. Thomas, R.A. Forber, L.A. Spinelli, M.S. Feld, *Phys. Rev. Lett.* **51**: 2194–2197 (1983).
113. M. Tanigawa, Y. Fukuda, T. Kohmoto, *et al.*, *Opt. Lett.* **8**: 620–622 (1983).
114. L.S. Vasilenko, N.N. Rubtsova, V.P. Chebotayev, *Pis'ma Zh. Eksp. Teor. Fiz.* **38**: 391–393 (1983).
115. R. Beach, B. Brody, S.R. Hartmann, *Photon Echoes with Angled Beams, in Laser Spectroscopy VI* (Berlin: Springer, 1983), pp. 48–52.
116. F. De Rougemont, J.-C. Keller, J.-L. Le Gouët, *Stimulated Photon Echoes in Ytterbium Vapour, in Laser Spectroscopy VI* (Berlin: Springer, 1983), pp. 53–55.
117. T.W. Mossberg, *Echo Spectroscopic Studies of Velocity-Changing Collisions, in Spectral Line Shapes* (Berlin: De Gruyter, 1983), Vol. 2, pp. 899–915.
118. R.A. Forber, J.E. Thomas, L. Spinelli, M.S. Feld, *Total Optical Coherence-Perturber Collision Cross Sections Using Velocity Selective Photon Echoes, in Spectral Line Shapes* (Berlin: De Gruyter, 1983), Vol. **2**, pp. 945–959.
119. R.G. Brewer, *New Phenomena in Coherent Optical Transients, in Advanced Laser Spectroscopy* (New York: Plenum, 1983), Vol. **95**, pp. 365–367.
120. J. L. Le Gouët, *Ann. Phys.* **8**: 169–174 (1983).
121. T.H. Jeys, F.B. Dunning, R.F. Stebblings, *Phys. Rev. A* **29**: 379–380 (1984).
122. K.P. Leung, T.W. Mossberg, S.R. Hartmann, *Phys. Rev. A* **29**: 225–229 (1984).
123. J.E. Thomas, R.A. Forber, *Opt. Lett.* **9**: 56–58 (1984).
124. J.-C. Keller, J.-L. Le Gouët, *Phys. Rev. Lett.* **52**: 2034–2037 (1984).
125. R. Beach, B. Brody, S.R. Hartmann, *J. Opt. Soc. Am. B* **1**: 189–194 (1984).
126. N.W. Carlson, W.R. Babbitt, Y.S. Bai, T.W. Mossberg, *Opt. Lett.* **9**: 232–234 (1984).
127. A.V. Durrant, J. Manners, *Opt. Commun.* **49**: 293–296 (1984).
128. R. Beach, S.R. Hartmann, *Phys. Rev. Lett.* **53**: 663–666 (1984).

129. A.G. Yodh, J.Golub, N.W. Carlson, T.W. Mossberg, *Phys. Rev. Lett.* **53**: 659–662 (1984).
130. V. Brückner, E.A.J.M. Bente, J. Langelaar, *et al.*, *Opt. Commun.* **51**: 49–52 (1984).
131. Y.S. Bai, W.R. Babbitt, N.W. Carlson, T.W. Mossberg, *Appl. Phys. Lett.* **45**: 714–716 (1984).
132. J.-C. Keller, J.-L. Le Gouët, *J. Opt. Soc. Am. B* **1**: 484–485 (1984).
133. N.W. Carlson, W.R. Babbitt, Y.S. Bai, T.W. Mossberg, *J. Opt. Soc. Am. B* **1**: 506–507 (1984).
134. E.T. Sleva, A.H. Zewail, *Chem. Phys. Lett.* **110**: 582–587 (1984).
135. M.A. Gubin, I.V. Yevseyev, V.A. Reshetov, *Fotonnoe ekho v gazakh: eskeprimental'nye metody formirovamiya I roaznovidnosti* (Photon Echo in Gases: Experimental Methods of Formation and Modifications) (Moscow: P.N. Lebedev Physical Institute, Preprint No. 214, 1984) (in Russian).
136. L.S. Vasilenko, I.D. Matveenko, N.N. Rubtsova, *Issledovanie vremennykh i spektral'nykh kharakteristik kogerentnykh perekhodnykh protsessov* (Investigation of Temporal and Spectral Characteristics of Coherent Transient Processes) (Novosibirsk: Sib. Div. USSR Acad. Sci., Preprint No. 114) (in Russian).
137. A.V. Durrant, J. Manners, *J. Phys. B* **17**: L701–L706 (1984).
138. L.S. Vasilenko, I.D. Matveenko, N.N. Rubtsova, *Investigation of Relaxation Processes in Gases by Means of the Photon Echo (PE) and Stimulated Photon Echo (SPE), in Kineticheskie I gazodinamicheskie protsessy v neravnovesnykh sredakh* (Kinetic and Gas-Dynamic Processes in Nonequilibrium Media) (Noscow: Moscow State Univ., 1984) (in Russian), pp. 26–27.
139. M.R. Woodworth, I.D. Abella, *Photon Echoes in Krypton and Xenon Discharges, in Coherence and Quantum Optics V* (New York: Plenum, 1984), pp. 287–292.
140. N.W. Carlson, A.G. Yodh, T.W. Mossberg, *Standing-Wave Induced Backward Photon Echoes in Gases, in Coherence and Quantum Optics V* (New York: Plenum, 1984), pp. 309–315.
141. J.D.W. Van Voorst, V. Brückner, E.A.J.M. Bente, *Proc. IX Int. Conf. Raman Spectrosc.* (Tokyo, 1984), pp. 516–517.
142. J.-C. Keller, J.-L. Le Gouët, *Stimulated Photon Echo for Elastic and Depolarizing Collision Studies, in Ultrafast Phenomena IV* (Berlin: Springer, 1984), pp. 236–238.
143. L.S. Vasilenko, N.N. Rubtsova, *Opt. Spektrosk.* **58**: 697–699 (1985).
144. L.S. Vasilenko, N.N. Rubtsova, *Opt. Spektrosk.* **59**: 52–56 (1985).
145. S. Tao-Heng, J.A. Kash, E.L. Hahn, *Acta Phys. Sinica* **34**: 359–367 (1985).
146. T.J. Chen, S.R. Hartmann, *Acta Phys. Sinica* **34**: 1034–1039 (1985).
147. M.R. Woodworth, I.D. Abella, *Bull. Am. Phys. Soc.* **30**: 146 (1985).
148. A.G. Yodh, J. Golub, T.W. Mossberg, *Phys. Rev. A* **32**: 844–853 (1985).
149. A.P. Ghosh, C.D. Nabors, M.A. Atilli, J.E. Thomas, *Phys. Rev. Lett.* **54**: 1794–1797 (1985).
150. J.C. Keller, J.L. Le Gouët, *Phys. Rev. A* **32**: 1624–1642 (1985).
151. L.S. Vasilenko, I.D. Matveyenko, N.N. Rubtsova, *Opt. Commun.* **53**: 371–374 (1985).
152. R. Beach, D. DeBeer, S.R. Hartmann, *Phys. Rev. A*, **32**: 3467–3474 (1985).
153. K.E. Drabe, A. De Groot, J.D.W. Van Voorst, *Chem. Phys.* **99**: 121–134 (1985).

154. K.E. Drabe, A. De Groot, J.D.W. Van Voorst, *Chem. Phys.* **99**: 135–153 (1985).
155. L.S. Vasilenko, N.N. Rubtsova, *Proc. III All-Union Symp. Optical Echo and Coherent Spectroscopy* (Khar'kov: Ukr. SSR Acad. Sci., 1985) (in Russian), p. 47.
156. J.M. Liang, L.A. Spinelli, R.W. Quinn, *et al.*, *Phys. Rev. Lett.* **55**: 2684–2687 (1985).
157. A.G. Yodh, J. Golub, T.W. Mossberg, *Colliding without Relaxing: the Suppression of Collisional Dephasing with Strong Optical Fields, in Laser Spectroscopy VII* (Berlin: Springer, 1985), Vol. **49**, pp. 296–297.
158. A.G. Yodh, J. Golub, T.W. Mossberg, *The Collisional Relaxation of Excited-State Zeeman Coherence in Atomic Ytterbium Vapor, in Electronic and Atomic Collisions* (Amsterdam: North-Holland, 1985), p. 352.
159. D. DeBeer, L.G. Van Wagenen, R. Beach, S.R. Hartmann, *Phys. Rev. Lett*, **56**: 1128–1131 (1986).
160. M. Defour, J.C. Keller, J.L. Le Gouët, *J. Opt. Soc. Am. B* **3**: 544–547 (1986).
161. Y.S. Bai, T.W. Mossberg, *Opt. Lett.* **11**: 30–32 (1986).
162. T.J. Chen, D. DeBeer, S.R. Hartmann, *J. Opt. Soc. Am. B* **3**: 493–496 (1986).
163. M. Defour, J.C. Keller, J.L. Le Gouët, *Ann. Phys. Colloq.* **11**: 229–230 (1986).
164. E.T. Sleva, I.M. Xavier, A.H. Zewail, *J. Opt. Soc. Am. B* **3**: 483–487 (1986).
165. J.M. Liang, R.R. Dasari, M.S. Feld, J.E. Thomas, *J. Opt. Soc. Am. B* **3**: 506–513 (1986).
166. E.Y. Xu, F. Moshary, S.R. Hartmann, *J. Opt. Soc. Am. B* **3**: 497–505 (1986).
167. J.E. Thomas, A.P. Ghosh, M.A. Attili, *Phys. Rev. A* **33**: 3029–3046 (1986).
168. T.J. Chen, D. DeBeer, S.R. Hartmann, *Am. Inst. Phys. Conf. Proc.* No. **146**: 437–438 (1986).
169. E.Y. Xu, F. Moshary, S.R. Hartmann, *Am. Inst. Phys. Conf. Proc.* No. **146**: 439–440 (1986).
170. D. DeBeer, L.G. Van Wagenen, R. Beach, S.R. Hartmann, *Am. Inst. Phys. Conf. Proc.* No. **146**: 592–593 (1986).
171. Y.S. Bai, A.G. Yodh, T.W. Mossberg, *Phys. Rev. A* **34**: 1222–1227 (1986).
172. J. Manners, A.V. Durrant, *Opt. Commun.* **58**: 389–394 (1986).
173. A.G. Yodh, T.W. Mossberg, J.E. Thomas, *Phys. Rev A* **34**: 5150–5153 (1986).
174. A.G. Yodh, T.W. Mossberg, J.E. Thomas, *J. Opt. Soc. Am. B* **3**: 200–201 (1986).
175. J.E. Thomas, J.M. Liang, R.R. Dasari, *J. Opt. Soc. Am. B* **3**: 202–203 (1986).
176. M. Defour, J.C. Keller, J.L. Le Gouët, *J. Opt. Soc. Am. B* **3**: 232–234 (1986).
177. L.S. Vasilenko, N.N. Rubtsova, *Proc. VII All-Union Symp. High- and Ultrahigh-Resolution Molecular Spectroscopy* (Tomsk: Sib. Div. USSR Acad. Sci., 1986), Part 2, pp. 162–165.
178. Y.S. Bai, W.R. Babbitt, T.W. Mossberg, *Opt. Lett.* **11**: 724–726 (1986).
179. I.D. Matvienko, *Development and Application of an Automated Laser Spectrometer for the Investigation of Coherent Transient Processes* (Novosibirsk: Candidate of Phys. Math. Science Dissertation, 1986) (in Russian).
180. W.R. Babbitt, Y.S. Bai, T.W. Mossberg, *Convolution, Correlation, and Storage of Optical Data in Inhomogeneously Broadened Absorbing Materials, in Optical Information Processing II* (New York: SPIE, 1986), pp. 240–247.

181. J.M. Liang, J.E. Thomas, R.F. Dasari, M.S. Feld, *Ind. J. Phys. B* **60**: 318–335 (1986).
182. L.S. Vasilenko, N.N. Rubtsova, *J. Sov. Laser Res.* **7**: 409–416 (1986).
183. N.S. Belousov, L.S. Vasilenko, I.D. Matveenko, N.N. Rubtsova, *Opt. Spektrosk.* **63**: 34–38 (1987).
184. T. Mishina, M. Tanigawa, Y. Fukuda, T. Hashi, *Opt. Commun.* **62**: 166–170 (1987).
185. V.A. Sautenkov, *Kratk. Soobshch. Fiz.* No. **8**: 23–25 (1987).
186. M. Defour, J.C. Keller, J.L. Le Gouët, *Phys. Rev. A* **36**: 5226–5234 (1987).
187. I.I. Popov, I.S. Bikbov, *Proc. III All-Union Conf. Comput. Optoelectronics "Optical Memory Problems"* (Erevan: ArmSSR Acad. Sci., 1987) (in Russian), Part **1**, pp. 141–142.
188. I.S. Bikbov, I.I. Popov, V.V. Samartsev, *Prib. Tekh. Eksp.* No. **5**: 172–174 (1988).
189. M. Saidi, J.-C. Keller, J.-L. Le Gouët, *J. Phys.* **49**: 1513–1527 (1988).
190. P. Tchenio, A. Debarre, J.-C. Keller, J.-L. Le Gouët, *Phys. Rev. A* **38**: 5235–5248 (1988).
191. A. Debarre, J.-C. Keller, J.-L. Le Gouët, P. Tchenio, *J. Phys. Colloq.* **49**: 259–262 (1988).
192. I.S. Bikbov, I.I. Popov, *Fotonnoe (svetovoe) ekho v parakh molekulyarnogo ioda* (Photon (Optical) Echo in Molecular Iodine Vapor) (Kazan: KF USSR Acad. Sci., Preprint No. 1, 1988) (in Russian), pp. 33–42.
193. P. Tchenio, A. Debarre, J.-C. Keller, J.-L. Le Gouët, *Phys. Rev. Lett.* **62**: 415–418 (1989).
194. I.S. Bikbov, I.V. Yevseyev, I.I. Popov, V.V. Samartsev, *Opt. Spektrosk.* **67**: 224–226 (1989).
195. A.V. Durrant, J. Manners, K.W. Godfrey, *J. Mod. Opt.* **36**: 1173–1184 (1989).
196. I.S. Bikbov, I.I. Popov, S.B. Chemodanov, *et al.*, *Proc. IV All-Union Symp Optical Echo and Methods of Its Practical Applications* (Kuibyshev: Kuibyshev State Univ., 1989) (in Russian), p. 36.
197. L.S. Vasilenko, N.M. Dyuba, N.N. Rubtsova, *Proc. IV All-Union Symp. Optical Echo and Methods of Its Practical Applications* (Kuibyshev: Kuibyshev State Univ., 1989) (in Russian), p. 45.
198. L.S. Vasilenko, N.N. Rubtsova, *Proc. IV All-Union Symp. Optical Echo and Methods of Its Practical Applications* (Kuibyshev: Kuibyshev State Univ., 1989) (in Russian), p. 46.
199. I.I. Popov, I.S. Bikbov, *Proc. IV All-Union Symp. Optical Echo and Methods of Its Practical Applications* (Kuibyshev: Kuibyshev State Univ., 1989) (in Russian), p. 137.
200. I.I. Popov, I.S. Bikbov, *Proc. IV All-Union Symp. Optical Echo and Methods of Its Practical Applications* (Kuibyshev: Kuibyshev State Univ., 1989) (in Russian), p. 138.
201. I.I. Popov, I.S. Bikbov, *Proc. IV All-Union Symp. Optical Echo and Methods of Its Practical Applications* (Kuibyshev: Kuibyshev State Univ., 1989) (in Russian), p. 139.
202. L.S. Vasilenko, N.N. Rubtsova, *Izv. Akad. Nauk SSSR, Ser. Fiz.* **53**: 2329–2333 (1989).
203. I.I. Popov, I.S. Bikbov, V.V. Samartsev, *Izv. Akad. Nauk SSSR, Ser. Fiz.* **53**: 2334–2339 (1989).
204. N.A. Kurnit, I.D. Abella, S.R. Hartmann, *Phys. Rev. Lett.* **13**: 567–570 (1964).

205. V.R. Nagibarov, V.V. Samartsev, *Prib. Tekh. Eksp.* No. **3**: 189–190 (1970).
206. I.A. Nagibarova, V.V. Samartsev, *Phys. Lett. A* **35**: 231–232 (1971).
207. A.H. Zewail, T.E. Orlowski, D.R. Dawson, *Chem. Phys. Lett.* **44**: 379–383 (1976).
208. L.S. Vasilenko, *Coherent Doppler-free Laser Spectroscopy of Gas Media* (Novosibirsk: Doctor of Phys. Math. Science Dissertation, 1987) (in Russian).
209. V.R. Nagibarov, V.V. Samartsev, *Optical and Sound Echo Excited with Standing Waves, in Nekotorye voprosy magnitnoi radiospektroskopii i kvantovoi akustiki* (Some Problems of Magnetic Radio Spectroscopy and Quantum Acoustics) (Kazan': KF USSR Acad. Sci., 1968) (in Russian), pp. 102–104.
210. V.R. Nagibarov, V.V. Samartsev, *Chem. Phys. Lett.* **5**: 61–63 (1970).
211. I.I. Popov, I.S. Bikbov, I.V. Yevseyev, V.V. Samartsev, *Zh. Prikl. Spektrosk.* **52**: 794–798 (1990).
212. A.I. Alekseyev, *Zh. Eksp. Teor. Fiz.* **9**: 472–475 (1969).
213. I.V. Yevseyev, V.M. Yermachenko, *Opt. Spektrosk.* **47**: 1139–1144 (1979).
214. I.V. Yevseyev, V.A. Reshetov, *Laser Phys.* **1**: 380–338 (1991).
215. I.S. Bikbov, I.I. Popov, V.V. Samartsev, I.V. Yevseyev, *Laser Phys.* **5**: 580–583 (1995).
216. I.I. Popov, *Optical Echo in Molecular Iodine Vapor and Its Application* (Kazan: Candidate of Phys. Math. Science Dissertation, 1990) (in Russian).
217. A.V. Yevseyev, I.V. Yevseyev, V.M. Yermachenko, *Fotonnoe ekho v gazakh: Vliyanie depolyarizuyushchikh stoknovenii* (Photon Echo in Gases: The Influence of Depolarizing Collisions) (Moscow: Inst. Atom. Energ., Preprint No. 3602/1, 1982) (in Russian).
218. V.R. Nagibarov, V.V. Samartsev, L.A. Nefediev, *Phys. Lett. A* **43**: 195–196 (1973).
219. E.L. McCall, E.L. Hahn, *Phys. Rev.* **183**: 457–485 (1969).
220. E. Kamke, *Spravochnik po obyknovennym differentsial'nym uravneniyam* (Handbook on Ordinary Differential Equations) (Moscow: Nauka, 1971) (in Russian).
221. I.A. Poluektov, Yu.M. Popov, V.S. Roitberg, *Kvantovaya Elektron.* **1**: 757–785 (1974).
222. C.K.N. Patel, R.E. Slusher, *Phys. Rev. Lett.* **19**: 1019–1022 (1967).
223. S.S. Alimpiev, *Trudy FIAN* **87**: 92–133 (1976).
224. V.R. Nagibarov, V.A. Pirozhkov, V.V. Samartsev, R.G. Usmanov, *Pis'ma Zh. Eksp. Teor. Fiz.* **19**: 391–394 (1974).
225. N.N. Kostin, V.A. Khodovoi, *Izv. Akad. Nauk SSSR, Ser. Fiz.* **37**: 2089–2092 (1973).
226. N.N. Kostin, V.A. Khodovoi, *Izv. Akad. Nauk SSSR, Ser. Fiz.* **37**: 2093–2098 (1973).
227. K.W. Smith, L. Allen, *Opt. Commun.* **8**: 166–170 (1973).

Chapter 4

POLARIZATION ECHO SPECTROSCOPY

As mentioned in Chapter 2, the phenomenon of spin echo is widely employed for spectroscopic investigations. The phenomenon of photon echo has a similar area of applications. Let us consider two examples to illustrate photon-echo applications.

The decay of the signal intensity I_e with the growth in the time interval τ between pump pulses in spin echo is described (e.g., see [1–4]) by formula (2.4.9), where $1/\gamma_{ab}^{(1)}$ is understood as the irreversible transverse relaxation time T_2. Therefore, varying in experiments the time interval τ and processing the results of these measurements with the use of formula (2.4.9), where the quantity $\gamma_{ab}^{(1)}$ is replaced by T_2^{-1}, one can extract the information concerning the irreversible transverse relaxation time T_2 by means of the spin echo. Experiments of this kind made it possible to obtain considerable amounts of data in both nuclear magnetic and electron paramagnetic resonance spectroscopy.

Similarly, if we characterize the irreversible relaxation of the polarization of a medium in terms of the transverse relaxation time T_2, then the decay of the photon echo intensity I_e with the growth in the time interval τ in optical echo spectroscopy is also described (e.g., see [5–7]) by formula (2.4.9), where $1/\gamma_{ab}^{(1)}$ should be understood as the time T_2. Consequently, performing appropriate measurements and processing the results of these measurements with the use of formula (2.4.9), where $\gamma_{ab}^{(1)}$ is replaced by T_2^{-1}, one can obtain the experimental information concerning the time T_2 by means of the photon echo.

Let us consider the second example. The stimulated spin echo is employed in spin-echo spectroscopy to determine the irreversible longitudinal relaxation time T_1. As the time interval τ_2 between the second and third pump pulses increases, the intensity I_{se} of the stimulated spin echo signal decays according to (2.5.16) [1 4]. Therefore, varying τ_2 and processing the results of measurements with the use of formula (2.5.16), one can extract the information concerning the irreversible longitudinal relaxation time T_1.

The stimulated photon echo, as mentioned above, is produced similar to the stimulated spin echo. If we characterize the decay of populations in resonant levels in terms of the irreversible longitudinal relaxation time T_1, then the dependence of the stimulated photon echo intensity I_{se} on the time interval τ_2 is described by formula (2.5.16) (e.g., see [6, 7]). Thus, the formulas governing the intensity of the stimulated echo signal as a function of the time interval τ_2 are the same in the case of the spin and photon echoes.

The analogy between spectroscopic studies using spin and photon echoes is not restricted to the examples considered above and can be extended to other situations.

In the case of the photon echo in solids, the description of irreversible relaxation in terms of the longitudinal and transverse relaxation times T_1 and T_2 is usually a good approximation [5–7]. Therefore, the use of formula (2.4.9), where the relaxation characteristic $\gamma_{ab}^{(1)}$ is replaced by T_2^{-1}, and formula (2.5.16) may be very fruitful for the spectroscopy of inhomogeneously broadened resonant transitions of impurity paramagnetic ions or molecules. Consequently, we can infer that the photon-echo spectroscopy of solids has much in common with spin-echo spectroscopy, and the methods of spin-echo spectroscopy can be employed for spectroscopic investigations of impurity paramagnetic ions or molecules in solids in the optical range without substantial changes.

We deal with a different situation in the case of gas media. The description of the irreversible relaxation of a resonant transition and resonant levels in gas media in terms of the longitudinal and transverse relaxation times T_1 and T_2 is a rather rough approximation. Indeed, elastic collisions of resonant atoms (molecules) with each other and with atoms (molecules) of impurity gases may play an important role in gas media. These collisions change the velocities of resonant atoms (molecules) and result in the redistribution of these particles in the Zeeman sublevels of degenerate resonant states. For gas pressures usually achieved in photon-echo experiments, collisions leading to the redistribution of resonant atoms or molecules in the Zeeman sublevels usually play a very significant role. Such collisions are usually referred to as elastic depolarizing collisions. The integral of elastic depolarizing collisions has been thoroughly investigated in Chapter 1.

An adequate description of elastic depolarizing collisions requires, as mentioned above, the introduction of a set of relaxation characteristics, including $\gamma_{ab}^{(k)}$ and $\Delta_{ab}^{(k)}$ ($|J_a - J_b| \le k \le J_a + J_b$), $\gamma_a^{(k)}$ ($0 \le k \le 2J_a$), and $\gamma_b^{(k)}$ ($0 \le k \le 2J_b$). Here, J_b and J_a are the angular momenta of the upper and lower levels b and a involved in the resonant transition, $\gamma_{ab}^{(k)}$ and $\Delta_{ab}^{(k)}$ are the relaxation characteristics of multipole moments of the $b \to a$ resonant transition, and $\gamma_a^{(k)}$ and $\gamma_b^{(k)}$ are the relaxation characteristics of multipole moments of the resonant levels a and b. Therefore, the intensity decay of photon echo and stimulated photon echo signals in gas media is no longer described by simple formulas (2.4.9) and (2.5.16). The only exception is associated with $0 \rightleftarrows 1$, $1/2 \to 1/2$, $1 \to 1$, and $1/2 \rightleftarrows 3/2$ transitions. The intensity of the photon echo signal as a function of the time interval τ between the pump pulses for such transitions is described by formula (2.4.9). The validity of this formula for $J_b = 1 \to J_a = 0$ transitions was demonstrated in Section 2.4. The applicability of this expression for other transitions will be discussed in Section 4.2. However, even for these elementary transitions, formula (2.5.16) cannot describe the decay of the intensity of the stimulated photon echo signal as a function of the time interval τ_2 between the second and third pump pulses. The fact that this formula fails for $J_b = 1 \to J_a = 0$ transitions was demonstrated in Section 2.5. For other transitions, this circumstance will be discussed in Section 4.8.

Thus, the intensities of photon echo and stimulated photon echo signals in gas media are generally described by rather complicated functions of a large number of relaxation characteristics. Therefore, it is impossible to obtain the information concerning each of these characteristics by studying the intensity I_e of photon echo signals as a function of the time interval τ and the intensity I_{se} of stimulated photon echo signals as a function of the time interval τ_2.

The above-discussed factors require the development of new approaches in the photon-echo spectroscopy of gas media that would differ from the methods applied in spin-echo spectroscopy. In other words, a more complicated theory of echo spectroscopy based on the polarization properties of the photon echo and its modifications should be developed for gas media. This theory is called the theory of polarization echo spectroscopy. In this chapter we present this theory along with the description of the available experimental data and the discussion of the possibility of new experimental studies in this direction.

4.1 Identification of Resonant Transitions

As mentioned in Chapter 2, beginning with studies [8, 9], the approach to theoretical and experimental investigations of polarization properties of the photon echo and its modifications was radically revised. It was proposed to perform such investigations with small areas of one or all pump pulses. This approach makes it possible to derive relatively simple formulas, allowing the processing of the experimental data and the extraction of the required spectroscopic data. In particular, this approach provides an opportunity to identify a resonant transition or the type of a resonant transition and to find out how many resonant (independent or adjacent) transitions were involved in the formation of the photon echo or its modifications.

We should note that the identification of resonant transitions involved in the formation of the photon echo or its modifications is one of the most urgent problems of echo spectroscopy.

First, we will consider the possibility of identifying a resonant transition or its type from the polarization properties of the photon echo signal produced through this transition. We will assume that the photon echo signal is produced under the action of linearly polarized small-area pump pulses with arbitrary waveforms. The polarization vectors of the pump pulses are assumed to make an angle ψ with respect to each other. Suppose that the electric field strengths in these pulses are defined by formulas (2.3.1) and (2.3.2), where the functions g_1 and g_2, which describe the waveforms of the pump pulses, are no longer given by formulas (2.3.3) and (2.3.4), but only satisfy the normalization condition

$$\frac{1}{\widetilde{T}_n}\int_0^{\widetilde{T}_n} g_n(s)ds = 1. \tag{4.1.1}$$

Here, $\widetilde{T}_n$ ($n = 1, 2$) is the effective duration of the n-th pump pulse. We assume also that the electric field strength in the photon echo signal can be calculated with the use of the collision integral involved in the right-hand side of equations (1.5.5)–(1.5.7). Note that the conditions when calculations can be carried out with such an approximation, which will be discussed in Section 4.6, require that the mass of the resonant atom (molecule) should be larger than or on the order of the mass of atoms (molecules) of the buffer gas. Recall that the collision integral involved in the right-hand side of equations (1.5.5)–(1.5.7) can be derived from the collision integral involved in the right-hand sides of equations (1.3.2)–(1.3.4) by averaging over the direction of the velocity of resonant atoms (molecules) in the gas.

In the case when the above-specified conditions are satisfied and the spectral line of an optically allowed transition involved in the formation of the photon echo signal is inhomogeneously broadened and the relevant resonant levels have arbitrary angular momenta, the signal is linearly polarized, and the electric field strength $\boldsymbol{E}^{\mathrm{e}}$ of this signal is described by formula (2.3.36) [10]. Under these conditions, the quantity S, which is involved in (2.3.36) and which characterizes the decay and the waveform of the photon echo signal, is written as

$$S = \int c_1(v_y)[c_2^*(v_y)]^2 \exp[i(kv_y - \Delta\omega(v)(t - t_e)] \times$$

$$\times \exp[-\gamma_{ab}^{(1)}(v)(t - y/c)] f(\mathbf{v}) d\mathbf{v}, \tag{4.1.2}$$

where

$$c_n(v_y) = \frac{1}{\widetilde{T}_n} \int_{-\widetilde{T}_n/2}^{\widetilde{T}_n/2} g_n(s + \widetilde{T}_n/2) \exp[i(kv_y - \omega + \omega_0)s] ds. \tag{4.1.3}$$

The quantities $\Delta\omega(v)$ and $\gamma_{ab}^{(1)}(v)$ are defined by formulas (2.3.13) and (2.3.14), respectively; the time interval t_e is given by formula (2.3.35); v_y is the projection of the velocity $\mathbf{v}$ of resonant atoms (molecules) on the Y axis; and $f(\mathbf{v})$ describes the Maxwellian distribution of resonant atoms (molecules) in the gas in velocities $\mathbf{v}$.

Finally, the nonvanishing components of the vector $\boldsymbol{e}^e$, which is involved in (2.3.36) and which characterizes the polarization properties of the photon echo signal, are given by

$$e_x^e = \frac{1}{2} B(J_b, J_a) \sin\psi, \quad e_z^e = \frac{1}{3} A(J_b, J_a) \cos\psi, \tag{4.1.4}$$

where

$$A(J,J) = 2(3J^2 + 3J - 1)[5J(J+1)(2J+1)]^{-1}, \tag{4.1.5}$$

$$B(J,J) = 4(J-1)(J+2)[15J(J+1)(2J+1)]^{-1}, \tag{4.1.6}$$

$$A(J+1,J) = A(J,J+1) =$$
$$= 2(4J^2 + 8J + 5)[5(J+1)(2J+1)(2J+3)]^{-1}, \tag{4.1.7}$$

$$B(J+1,J) = B(J,J+1) =$$

$$= -8J(J+2)[15(J+1)(2J+1)(2J+3)]^{-1} . \quad (4.1.8)$$

Similar to Section 2.3, the Z axis in formulas (4.1.4)–(4.1.8) is directed along the polarization vector of the second pump pulse, and the polarization vector of the first pulse makes an angle ψ (in the clockwise direction if we look at these vectors along the Y axis) with the polarization vector of the second pulse.

Let us introduce an angle φ between the polarization vector of the photon echo signal and the polarization vector of the second pump pulse. The angle φ is positive when it is measured in the clockwise direction from the polarization vector of the second pump pulse, and we look at the polarization vectors along the Y axis. Otherwise, the angle φ is negative. As can be seen from (4.1.4)–(4.1.8), the angle φ is given by

$$\tan\varphi = (J-1)(J+2)(3J^2+3J-1)^{-1}\tan\psi \quad (4.1.9)$$

for $J \to J$ transitions and

$$\tan\varphi = -2J(J+2)(4J^2+8J+5)^{-1}\tan\psi \quad (4.1.10)$$

for $J \rightleftarrows J+1$ transitions.

Let us point out the specific features of formulas (4.1.9) and (4.1.10): (1) these formulas are valid for any relation between the Doppler width k_0u of a resonant spectral line and the inverse durations $1/\widetilde{T}_n$ ($n = 1, 2$) of the pump pulses; (2) for transitions involving resonant levels with low angular momenta ($0 \rightleftarrows 1$, $1 \to 1$, $1/2 \to 1/2$, $1/2 \rightleftarrows 3/2$) the polarization properties of the photon echo governed by (4.1.9) and (4.1.10) coincide with the polarization properties predicted earlier in [11–14] without the assumption of the smallness of pump-pulse areas; and (3) when a photon echo signal is produced by small-area pulses, the polarization vector of this signal lies within the angle ψ between the polarization vectors of the pump pulses in the case of $J \to J$ transitions ($J \geq 3/2$) and falls outside this angle in the case of $J \rightleftarrows J+1$ transitions ($J \geq 1/2$).

Formulas (4.1.9) and (4.1.10) allow resonant transitions involving levels with relatively low angular momenta ($J \leq 10$) to be identified by means of the photon echo. When transitions between levels with high angular momenta are involved in the formation of the photon echo, these formulas permit the type of the resonant transition ($J \to J$ or $J \rightleftarrows J + 1$) to be identified. In the corresponding limiting case ($J >> 1$), formulas (4.1.9) and (4.1.10) are reduced to

$$\tan\varphi = (1/3)\tan\psi \tag{4.1.11}$$

for $J \to J$ transitions and

$$\tan\varphi = -(1/2)\tan\psi \tag{4.1.12}$$

for $J \rightleftarrows J + 1$ transitions. Hence, the polarization vector of the photon echo signal lies within the angle ψ between the polarization vectors of the pump pulses in the case of $J \to J$ transitions ($J >> 1$), when the relation $\tan\varphi/\tan\psi = 1/3$ is satisfied, and falls outside this angle in the case of $J \rightleftarrows J + 1$ transitions ($J >> 1$), when the relationship $\tan\varphi/\tan\psi = -1/2$ holds true.

Formulas (4.1.9)–(4.1.12) are illustrated by Fig. 4.1, where curves 8, 5, 1, 2, 6, 1, 7, 3, and 4 represent the dependences $\varphi(\psi)$ for $J \rightleftarrows J + 1$ ($J >> 1$), $J \to J$ ($J >> 1$), $0 \rightleftarrows 1$, $1/2 \to 1/2$, $1/2 \rightleftarrows 3/2$, $1 \to 1$, $1 \rightleftarrows 2$, $3/2 \to 3/2$, and $2 \to 2$ transitions, respectively.

We emphasize that formulas (4.1.9) and (4.1.10) do not involve the areas θ_1 and θ_2 of the pump pulses. Therefore, the areas θ_1 and θ_2 in experiments should be gradually decreased from their optimal values, corresponding to the maximum intensity of the echo signal, until the polarization properties of the echo signal become independent of θ_1 and θ_2.

Due to their simplicity and usefulness for the processing of the experimental data, formulas (4.1.9)–(4.1.12) immediately attracted the attention of researchers working the echo spectroscopy. Recall that these formulas were derived for the first time in [8, 9] for the cases when the spectral line involved in the formation of the photon echo signal is either narrow or broad. The following assumptions were employed in the derivation of these formulas: (1) the carrier frequency ω of the pump pulses is exactly resonant to the frequency ω_0 of the resonant transition; (2) the collision integral in the model of elastic depolarizing collisions can be averaged not only in the direction, but also in the modulus of the velocity $\mathbf{v}$ of resonant atoms

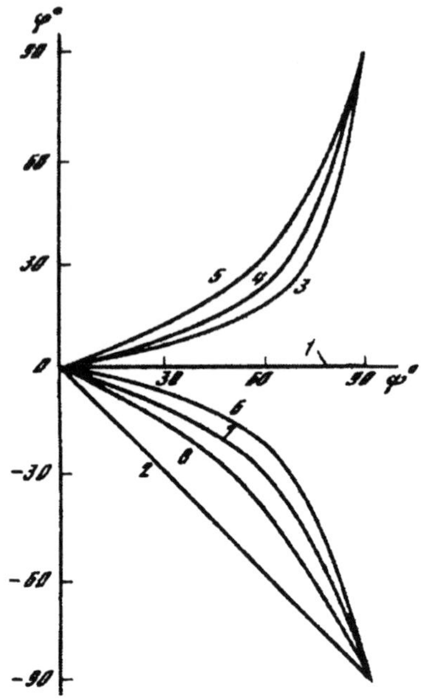

Fig. 4.1. Dependence of the angle φ on the angle ψ for different resonant transitions in the approximation of small-area pump pulses.

(molecules); and (3) the pump pulses have a rectangular shape. Note that these restrictions were subsequently lifted. The first restriction was lifted in [15], where it was demonstrated that formulas (4.1.9)–(4.1.12) hold true also away from the exact resonance. The authors of [15] have also shown that these expressions remain valid when one employs the collision integral involved in the right-hand sides of equations (1.5.5)–(1.5.7). Finally, Yevseyev and Reshetov lifted the third restriction, by demonstrating that formulas (4.1.9)–(4.1.12) hold true for pump pulses with arbitrary waveforms.

The reliability of formulas (4.1.9)–(4.1.12) has been verified many times. In particular, the authors of [16] have demonstrated that the limiting transition to small-area pump pulses performed in the Wang formulas [17], which are applicable only for the case when the spectral line involved in the formation of the photon echo signal is narrow, yields formulas (4.1.9) and (4.1.10). Alekseev and Beloborodov [18] have tested these formulas by letting the areas of the pump pulses be small in expressions governing the polarization properties of the photon echo signal in the case of optically allowed transitions with high angular momenta ($J >> 1$) of the resonant levels.

As mentioned above, expressions (4.1.9) and (4.1.10) were derived in the limiting case when both pump pulses have small areas. However, the applicability range of these formulas is apparently much broader. We can arrive at such a conclusion by analyzing the results presented in [19, 20]. In particular, Yevseyev and Reshetov [19] considered the case when the area θ_1 of the first pump pulse is small, while the second pump pulse has an arbitrary area θ_2. The authors of [20] analyzed a situation when the first pump pulse has an arbitrary area, while the area of the second pump pulse is small. These studies have shown that the considered formulas remain valid up to the areas of the pump pulses on the order of unity. A comprehensive discussion of these results is provided in [10].

Formulas (4.1.9)–(4.1.12) have been already employed for the identification of resonant transitions. Figure 4.2 presents the results obtained by Vasilenko and Rubtsova [21], who applied formulas (4.1.9)–(4.1.12) to identify resonant transitions in SF_6 molecules excited by CO_2-laser pulses. The solid curves in Fig. 4.2 represent the results of calculations carried out with the use of formulas (4.1.11) and (4.1.12), while the crosses and circles correspond to the experimental data obtained for $P(16)$ and $P(18)$ oscillation lines of a CO_2-laser. Comparison of the experimental data and theoretical predictions allowed the authors of [21] to infer that the absorption line of SF_6 molecules near the $P(16)$ oscillation line of a CO_2-laser is due to a transition with ($J >> 1$) belonging to the Q-branch. The authors of [21] failed to unambiguously identify the transition responsible for the absorption line corresponding to the $P(18)$ oscillation line of a CO_2-laser. We emphasize that, in contrast to the paper by Alimpiev and Karlov [22], which was already mentioned in Chapter 2, the authors of [21] identified transitions without invoking an *a priori* information concerning large values of rotational quantum numbers related to resonant transitions, because they employed formulas (4.1.9)–(4.1.12) to process their experimental data.

Note that resonant transitions can be identified not only by analyzing the angle φ between the polarization vector of the photon echo signal and the polarization vector of the second pump pulse as a function of the angle ψ between the polarization vectors of the pump pulses with subsequent processing of this dependence with the use of formulas (4.1.9)–(4.1.12), but also by experimentally studying the dependence of the photon echo intensity on the angle ψ. Indeed, formulas (2.3.36) and (4.1.2)–(4.1.8) show that the variation in the angle ψ between the polarization vectors of the pump pulses changes the intensity of the photon echo signal. For example, in the

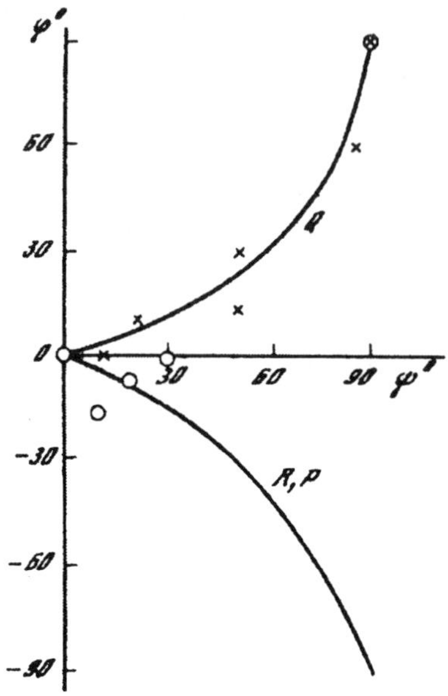

Fig. 4.2. Dependence of the angle φ between the polarization vector of the photon echo and the polarization vector of the second pump pulse on the angle ψ between polarization vectors of the pump pulses.

limiting case when ($J >> 1$), formulas (2.3.36) and (4.1.2)–(4.1.8) yield the following expressions for the dimensionless intensity:

$$I_e(\psi)/I_e(0) = (1/9)(5 + 4\cos 2\psi) \tag{4.1.13}$$

for $J \rightarrow J$ transitions and

$$I_e(\psi)/I_e(0) = (1/8)(5 + 3\cos 2\psi) \tag{4.1.14}$$

for $J \rightleftarrows J + 1$ transitions. As can be seen from these expressions, the curves representing the angular dependence of the photon echo intensity start with 1 for $\psi = 0$ and then converge, reaching the values of 1/9 and 1/4 at $\psi = \pi/2$. Thus, expressions (4.1.13) and (4.1.14) can be also employed for the identification of the type ($J \rightarrow J$ or $J \rightleftarrows J + 1$) of resonant transitions.

The identification of resonant transitions can be performed with the use of polarization properties of photon echo signals produced by elliptically polarized small-area pump pulses. This scheme of the photon echo was investigated for the first time by the authors of [23] and is discussed in detail in [10].

In the context of photon-echo experiments performed with vibrational-rotational transitions in molecular gases, the case of resonant levels with high angular momenta is of special interest. As mentioned above, the method of photon echo allows one in this case to determine the type of resonant transitions, or, in other words, to identify the branch (Q or $P(R)$) of the transition. However, we should take into account that two or several transitions may be involved in the formation of the photon echo signal in a molecular gas. These transitions may be either independent or adjacent to each other. In this context, Yevseyev and Yermachenko [24] considered the formation of the photon echo signal by small-area pump pulses in the case of two independent or adjacent transitions. Analysis performed in [24] shows that one can identify the type of the manifold of resonant levels involved in the formation of the photon echo.

Let us first discuss the results of [24] related to the case when independent transitions are involved in the formation of a photon echo signal in the presence of two small-area pump pulses. Let the electric field strengths of the pump pulses be defined by expressions (2.3.1) and (2.3.2), where the function g_n is normalized in accordance with (4.1.1), J_a and J_b are the angular momenta of the lower and upper levels a and b of one of the independent optically allowed transition ($b \to a$ transition), and $J_{a'}$ and $J_{b'}$ are the angular momenta related to the other independent optically allowed transition ($b' \to a'$ transition). We introduce N_0' to denote the density of the population difference for the Zeeman sublevels of resonant levels b' and a' before the moment of time when the first pump pulse reaches the point y of the gas medium. Suppose also that ω_0' and d' are the frequency and the reduced matrix element of the dipole moment operator for the $b' \to a'$ transition, respectively. Finally, the relaxation characteristics of the dipole moment of the $b' \to a'$ resonant transition will be denoted as $\gamma^{(1)}_{a'b'}(v)$ and $\Delta^{(1)}_{a'b'}(v)$. Analogous quantities for the $b \to a$ resonant transition will be denoted as N_0, ω_0, d, $\gamma^{(1)}_{ab}(v)$, and $\Delta^{(1)}_{ab}(v)$, as before. Then, the electric field strength in the photon echo signal corresponding to the optically allowed $b \to a$ transition is described by formulas (2.3.36) and (4.1.2)–(4.1.8). Suppose that $N_0' = N_0$, $|d'| = |d|$, $|\omega_0' - \omega_0|\tau << 1$, and the inequalities $|\gamma^{(1)}_{a'b'}(v) - \gamma^{(1)}_{ab}(v)|\tau << 1$ and $|\Delta^{(1)}_{a'b'}(v) - \Delta^{(1)}_{ab}(v)|\tau << 1$ are satisfied for the characteristics values of v. Then, the electric field strength in the photon echo signal produced through the optically allowed $b' \to a'$ transition is also described by formulas (2.3.36) and (4.1.2)–(4.1.8) if the replacements $J_a \to J_{a'}$ and $J_b \to J_{b'}$ are made in these formulas. Thus, the electric field strength of the photon echo signal produced through both independent transitions $b \to a$ and $b' \to a'$ is described in the case under study by formulas (2.3.36), (4.1.2), and (4.1.3), where the vector $\boldsymbol{e}^e$, which characterizes the polarization properties of the photon echo, has the following nonvanishing components:

$$e_x^e = \frac{1}{2}[B(J_b, J_a) + B(J_{b'}, J_{a'})]\sin\psi\,, \tag{4.1.15}$$

$$e_z^e = \frac{1}{3}[A(J_b, J_a) + A(J_{b'}, J_{a'})]\cos\psi\,. \tag{4.1.16}$$

Here, the quantities $A(J_b, J_a)$ and $B(J_b, J_a)$ are defined by formulas (4.1.5)–(4.1.8), and the quantities $A(J_{b'}, J_{a'})$ and $B(J_{b'}, J_{a'})$ can be obtained from $A(J_b, J_a)$ and $B(J_b, J_a)$ with the replacements $J_b \to J_{b'}$ and $J_a \to J_{a'}$.

Below, we will consider vibrational-rotational transitions with high rotational quantum numbers. In other words, we assume that the angular momenta of the resonant levels a, b, a', and b' are high. Formulas (4.1.5)–(4.1.8) yield in this case

$$A(J, J) = 3(5J)^{-1}, \quad B(J, J) = 2(15J)^{-1}, \tag{4.1.17}$$

$$A(J, J+1) = A(J+1, J) = 2(5J)^{-1},$$

$$B(J, J+1) = B(J+1, J) = -2(15J)^{-1}. \tag{4.1.18}$$

Let us consider three possible schemes of photon-echo formation with pump pulses resonant to two independent vibrational-rotational transitions with high rotational quantum numbers. Suppose that each of the independent transitions belongs to the Q branch in the first case. Then, using J and J' to denote the rotational quantum numbers of the levels involved in resonant transitions $b \to a$ and $b' \to a'$, we can apply (4.1.15)–(4.1.18) to derive

$$e_x^e = \frac{1}{15}[J^{-1} + (J')^{-1}]\sin\psi ,$$

$$e_z^e = \frac{1}{5}[J^{-1} + (J')^{-1}]\cos\psi . \tag{4.1.19}$$

Thus, the tilt angle φ of the polarization vector of the photon echo signal relative to the polarization vector of the second pump pulse in this case can be found from the equation

$$\tan\varphi = (1/3)\tan\psi.$$

This equation coincides with relation (4.1.11). Consequently, the case when the photon echo signal is produced through a single resonant transition belonging to the *Q* branch with high rotational quantum numbers is indistinguishable in the polarization properties of the photon echo signal from the case when the photon echo signal is produced through two independent transitions belonging to the *Q* branches with high rotational quantum numbers.

Suppose that, in the second case, the photon echo signal is produced through two independent transitions, and each of these transitions belongs to the *P* (*R*) branch or one of these transitions belongs to the *P* branch, while the other one belongs to the *R* branch. Then, assuming that the rotational quantum numbers of these transitions are high, we can apply (4.1.15)–(4.1.18) to find that

$$e_x^e = -\frac{1}{15}[J^{-1} + (J')^{-1}]\sin\psi ,$$

$$e_z^e = \frac{2}{15}[J^{-1} + (J')^{-1}]\cos\psi . \tag{4.1.20}$$

Here, J is the rotational quantum number of the levels involved in the $b \rightarrow a$ transition, and J' is the rotational quantum number of the levels involved in the $b' \to a'$ transition. Thus, the angle φ in this case can be found from the equation

$$\tan\varphi = -(1/2)\tan\psi,$$

which coincides with relation (4.1.12). Consequently, the considered case of photon-echo formation is indistinguishable in the polarization properties of the photon echo signal from the case when the photon echo signal is produced through a single transition belonging to the *P* (*R*) branch with high rotational quantum numbers.

Finally, suppose that the photon echo signal is produced through two independent transitions with high rotational quantum numbers and one of these transitions belongs to the *Q* branch, while the other one belongs to the *P* (*R*) branch. Formulas (4.1.15)–(4.1.18) yield in this case

$$e_x^e = \frac{1}{15}[J^{-1} - (J')^{-1}]\sin\psi, \tag{4.1.21}$$

$$e_z^e = \frac{1}{15}[3J^{-1} + 2(J')^{-1}]\cos\psi. \tag{4.1.22}$$

Here, as before, J is the rotational quantum number of the levels involved in the $b \to a$ transition, and J' is the rotational quantum number of the levels involved in the $b' \to a'$ transition. Thus, the angle φ in this case can be found from the equation

$$\tan\varphi = (J' - J)(3J' + 2J)^{-1}\tan\psi. \tag{4.1.23}$$

If $J' >> J$, then equation (4.1.23) gives equation (4.1.11). Provided that $J' << J$, equation (4.1.23) is reduced to (4.1.12). Finally, in the case when $J' \approx J$, which is of special interest for practical applications, equation (4.1.23) yields

$$\tan\varphi = 0. \tag{4.1.24}$$

Consequently, if one of the independent transitions belongs to the Q branch, the other one belongs to the P (R) branch, and the rotational quantum numbers of these transitions are high and approximately equal to each other, then the polarization vector of the photon echo signal, as it follows from (4.1.24), coincides with the polarization vector of the second pump pulse. This circumstance provides an opportunity to easily distinguish this case from the cases when the photon echo signal is produced through a single vibrational–rotational transition with high rotational quantum numbers.

We should note that Vasilenko and Rubtsova [25] have already applied formula (4.1.23) for the processing of the results of experiments devoted to the identification of the type of resonant transitions involved in the formation of the photon echo signal. The use of this formula allowed the authors of [25] to identify a background of a Q-type line in the absorption corresponding to the $P(33)$ transition in the $0 \to 1$ band of the ν_3 mode of SF_6 molecules.

Finally, let us consider the results of [24] related to the case when two adjacent optically allowed transitions are involved in the formation of a photon echo signal in the presence of two small-area pump pulses. A V-system was considered in [24] as a system of two adjacent optically allowed transitions. Suppose that the angular momentum of the lower resonant level a is equal to J_a, and the angular momenta of the upper resonant levels b and c are equal to J_b and J_c, respectively. We assume that $c \to a$ and $b \to a$ transitions are optically allowed, while the $c \to b$ transition is optically forbidden ($E_c > E_b > E_a$). The strengths of the electric field in the pump pulses are described by formulas (2.3.1) and (2.3.2), where the function g_n is normalized in accordance with condition (4.1.1), and the carrier frequency ω of the pump pulses is resonant to the frequencies of optically allowed $c \to a$ and $b \to a$ transitions. Although calculations in [24] have been performed for arbitrary angular momenta of resonant levels, in what follows, we will restrict our consideration to the case when $J >> 1$.

As pointed out in [24], the three-level systems of the considered type can be divided into four groups. The first group includes systems where both optically allowed resonant transitions belong to the Q branch. The second group includes systems where both optically allowed resonant transitions belong to the P (R) branch. The third group includes systems where one of the optically allowed transitions belongs to the Q branch, while the second one belongs to the P (R) branch. Finally, the fourth group includes systems where one of the optically allowed transitions belongs to the P branch, while the second one belongs to the R branch.

Suppose that the population differences for the Zeeman sublevels of the resonant levels c and a, as well as b and a, are equal to each other. We also assume that the reduced matrix elements of optically allowed $c \to a$ and $b \to a$ transitions are equal to each other, and the splitting frequency $\Delta = (E_c - E_b)\hbar^{-1}$ is small as compared with the inverse pump pulse durations $1/\widetilde{T}_n$ ($n = 1, 2$). In this case, when narrow spectral lines of $c \to a$ and $b \to a$ transitions are involved in the formation of the photon echo signal, the maximum of the signal is characterized by a linear polarization, whereas in the case when broad spectral lines of $c \to a$ and $b \to a$ transitions are involved in the formation of the photon echo signal, the entire signal is linearly polarized. Equations that govern the tilt angle φ of the polarization vector corresponding to the maximum of the photon echo signal (the entire photon echo signal) with respect to the polarization vector of the second pump pulse for the above-specified four groups considerably differ from each other. In particular, for the first group, we have the relation [24] $\tan\varphi = (1/3)\tan\psi$, which coincides with equation (4.1.11). This result indicates that the case when two adjacent transitions with high rotational quantum numbers belonging to the Q branch are involved in the formation of the photon echo signal cannot be distinguished in the polarization properties of the photon echo signal from the case when a single transition with high rotational quantum numbers belonging to the Q branch is involved in the formation of the photon echo signal.

For the second group of systems, we find that, when two adjacent optically allowed transitions with high rotational quantum numbers belonging to the P (R) branch are involved in the formation of the photon echo signal, the angle φ can be found from the equation $\tan\varphi = -(1/2)\tan\psi$.

This relation coincides with equation (4.1.12). Therefore, the case when two adjacent optically allowed resonant transitions belonging to the second group are involved in the formation of the photon echo signal cannot be distinguished from the situation when the photon echo is produced through a transition that belongs to the P (R) branch with high rotational quantum numbers.

For systems that belong to the third group, we have [24]

$$\tan\varphi = -\cos(\Delta\tau)[5 + 2\cos(\Delta\tau)]^{-1}\tan\psi. \tag{4.1.25}$$

In this case, the polarization plane of the photon echo signal displays quantum beats as a function of the time interval τ between the pump pulses.

With $\Delta\tau << 1$, such beats are not observed, and the angle φ can be found from the equation

$$\tan\varphi = -(1/7)\tan\psi. \tag{4.1.26}$$

Thus, for systems that belong to the third group with $\Delta\tau << 1$, the polarization vector of the photon echo signal falls outside the angle ψ between the polarization vectors of pump pulses. Note that equation (4.1.26) considerably differs from equations (4.1.11), (4.1.12), and (4.1.24). Therefore, the case when two adjacent optically allowed transitions belonging to the third group are involved in the formation of the photon echo signal can be identified through the analysis of the polarization of the photon echo signal as a function of the angle ψ.

For systems including two adjacent optically allowed transitions belonging to the fourth group, the angle φ can be determined [24] from the following equation:

$$\tan\varphi = -(1/4)[2 - \cos(\Delta\tau)][1 + \cos(\Delta\tau)]^{-1}\tan\psi. \tag{4.1.27}$$

Consequently, similar to the systems that belong to the third group, the polarization plane of the photon echo signal in the systems that belong to the fourth group displays quantum beats as a function of the time interval τ between the pump pulses. With $\Delta\tau << 1$, formula (4.1.27) yields

$$\tan\varphi = (1/8)\tan\psi. \tag{4.1.28}$$

Thus, the polarization vector of the photon echo signal lies within the angle ψ for the systems that belong to the fourth group with $\Delta\tau << 1$. As can be seen from (4.1.11), (4.1.12), (4.1.24), (4.1.26), and (4.1.28), the case when a system of adjacent optically allowed resonant transitions belonging to the fourth group is involved in the formation of the photon echo signal can be identified by analyzing the polarization of the photon echo signal as a function of the angle ψ between the polarization vectors of the pump pulses.

Closing this section, we will consider the possibility of using the non-Faraday rotation of the polarization vector of the photon echo signal in a permanent uniform longitudinal magnetic field for the identification of resonant transitions. This method of identification has been already mentioned in Chapter 2. Our analysis of this approach is based on the results of [26, 27], which have been obtained for the case when the photon echo signal is produced by small-area pump pulses of an arbitrary shape through optically allowed transitions with arbitrary angular momenta J_a and J_b of resonant levels. The case of resonant levels with close g-factors,

$$(J_a + J_b)|\varepsilon_a - \varepsilon_b|\tau << 1, \tag{4.1.29}$$

was considered in [26], while the case of resonant levels with an arbitrary relation between the g-factors was analyzed in [27]. The quantities ε_a, ε_b in (4.1.29) stand for the frequencies of transitions between the neighboring Zeeman sublevels of the considered states:

$$\varepsilon_{a,\,b} = \mu_0 g_{a,\,b} H\hbar^{-1}, \tag{4.1.30}$$

where μ_0 is the Bohr magneton, $\boldsymbol{H}$ is the strength of the magnetic field, and g_a and g_b are the g-factors of the resonant levels.

Let us consider the formation of a photon echo signal in a gas medium in the presence of a permanent uniform magnetic field with a strength $\boldsymbol{H}$. In this and three next sections, we assume that the electric field strength in the photon echo signal can be calculated with the use of the collision integral determined in the model of elastic depolarizing collisions and averaged not only in the direction, but also in the modulus of the velocity $\mathbf{v}$ of resonant atoms (molecules).

The lower and upper degenerate levels a and b with energies E_a and E_b and total angular momenta J_a and J_b which are resonant to the frequency ω of pump pulses, are transformed in the presence of a magnetic field into two groups of Zeeman sublevels with energies E_{m_a} and E_{m_b}. In the approximation linear in H, the quantities E_{m_a} and E_{m_b} are given by

$$E_{m_a} = E_a + \mu_0 g_a H m_a, \qquad E_{m_b} = E_b + \mu_0 g_b H m_b,$$

where the indices m_a and m_b characterize the projection of the total angular momenta of the lower and upper resonant levels on the quantization axis, which is chosen along the Z axis directed along the vector $\boldsymbol{H}$. The Zeeman splitting is assumed to be small as compared with $\hbar\omega$.

Since we are searching for the electric field strength in the photon echo signal in a longitudinal magnetic field directed along the Z axis, the electric field strengths in the pump pulses are given by formulas (2.3.1) and (2.3.2), where the replacement $y \to z$ should be made, and the quantity ξ should be understood as $(t - z/c)$. We assume that the pump pulses are linearly polarized in the same plane ($\boldsymbol{l}_1 = (0, 1, 0)$ and $\boldsymbol{l}_2 = (0, 1, 0)$) and the function g_n, which describes the waveform of the n-th pump pulse, is normalized in accordance with (4.1.1). Then, we can find that electric field strength in the photon echo signal using the set of equations (2.3.8)–(2.3.11), where the replacements $y \to z$ and $v_y \to v_z$ should be made and some additional terms should be introduced. For equations (2.3.9)–(2.3.11), these additional terms are written as

$$- iq\varepsilon_a \widetilde{\psi}_q^{(k)} + i^{1/2}(\varepsilon_a - \varepsilon_b)\sum_{k'} Q_q^{kk'} \widetilde{\psi}_q^{(k')}, \tag{4.1.31}$$

$$- iq\varepsilon_b f_q^{(k)}, \tag{4.1.32}$$

$$- iq\varepsilon_a \varphi_q^{(k)}. \tag{4.1.33}$$

These terms should be introduced into the left-hand sides of the corresponding equations. The matrix operator

$$Q_q^{kk'} = (-1)^{J_a + J_b + q}[2J_b(2J_b + 1)(2J_b + 2)]^{1/2}(2k' + 1) \times \\ \times \begin{pmatrix} 1 & k & k' \\ 0 & -q & q \end{pmatrix} \begin{Bmatrix} 1 & k & k' \\ J_a & J_b & J_b \end{Bmatrix} \tag{4.1.34}$$

determines the degree of coupling between various components of the optical coherence matrix due to the difference in the g-factors of resonant levels.

In our calculations, we will neglect the influence of the magnetic field during the propagation of the pump pulses through the gas medium, since the pump pulses are assumed to be sufficiently short:

$$\varepsilon_a \widetilde{T}_n << 1, \varepsilon_b \widetilde{T}_n << 1, \quad n = 1, 2. \tag{4.1.35}$$

In addition, similar to calculations performed above, we assume that the durations $\widetilde{T}_n$ ($n = 1, 2$) of the pump pulses are small as compared with the time interval τ between these pulses and irreversible relaxation times.

The method of determining the electric field strength in the photon echo signal in the case under study is similar to the procedure described in Section 2.3. Since inequalities (4.1.35) are satisfied, the field is subject to changes only within the time intervals between the first and second pump pulses and between the second pump pulse and the photon echo signal. The slowly varying amplitude $\widetilde{\psi}_q^{(k)}$ within these intervals, along with the irreversible relaxation governed by $\gamma_{ab}^{(k)}$ and $\Delta_{ab}^{(k)} - \Delta_{ab}^{(1)}$, displays changes due to the presence of a longitudinal magnetic field, which is described by (4.1.31). Such calculations have been carried out in [26, 27]. Therefore, we present only the results of these calculations here. To ensure the uniformity of our consideration, we present the results of these calculations in such a way as to permit a convenient comparison of these results with the formulas presented in the previous and following sections. For this purpose, we make the replacements $z \to y$ and $y \to z$ in the final results, which implies that the pump pulse again propagate along the Y axis, which is directed along the magnetic field, and the polarization vectors of the first and second pump pulses are directed along the Z axis.

In the case when the resonant levels have close g factors [26], the electric field strength in the photon echo signal is determined by formula (2.3.36), where the quantity S is given by

$$S = \exp[-\gamma_{ab}^{(1)}(t - y/c)] \int c_1(v_y)[c_2^*(v_y)]^2 f(v_y) \times$$

$$\times \exp[i(kv_y - \omega + \omega_0 + \Delta_{ab}^{(1)})(t - t_e)] dv_y. \tag{4.1.36}$$

Here, the function $c_n(v_y)$ is defined by formula (4.1.3), and the quantity t_e is given by formula (2.3.35). The vector $\boldsymbol{e}^e$ involved in (2.3.36) should be replaced by a vector $\boldsymbol{e}^e(t')$ characterizing polarization properties of the photon echo signal and having the following nonvanishing components:

$$e_x^e(t') = (1/3)A(J_b, J_a)\cos(\varepsilon\tau)\sin(\varepsilon t') + \\ + (1/2)B(J_b, J_a)\sin(\varepsilon\tau)\cos(\varepsilon t'), \tag{4.1.37}$$

$$e_z^e(t') = (1/3)A(J_b, J_a)\cos(\varepsilon\tau)\cos(\varepsilon t') - \\ - (1/2)B(J_b, J_a)\sin(\varepsilon\tau)\sin(\varepsilon t'). \tag{4.1.38}$$

Here, $\varepsilon = (\varepsilon_a + \varepsilon_b)$, $A(J_b, J_a)$, and $A(J_b, J_a)$ are determined by formulas (4.1.5)–(4.1.8), and t' is given by expression (2.3.33).

Analysis of the quantity S, determined by formula (4.1.36), shows (e.g., see [10]) that the vector $\boldsymbol{e}^e(\tau)$ describes polarization properties of the maximum of the photon echo intensity (in the case when the spectral line involved in the formation of the photon echo is narrow for both pump pulses). By virtue of inequalities (4.1.35), this vector also describes polarization properties of the entire signal (in the case when the spectral line involved in the formation of the photon echo is broad for both pump pulses). Applying (4.1.37) and (4.1.38) with $t' = \tau$, we derive the following expressions for the nonvanishing components of the vector $\boldsymbol{e}^e(\tau)$:

$$e_x^e(\tau) = (1/2)[(1/3)A(J_b, J_a) + (1/2)B(J_b, J_a)]\sin(2\varepsilon\tau), \tag{4.1.39}$$

$$e_z^e(\tau) = (1/2)[(1/3)A(J_b, J_a) - (1/2)B(J_b, J_a) + \\ + ((1/3)A(J_b, J_a) + (1/2)B(J_b, J_a))\cos(2\varepsilon\tau)]. \tag{4.1.40}$$

The angle φ between the polarization vector of the photon echo signal and the polarization vectors of the pump pulses can be determined from the equation

$$\tan\varphi = e_x^e(\tau)/e_z^e(\tau). \tag{4.1.41}$$

As it follows from (4.1.39)–(4.1.41) and (4.1.5)–(4.1.8), the clockwise rotation of the polarization vector of the photon echo signal around $\boldsymbol{H}$ occurs for all the optically allowed transitions, except for the $J_b = 1/2 \rightarrow J_a = 1/2$ transition. This effect can be explained in the following way.

Suppose for simplicity that $\varepsilon\tau < \pi/2$. The first pump pulse passing through a medium induces polarization in this medium. Under these conditions, the polarization vector $\boldsymbol{P}$ of a group of atoms (molecules) moving with a velocity $\mathbf{v}$ is directed along the polarization vector of the first pump pulse at the moment of time when this pulse passes through the point y of the gas medium (which is a consequence of the constant-field approximation). Within the time interval between the first and second pump pulses, the vector $\boldsymbol{P}$ displays clockwise precession around $\boldsymbol{H}$ with a frequency ε in such a way that the angle of the clockwise rotation of the vector $\boldsymbol{P}$ around $\boldsymbol{H}$ at the moment of time when the second pump pulse reaches the point y of the gas medium is equal to $\varepsilon\tau$. The second pump pulse changes the sign of the phase of some terms in the expression for the vector $\boldsymbol{P}$. As demonstrated in Section 2.3, these terms contribute to the electric field strength of the photon echo signal. These terms in the expression for $\boldsymbol{P}$ will be denoted as $\boldsymbol{P}^e$. Note that, at the moment of time when the second pump pulse leaves the point y of the gas medium, the vector $\boldsymbol{P}^e$ has different orientations for different transitions. In particular, for $J_b = 1/2 \rightarrow J_a = 1/2$ transition, this vector is turned counterclockwise by the angle $\varepsilon\tau$ around the vector $\boldsymbol{H}$, while for $J \rightleftarrows J + 1$ transitions ($J \geq 1/2$), the rotation angle is less than $\varepsilon\tau$. For $J_b = 1 \rightarrow J_a = 0$, $J_b = 0 \rightarrow J_a = 1$, and $J_b = 1 \rightarrow J_a = 1$ transitions, the vector $\boldsymbol{P}^e$ is directed along the polarization vectors of the pump pulses. Finally, for $J \rightarrow J$ ($J \geq 3/2$) transitions, the angle of the clockwise rotation of the vector $\boldsymbol{P}^e$ around $\boldsymbol{H}$ is less than $\varepsilon\tau$. Within the time interval between the second pump pulse and the generation of the photon echo signal, the vector $\boldsymbol{P}^e$ displays a clockwise precession around the vector $\boldsymbol{H}$ with a frequency ε again. Thus, at the moment of time $t \approx 2\tau + y/c$, when the emitting particles have the same phase, the vector $\boldsymbol{P}^e$ is turned clockwise around the vector $\boldsymbol{H}$ for all the transitions, except for $1/2 \rightarrow 1/2$ transition. This effect gives rise to the rotation of the polarization vector of the photon echo signal with respect to the polarization vectors of pump pulses. Note that the rotation angle in this case is sensitive to the angular momenta of the levels involved in the optically allowed resonant transition.

Expressions (4.1.39)–(4.1.41) make it possible to employ the results of measurements of the angle φ as a function of the parameter $\varepsilon\tau$ to find the angular momenta of the levels involved in a resonant transition. These quantities can be determined with a reasonable reliability only when they are sufficiently low. Note that the parameter $\varepsilon\tau$ can be changed in experiments by varying the delay time between the pump pulses or by changing the strength $\boldsymbol{H}$ of the longitudinal magnetic field.

As mentioned above, the case of high angular momenta ($J >> 1$) is especially important for the experimental investigation of the photon echo in molecular gases. In this case, formulas (4.1.39)–(4.1.41) yield

$$\tan\varphi = 2\sin(2\varepsilon\tau)[1 + 2\cos(2\varepsilon\tau)]^{-1} \qquad (4.1.42)$$

for $J \rightarrow J$ transitions and

$$\tan\varphi = \sin(2\varepsilon\tau)[3 + \cos(2\varepsilon\tau)]^{-1} \qquad (4.1.43)$$

for $J \rightleftarrows J + 1$ transitions.

As can be seen from formulas (4.1.42) and (4.1.43), the rotation angle of the polarization plane of the photon echo signal may depend in a different way on $\varepsilon\tau$ for $J \rightarrow J$ and $J \rightleftarrows J + 1$ transitions with $J >> 1$. Indeed, for $J \rightarrow J$ transitions ($J >> 1$), we can choose the parameter $\varepsilon\tau$ in such a way that the photon echo signal is polarized perpendicular to the polarization plane of the pump pulses. For $J \rightleftarrows J + 1$ transitions ($J >> 1$), the maximum rotation angle $\arctan(2)^{-3/2}$ of the polarization vector of the photon echo signal is comparatively small. Thus, expressions (4.1.42) and (4.1.43) can be employed to experimentally identify the type of a resonant transition ($J \rightarrow J$ or $J \rightleftarrows J + 1$) involved in the formation of a photon echo signal. Note that such an identification has been recently carried out by Popov *et al.* [28], who investigated the generation of a photon echo signal in the presence of a permanent uniform longitudinal magnetic field in a molecular iodine vapor.

A variation in the parameter $\varepsilon\tau$ changes also the intensity of the photon echo signal. Specifically, in the limiting case when $J >> 1$, we can apply (2.3.36), (4.1.39), and (4.1.40) to derive the following expressions for the normalized photon echo intensities:

$$I_e(\varepsilon\tau)/I_e(0) = (1/9)[5 + 4\cos(2\varepsilon\tau)] \qquad (4.1.44)$$

for $J \rightarrow J$ transitions, and

$$I_e(\varepsilon\tau)/I_e(0) = (1/8)[5 + 3\cos(2\varepsilon\tau)] \qquad (4.1.45)$$

for $J \rightleftarrows J+1$ transitions. These formulas can also be employed for the identification of the type of resonant transitions.

In the case of an arbitrary relation between the g factors of resonant levels, the expression for the electric field strength in the photon echo signal in the presence of a permanent uniform longitudinal magnetic field can be derived only for resonant transitions with relaxation characteristics meeting the following inequalities:

$$|\gamma_{ab}^{(k)} - \gamma_{ab}^{(1)}|\tau << 1, \qquad |\Delta_{ab}^{(k)} - \Delta_{ab}^{(1)}|\tau << 1. \tag{4.1.46}$$

The formulas describing the photon echo signal in this case were derived in [27]. Similar to expressions (2.3.36) and (4.1.36)–(4.1.38), these formulas make it possible to identify a resonant transition or the type of a resonant transition through the experimental investigation of the rotation angle of the polarization vector of the photon echo signal as a function of the magnetic field strength $\boldsymbol{H}$ or the time interval τ between the pump pulses. In particular, similar to the case when inequality (4.1.29) is satisfied, the magnetic field (the time interval τ) can be chosen in such a way in the case of resonant levels with arbitrary g factors that the photon echo signal is polarized perpendicular to the polarization plane of the pump pulses for $J \to J$ transitions ($J >> 1$) or the maximum rotation angle of the polarization plane of the photon echo signal depends on the relation between the g factors of the upper and lower resonant levels for $J \rightleftarrows J+1$ transitions ($J >> 1$). In any case, this angle cannot exceed arctan(1/2).

Thus, we have described in this section how resonant transitions (or the type of resonant transitions) can be identified by means of the photon echo. Such an identification requires the investigation of the polarization of the photon echo signal produced by small-area pump pulses as a function of the polarizations of the pump pulses, the magnitude of the applied longitudinal magnetic field, or the time interval between the pump pulses.

4.2 Conditions Imposed on the Parameters of Pump Pulse for Measuring the Homogeneous Half-Width of a Resonant Spectral Line

In this section, we assume that we can employ the collision integral calculated in the model of elastic depolarizing collisions and averaged not only in the direction, but also in the modulus of the velocity $\mathbf{v}$ of resonant atoms (molecules). In this case, the irreversible relaxation of the optical coherence matrix is described by the quantities $\gamma_{ab}^{(k)}$ and $\Delta_{ab}^{(k)}$ determined by formulas similar to (2.3.14) and (2.3.15), but, unlike formulas (2.3.14) and (2.3.15), independent of v.

As mentioned in Chapter 2, the electric field strength $\boldsymbol{E}^e$ of the photon echo signal produced by pump pulses (2.3.1)–(2.3.4) through optically allowed transitions $b \to a$ involving resonant levels with arbitrary angular momenta generally depends on the quantities $\gamma_{ab}^{(k)}$ and $\Delta_{ab}^{(k)}$ with odd k [17, 19]. Therefore, we should specify a method for the independent measurement of these relaxation characteristics with the use of the photon echo.

In this section, we discuss conditions when photon-echo experiments make it possible to extract the spectroscopic information concerning the homogeneous half-width $\gamma_{ab}^{(1)}$ of a spectral line corresponding to an inhomogeneously broadened resonant transition involved in the formation of a photon echo signal.

Such experimental conditions were defined for the first time by Yevseyev and Yermachenko [9], who proposed to employ the transition under study to generate a photon echo signal in the field of small-area pump pulses. Calculations in this paper were performed fort he case of rectangular pump pulses (2.3.1)–(2.3.4). Yevseyev and Reshetov [16] have demonstrated that, to obtain the spectroscopic information concerning the homogeneous half-width $\gamma_{ab}^{(1)}$ of an inhomogeneously broadened spectral line of a resonant transition, one can employ pump pulses with an arbitrary waveform, but with a small area. Indeed, if the pump pulses have arbitrary waveforms and small areas, then the electric field strength in the photon echo signal produced through optically allowed transitions involving resonant levels with arbitrary angular momenta is described by formulas (2.3.36), (4.1.4), and (4.1.36). Under these conditions, the quantity S, which is defined by formula (4.1.36) and which characterizes the decay and the waveform of the photon echo signal, depends only on the relaxation parameter $\gamma_{ab}^{(k)}$ with $k = 1$.

Expression (4.1.36) can be analyzed in a way similar to the consideration performed in Section 2.4. Specifically, when the photon echo signal is produced through a narrow (2.3.24) spectral line corresponding to an inhomogeneously broadened resonant transition, the maximum of the echo signal reaches the observation point y at the moment of time $t = 2\tau + y/c$, and the duration of this signal is on the order of the reversible transverse relaxation time T_2*. Therefore, taking into account inequality (2.3.29) and using formulas (2.3.36), (4.1.4), and (4.1.36), we find that the intensity I_e of the photon echo signal decays with the growth in the time interval τ between the pump pulses in accordance with formula (2.4.9), which makes it possible to employ the data of photon-echo experiments in gases to extract the spectroscopic information on $\gamma_{ab}^{(1)}$. When the photon echo signal is produced through a broad (2.3.25) spectral line corresponding to an inhomogeneously broadened resonant transition, the maximum of the echo signal reaches the observation point y at the moment of time shifted with respect to the instant of time $t = 2\tau + y/c$ by the time approximately equal to the total duration of the pump pulses. The duration of the photon echo signal in this case is on the order of the maximum duration of the pump pulse. By virtue of a natural assumption that the durations of the pump pulses are small as compared with irreversible relaxation times and the time interval τ, we find that the intensity I_e of the photon echo signal decays with the growth in τ in accordance with formula (2.4.9) again. Thus, formula (2.4.9) describes the decay of the photon echo intensity with the growth in τ when the photon echo is produced through either a narrow or a broad spectral line corresponding to an inhomogeneously broadened optically allowed resonant transition involving resonant levels with arbitrary angular momenta if the photon echo signal is generated in the field of small-area pump pulses.

The aim of the analysis performed in [19] was to expand the applicability area of (2.4.9). This study has demonstrated that formula (2.4.9) can be employed in the case when the area of the first pump pulse is small, and the area of the second pump pulse is arbitrary (e.g., optimal). Indeed, the electric field strength $\boldsymbol{E}^e$ of the photon echo signal produced by pump pulses (2.3.1)–(2.3.4) with arbitrary areas through an optically allowed transition $b \to a$ involving resonant levels with arbitrary angular momenta is given by [10, 19]

$$\boldsymbol{E}^e = -\pi\hbar^{-3}\omega L c^{-1}|d|^4 e^{(1)}\,\tilde{T}_1\, e^{(2)\,2}\,\tilde{T}_2^{\,2} N_0 \boldsymbol{\varepsilon}^e \exp[i(\omega t - ky + 2\Phi_2 - \\ - \Phi_1) - \gamma_{ab}^{(1)}(t - y/c)] + \text{c.c.} \qquad (4.2.1)$$

Here, the nonvanishing components of the vector $\boldsymbol{\varepsilon}^e$, characterizing the polarization properties and the waveform of the photon echo signal, are equal to

$$\varepsilon_x^e = 2\sqrt{2}(3)^{-1} \sum_{k,\mu,m} [4\pi(2k+1)]^{1/2} Y_{k1}(\psi,0) \times$$

$$\times \begin{pmatrix} J_b & J_a & k \\ \mu & -m & 1 \end{pmatrix} \begin{pmatrix} J_a & J_b & 1 \\ \mu & -m & 1 \end{pmatrix} a_\mu a_m \times$$

$$\times \sum_\eta [a_\eta \begin{pmatrix} J_a & J_b & k \\ \eta & -\eta & 0 \end{pmatrix} T^{k\eta}_{\mu m}], \tag{4.2.2}$$

$$\varepsilon^e_z = 2(3)^{-1} \sum_{k,\mu} [4\pi(2k+1)]^{1/2} Y_{k0}(\psi, 0) a^3_\mu \times$$

$$\times \begin{pmatrix} J_a & 1 & J_b \\ -\eta & 0 & \eta \end{pmatrix} \sum_\eta [a_\eta \begin{pmatrix} J_a & J_b & k \\ \eta & -\eta & 0 \end{pmatrix} T^{k\eta}_{\mu\mu}], \tag{4.2.3}$$

where

$$T^{k\eta}_{\mu m} = 12\exp\{-[\gamma^{(k)}_{ab} - \gamma^{(1)}_{ab} + i(\Delta^{(k)}_{ab} - \Delta^{(1)}_{ab})]\tau\}\times$$

$$\times \int dv_y\, f(v_y) \exp[i\delta(t' - \tau)](U_\eta - iW_\eta)(V_2)_{\mu m}, \tag{4.2.4}$$

$$U_\eta = (\Omega^{(1)}_\eta \widetilde{T}_1)^{-1} \sin(\Omega^{(1)}_\eta \widetilde{T}_1), \tag{4.2.5}$$

$$W_\eta = \delta(\Omega^{(1)}_\eta)^{-2} \widetilde{T}_1^{\,-1}[1 - \cos(\Omega^{(1)}_\eta \widetilde{T}_1)], \tag{4.2.6}$$

$$(V_2)_{\mu m} = (\Omega^{(2)}_\mu \Omega^{(2)}_m \widetilde{T}_2^{\,2})^{-1} \sin(\tfrac{1}{2}\Omega^{(2)}_\mu \widetilde{T}_2) \sin(\tfrac{1}{2}\Omega^{(2)}_m \widetilde{T}_2), \tag{4.2.7}$$

$$(\Omega^{(n)}_\eta \widetilde{T}_n)^2 = (\delta \widetilde{T}_n)^2 + \theta^2_{n\eta}, \tag{4.2.8}$$

$$\delta = kv_y - \omega + \omega_0 + \Delta^{(1)}_{ab}, \tag{4.2.9}$$

$$\theta_{n\eta} = 2e^{(n)}|d|\widetilde{T}_n \hbar^{-1} a_\eta, \quad a_\eta = \begin{pmatrix} J_a & 1 & J_b \\ -\eta & 0 & \eta \end{pmatrix}. \tag{4.2.10}$$

Here, L is the length of the gas medium, d is the reduced matrix element of the dipole moment operator related to the resonant transition $b \to a$, N_0 is defined by formula (2.3.18), $Y_{lm}(\theta,\varphi)$ is the spherical function, $f(v_y)$ describes the Maxwell distribution in projections v_y of the velocity of resonant atoms (molecules) on the Y axis, the quantity t' is given by formula (2.3.33), and the other quantities are the same as in expressions (2.3.1)–(2.3.4).

In the general case, the photon echo signal described by (4.2.1)–(4.2.10) is elliptically polarized and propagates with a carrier frequency ω in the direction that coincides with the direction of incidence of the pump pulses (2.3.1)–(2.3.4).

Now, let us analyze formula (4.2.4) in the case when a photon echo signal is produced through narrow and broad spectral lines corresponding to inhomogeneously broadened resonant transitions. For simplicity, we restrict our consideration to the case of an exact resonance, $\omega = \omega_0 + \Delta_{ab}^{(1)}$, where ω is the carrier frequency of the pump pulses, ω_0 is the frequency of the resonant transition $b \to a$, and $\Delta_{ab}^{(1)}$ is the shift of the spectral line of the resonant transition due to elastic depolarizing collisions. When the spectral line involved in the formation of the photon echo signal is narrow for both pump pulses and we deal with the strong-field limit (2.4.4), formula (4.2.4) yields

$$T_{\mu m}^{k\eta} = 12\exp\{-[\gamma_{ab}^{(k)} - \gamma_{ab}^{(1)} + i(\Delta_{ab}^{(k)} - \Delta_{ab}^{(1)})]\tau\}(\theta_{1\eta}\theta_{2\mu}\theta_{2m})^{-1}\times$$
$$\times \sin(\theta_{1\eta})\sin(\tfrac{1}{2}\theta_{2\mu})\sin(\tfrac{1}{2}\theta_{2m})\exp[-\tfrac{1}{4}(T_2^*)^{-2}(t-2\tau-y/c)^2]. \tag{4.2.11}$$

In the case under consideration, the maximum of the photon echo signal reaches the observation point y at the moment of time $t = 2\tau + y/c$, and the signal width is on the order of the reversible transverse relaxation time T_2^*. Such a behavior of the photon echo signal is similar to the behavior of the photon echo produced by small-area pump pulses through a spectral line that is narrow for both pump pulses. However, in the case of pump pulses with arbitrary areas, as can be seen from (4.2.1)–(4.2.3) and (4.2.11), the intensity of the photon echo signal as a function of the time interval τ cannot be described by a simple formula (2.4.9) even when the spectral line involved in the formation of the photon echo signal is narrow for both pump pulses. The intensity of the photon echo signal (4.2.1)–(4.2.3) and (4.2.11) depends on a large number of relaxation parameters $\gamma_{ab}^{(k)}$ and $\Delta_{ab}^{(k)}$ with odd k ($|J_a - J_b| \le k \le J_a + J_b$). The only exception, as mentioned above, is associated with transitions involving resonant levels with low angular momenta ($0 \rightleftarrows 1$, $1 \to 1$, $1/2 \to 1/2$, and $1/2 \rightleftarrows 3/2$). The intensity of the photon echo signal in the case when the spectral line involved in echo formation is narrow for both pump pulses decays with the increase in the time interval τ in accordance with formula (2.4.9).

In the case of a broad spectral line with $\theta_1 \geq 1$ and $\theta_2 \geq 1$, integration in (4.2.4) can be performed only numerically. Such an integration for particular values of angular momenta of the levels involved in the resonant transition was carried out in [14], where it was demonstrated that the maximum of the photon echo signal is observed at the moment of time that approximately corresponds to the time t_e determined by formula (2.3.35), and the duration of the echo signal is on the order of the maximum duration of the pump pulses. Therefore, similar to the case of a narrow spectral line, the intensity of a photon echo signal produced by pump pulses with arbitrary areas through a broad spectral line depends on the relaxation parameters $\gamma_{ab}^{(k)}$ and $\Delta_{ab}^{(k)}$ with odd k.

Thus, as it follows from (4.2.1)–(4.2.10), in the case of optimal areas of the pump pulses, when the intensity of the echo signal reaches its maximum, and for low angular momenta of the levels involved in the resonant transition ($0 \rightleftarrows 1$, $1 \to 1$, $1/2 \to 1/2$, and $1/2 \rightleftarrows 3/2$), the intensity of the photon echo signal as a function of the time interval τ is governed by formula (2.4.9). For resonant transitions involving levels with high angular momenta and pump pulses with optimal areas, the intensity of the photon echo signal as a function of the time interval τ depends on a large number of relaxation parameters $\gamma_{ab}^{(k)}$ and $\Delta_{ab}^{(k)}$ with odd k.

An exception from this rule was considered in [19], where the area of the first pump pulse was assumed to be small ($\theta_1 << 1$), while the area of the second pump pulse was assumed to be arbitrary. Then, as it follows from (4.2.4), the quantity $T_{\mu m}^{k\eta}$ becomes independent of η, which makes it possible to perform summation in η for the functions in square brackets in the right-hand sides of formulas (4.2.2) and (4.2.3). This procedure yields a quantity proportional to $\delta_{k,1}$. Therefore, only the term with $k = 1$ survives when summation in k is performed in expressions (4.2.2) and (4.2.3). Consequently, the intensity of the photon echo signal in this case, similar to the situation when the areas of both pump pulses are small, decays with the growth in the time interval τ in accordance with formula (2.4.9) for arbitrary angular momenta of the levels involved in the optically allowed resonant transition. This finding makes constraints on the areas of the pump pulses less stringent, showing that the conditions of the detection of the photon echo signal can be improved through the optimization of the area of the second pump pulse.

Closing this section, we will discuss the experimental consequences of formula (2.4.9). This formula makes it possible to extract the experimental information concerning the homogeneous half-width $\gamma_{ab}^{(1)}$ of an inhomogeneously broadened spectral line involved in the formation of the photon echo. Indeed, varying the time interval τ between the pump pulses, keeping the pressure p of the resonant or buffer gas constant, and processing the experimental data with the use of formula (2.4.9), one can measure the homogeneous half-width $\gamma_{ab}^{(1)}$ of the relevant resonant spectral line (see Chapter 3). This experimental approach can be slightly modified. Changing the pressure p of the buffer gas, keeping the time interval constant, and applying formula (2.4.9), one can obtain the experimental information on $d\gamma_{ab}^{(1)}/dp$. However, using formula (2.4.9) for processing the experimental data, one has to take into consideration the area of applicability of this formula, which was described in this section. We emphasize once again that for resonant transitions involving resonant levels with low angular momenta ($0 \rightleftarrows 1$, $1 \to 1$, $1/2 \to 1/2$, and $1/2 \rightleftarrows 3/2$), formula (2.4.9) is applicable for arbitrary (including optimal) areas of the pump pulses. For resonant transitions involving levels with higher angular momenta, formula (2.4.9) is applicable only in the case when the area of the first pump pulse is small. In this case, we can specify which areas θ_1 of the first pump pulse can be considered as small in photon-echo experiments. As can be seen from (4.2.1)–(4.2.10), polarization properties of the photon echo signal in the limiting case $\theta_1 << 1$ are independent of θ_1. Therefore, the quantity θ_1 should be gradually decreased in experiments starting with its optimal value, corresponding to the maximum intensity of the photon echo signal, until the polarization properties of the echo signal become independent of θ_1. With such θ_1, the experimental data can be processed with the use of formula (2.4.9).

4.3 The Possibility of Measuring the Relaxation Parameters of the Octupole Moment of a Resonant Transition

As it was shown above, the intensity and polarization of the photon echo signal produced by linearly polarized pump pulses through $J \to J$ ($J \geq 3/2$) and $J \rightleftarrows J + 1$ ($J \geq 1$) transitions generally depend not only on $\gamma_{ab}^{(1)}$ and $\Delta_{ab}^{(1)}$, but also on $\gamma_{ab}^{(k)}$ and $\Delta_{ab}^{(k)}$ with $k = 3, 5, \ldots$. Let us show how such transitions can be employed to obtain the spectroscopic information concerning the relaxation parameters $\gamma_{ab}^{(3)}$ and $\Delta_{ab}^{(3)}$ of the octupole moment of a resonant transition $b \to a$. For this purpose, we will consider the results obtained in [20], which make it possible to apply the photon echo to measure these parameters.

The authors of [20] were the first to demonstrate that the area of the second pump pulse should be small, while the area of the first pump pulse should not be small to provide the required spectroscopic information on the relaxation characteristics of the octupole moment of a resonant transition. Indeed, as it follows from expressions (4.2.1)–(4.2.10), the electric field strength of the photon echo signal in the case under study is determined by formula (4.2.1), where the nonvanishing components of the vector ε^{e}, characterizing polarization properties and the waveform of the echo signal, are written as

$$\varepsilon_x^e = (1/2)\{B(J_b, J_a)S_1(J_b, J_a) + (1/3)(5\cos^2\psi - 1)S_3(J_b, J_a) \times$$
$$\times \exp[-(\gamma_{ab}^{(3)} - \gamma_{ab}^{(1)})\tau - i(\Delta_{ab}^{(3)} - \Delta_{ab}^{(1)})\tau]\}\sin\psi, \qquad (4.3.1)$$

$$\varepsilon_z^e = (1/3)\{A(J_b, J_a)S_1(J_b, J_a) + (1/2)(5\cos^2\psi - 3)S_3(J_b, J_a) \times$$
$$\times \exp[-(\gamma_{ab}^{(3)} - \gamma_{ab}^{(1)})\tau - i(\Delta_{ab}^{(3)} - \Delta_{ab}^{(1)})\tau]\}\cos\psi. \qquad (4.3.2)$$

Here, the quantities $A(J_b, J_a)$ and $B(J_b, J_a)$ are defined by formulas (4.1.5)–(4.1.8), while the quantities $S_1(J_b, J_a)$ and $S_3(J_b, J_a)$ are given by

$$S_1(J, J) = a(J) \sum_{m=-J}^{J} m^2 V_m,$$

$$S_3(J, J) = (2/5)a^2(J) \sum_{m=-J}^{J} [5m^4 - (3J^2 + 3J - 1)m^2] V_m,$$

$$S_1(J, J+1) = S_1(J+1, J) = b(J) \sum_{m=-J}^{J} [(J+1)^2 - m^2] V_m,$$

$$S_3(J, J+1) - S_3(J+1, J) = (2/5)b^2(J) \sum_{m=-J}^{J} [(J+1)^2 - 5m^2 - 1][(J+1)^2 -$$
$$- m^2] V_m,$$

where

$$a(J) = [J(J+1)(2J+1)]^{-1}, \qquad (4.3.3)$$

$$b(J) = [(J+1)(2J+1)(2J+3)]^{-1}, \tag{4.3.4}$$

$$V_m = 6\widetilde{T}_2^{-2} \int dv_y\, f(v_y) \exp[i\delta(t'-\tau)](U_m - iW_m)\delta^{-2}(1-\cos\delta\widetilde{T}_2). \tag{4.3.5}$$

The quantities U_m and W_m in formula (4.3.5) can be obtained from (4.2.5) and (4.2.6), respectively, by the replacement of indices $\eta \to m$; the quantities δ and t' are defined by formulas (4.2.9) and (2.3.33), respectively; $\widetilde{T}_2$ is the duration of the second pump pulse; and $f(v_y)$ describes the Maxwell distribution of resonant atoms (molecules) in projections v_y of their velocities on the Y axis.

Expression (4.3.5) with $\omega = \omega_0 + \Delta_{ab}^{(1)}$ can be analyzed similar to formula (4.2.4), which was investigated in the previous section. Therefore, omitting the discussion for brevity, we will employ the results obtained in the previous section. As it follows from the analysis of expression (4.3.5), the dependence of the electric field strength in the photon echo signal (4.2.1) and (4.3.1)–(4.3.5) on the time interval τ can be derived from (4.2.1) and (4.3.1)–(4.3.5) through the replacement of $t - y/c$ by 2τ. Note that formulas (4.2.1), (4.3.1), and (4.3.2) where $t - y/c$ is replaced by 2τ can be employed to extract the experimental information on the relaxation parameters $\gamma_{ab}^{(1)}$, $\gamma_{ab}^{(3)}$, and $\Delta_{ab}^{(3)} - \Delta_{ab}^{(1)}$. Specifically, in the case when $\psi = 0$, i.e., when the pump pulses are polarized in the same plane, the intensity of the photon echo signal as a function of the time interval τ between the pump pulses displays beats [29] with a frequency $|\Delta_{ab}^{(3)} - \Delta_{ab}^{(1)}|$. This effect can be employed to measure the quantity $|\Delta_{ab}^{(3)} - \Delta_{ab}^{(1)}|$. In the case when $\cos\psi = 5^{-1/2}$, the analysis of the decay of the component E_x^e of the electric field strength in the photon echo signal with the growth in τ makes it possible to measure the homogeneous half-width $\gamma_{ab}^{(1)}$ of the spectral line of a resonant transition. Finally, when the relaxation parameter $\gamma_{ab}^{(1)}$ is known and $|\Delta_{ab}^{(3)} - \Delta_{ab}^{(1)}|\tau \ll 1$, the analysis of the dependence of the tilt angle φ of the polarization vector of the photon echo signal with respect to the polarization vector of the second pump pulse provides an opportunity to extract the spectroscopic information on the relaxation time $1/\gamma_{ab}^{(3)}$ of the octupole moment of a resonant transition. Such measurements can be performed because, as it follows from (4.2.1), (4.3.1), and (4.3.2), the quantity

$$\tan\varphi = E_x^e / E_z^e \tag{4.3.6}$$

as a function of the time interval τ depends only on the difference $\gamma_{ab}^{(3)} - \gamma_{ab}^{(1)}$. Note that formulas (4.2.1) and (4.3.1)–(4.3.5) make it possible to measure the quantities $d\gamma_{ab}^{(1)}/dp$, $d\gamma_{ab}^{(3)}/dp$, and $d|\Delta_{ab}^{(3)} - \Delta_{ab}^{(1)}|/dp$ by varying the pressure p of a buffer gas.

Thus, we can make the following conclusions concerning the possibility of measuring the quantities $\gamma_{ab}^{(k)}$ and $\Delta_{ab}^{(k)} - \Delta_{ab}^{(1)}$ with $k = 3, 5, \ldots$. Generally, the photon echo signal is determined by the whole set of these relaxation parameters, and it is impossible to extract the information concerning each of these components. However, the information on the quantities $\gamma_{ab}^{(3)}$ and $\Delta_{ab}^{(3)} - \Delta_{ab}^{(1)}$ can be extracted even in the general case. For this purpose, one should perform measurements with a small-area second pump pulse taking into account the criterion that allows the area θ_2 of the second pump pulse to be considered as small for a given experiment. As it follows from (4.3.6), polarization properties of the photon echo signal, described by (4.2.1) and (4.3.1)–(4.3.5), are independent of the area of the second pump pulse. Therefore, one should gradually decrease the area of the second pump pulse in experiments starting with its optimal value, which corresponds to the maximum intensity of the echo signal, until polarization properties of the echo signal become independent of θ_2.

4.4 The Possibility of Measuring the Relaxation Parameters of the Quadrupole Moment of a Resonant Transition

In the two previous sections, we discussed the possibility of applying the photon echo method for extracting the spectroscopic information on the relaxation parameters $\gamma_{ab}^{(k)}$ and $\Delta_{ab}^{(k)} - \Delta_{ab}^{(1)}$ with odd k. As it follows from (4.2.1)–(4.2.10), the photon echo signal produced by linearly polarized pump pulses is insensitive to the parameters $\gamma_{ab}^{(k)}$ and $\Delta_{ab}^{(k)} - \Delta_{ab}^{(1)}$ with even k.

Application of a uniform longitudinal magnetic field to a medium gives rise to the dependence of the electric field strength in the photon echo signal on the parameters $\gamma_{ab}^{(k)}$ and $\Delta_{ab}^{(k)} - \Delta_{ab}^{(1)}$ not only with odd, but also with even k. This fact was highlighted for the first time by Bakaev *et. al.* [30], who demonstrated that the dependence of the tilt angle of the polarization vector of the photon echo signal on the strength $\boldsymbol{H}$ of the magnetic field permits the quantity H to be chosen in such a way for $1 \to 1$ and $1/2 \rightleftarrows 3/2$ transitions that the difference of the direction of the polarization vector of the echo signal from the direction of polarization vectors of the pump pulses is completely determined by elastic depolarizing collisions. This effect can be employed for the experimental determination of the relaxation time $1/\gamma_{ab}^{(2)}$ of the quadrupole moment of a resonant transition.

Let us consider the formation of a photon echo signal in a gas medium in the presence of a permanent uniform magnetic field with a strength $\boldsymbol{H}$ under conditions when the time interval τ between the pump pulses is comparable with irreversible relaxation times. Then, to find the electric field strength in the photon echo signal, we should solve equations (2.3.8)–(2.3.11) with additional terms (4.1.31)–(4.1.33). The technique of such calculations was discussed in detail in Section 2.3. Therefore, below, we will present only the final result for the electric field strength in the photon echo signal produced by pump pulses (2.3.1)–(2.3.4) through $J_b = 1 \rightarrow J_a = 1$ transition. This field strength is given [30, 31] by formula (4.2.1), where the vector ε^e should be replaced by a vector $\varepsilon^e(t')$ with the following nonvanishing components:

$$\varepsilon_x^e(t') = -(1/9)V(t')[a(t')\cos(\varepsilon\tau)\sin(\varepsilon t') - b(t')\sin(\varepsilon\tau)\cos(\varepsilon t')], \tag{4.4.1}$$

$$\varepsilon_z^e(t') = (1/9)V(t')[a(t')\cos(\varepsilon\tau)\cos(\varepsilon t') + b(t')\sin(\varepsilon\tau)\sin(\varepsilon t')]. \tag{4.4.2}$$

Here, the quantity t' is defined by formula (2.3.33); $\varepsilon = (\varepsilon_a + \varepsilon_b)/2$; the frequencies ε_a and ε_b are given by expression (4.1.30); and the quantity $V(t')$, which characterizes the waveform of the photon echo signal in the case of an exact resonance ($\omega = \omega_0 + \Delta_{ab}^{(1)}$), is defined as

$$V(t') = 2\tilde{T}_1^{-1}\tilde{T}_2^{-2}\int dv_y\, f(v_y)\Omega_1^{-1}\Omega_2^{-2}\{\sin(\Omega_1\tilde{T}_1)\cos[kv_y(t'-\tau)] +$$

$$+ kv_y\Omega_1^{-1}[1-\cos(\Omega_1\tilde{T}_1)]\sin[kv_y(t'-\tau)]\}[1-\cos(\Omega_2\tilde{T}_2)], \tag{4.4.3}$$

$$\Omega_n^2 = (kv_y)^2 + (\theta_n/\tilde{T}_n)^2, \qquad \theta_n = (2/3)^{1/2}e^{(n)}|d|\tilde{T}_n\hbar^{-1}. \tag{4.4.4}$$

The complex quantities $a(t')$ and $b(t')$ involved in (4.4.1) and (4.4.2) include the influence of the differences $|\gamma_{ab}^{(2)} - \gamma_{ab}^{(1)}|$ and $|\Delta_{ab}^{(2)} - \Delta_{ab}^{(1)}|$ on the photon echo signal. The expressions for these quantities are presented in [30]. Note that $a(t')$ and $b(t')$ become real when $t' = \tau$.

For simplicity, we will present expressions for the quantities $a(\tau)$ and $b(\tau)$ only for the particular case when $|\gamma_{ab}^{(2)} - \gamma_{ab}^{(1)}|\tau << 1$, $|\Delta_{ab}^{(2)} - \Delta_{ab}^{(1)}|\tau << 1$, and $|\varepsilon_a - \varepsilon_b|\tau \geq 1$, which is considered below. Then, employing the results obtained in [30], we find that

$$a(\tau) = \{\cos^2[\tfrac{1}{2}(\varepsilon_a - \varepsilon_b)\tau] - (\gamma_{ab}^{(2)} - \gamma_{ab}^{(1)})(\varepsilon_b - \varepsilon_a)^{-1}\sin[(\varepsilon_a - \varepsilon_b)\tau]\} \times$$

$$\times [1 - (\gamma_{ab}^{(2)} - \gamma_{ab}^{(1)})\tau], \tag{4.4.5}$$

$$b(\tau) = \sin^2[\tfrac{1}{2}(\varepsilon_a - \varepsilon_b)\tau][1 - (\gamma_{ab}^{(2)} - \gamma_{ab}^{(1)})\tau]. \tag{4.4.6}$$

The photon echo signal described by (4.2.1) and (4.4.1)–(4.4.4) is generally elliptically polarized. Expression (4.4.3) for the quantity $V(t')$ can be analyzed similarly to formula (4.2.4), which was investigated in Section 4.2. This analysis shows that the vector $\varepsilon^e(\tau)$ describes polarization properties of the maximum of the photon echo signal produced through a narrow spectral line. By virtue of inequalities (4.1.35), this vector also describes the entire echo signal in the case when a broad spectral line is involved in the formation of the echo signal. Thus, a photon echo signal described by (4.2.1) and (4.4.1)–(4.4.4) is linearly polarized in the case of a broad spectral line. We emphasize that, with $t' = \tau$, the echo signal is always linearly polarized.

For low pressures, when depolarizing collisions do not play a significant role ($\gamma_{ab}^{(2)} = \gamma_{ab}^{(1)}$), we can employ (4.4.5) and (4.4.6) to rewrite expressions for $\varepsilon_x^e(\tau)$ and $\varepsilon_z^e(\tau)$ as

$$\varepsilon_x^e(\tau) \sim \sin(2\varepsilon\tau)\cos[(\varepsilon_a - \varepsilon_b)\tau], \tag{4.4.7}$$

$$\varepsilon_z^e(\tau) \sim 1 + \cos(2\varepsilon\tau)\cos[(\varepsilon_a - \varepsilon_b)\tau]. \tag{4.4.8}$$

As can be seen from (4.4.7) and (4.4.8), the magnetic field strengths can be chosen in such a way for a given time interval τ between the pump pulses that the echo signal is polarized along the polarization vectors of the pump pulses at the moment of time $t' = \tau$. Suppose, for example, that $(\varepsilon_a - \varepsilon_b)\tau = \pi/2$. Then, increasing the pressure of the buffer or the resonant gas, we can observe at the moment of time $t' = \tau$ the appearance of the electric field component in the photon echo signal perpendicular to the polarization plane of the pump pulses. This component of the electric field is exclusively due to elastic depolarizing collisions.

Finally, taking into account (4.4.5) and (4.4.6), we derive the following expressions for $\varepsilon_x^e(\tau)$ and $\varepsilon_z^e(\tau)$:

$$\varepsilon_x^e(\tau) \sim 2\pi^{-1}(\gamma_{ab}^{(2)} - \gamma_{ab}^{(1)})\tau \sin[2^{-1}\pi(g_a + g_b)(g_a - g_b)^{-1}] \times$$

$$\times \exp[-(\gamma_{ab}^{(2)} + \gamma_{ab}^{(1)})\tau], \tag{4.4.9}$$

$$\varepsilon_z^e(\tau) \sim \exp[-(\gamma_{ab}^{(2)} + \gamma_{ab}^{(1)})\tau]. \tag{4.4.10}$$

If the g factors of the resonant levels are known, measuring the tangent of the tilt angle φ of the polarization vector of the photon echo signal relative to the polarization vectors of the pump pulses, $\tan\varphi = \varepsilon_x^e(\tau)/\varepsilon_z^e(\tau)$, we can employ (4.4.9) and (4.4.10) to determine the difference $\gamma_{ab}^{(2)} - \gamma_{ab}^{(1)}$. The relaxation parameter $\gamma_{ab}^{(1)}$ can be determined in this case in a usual way with the use of formula (2.4.9), which describes the decay of the photon echo intensity with the growth in the time interval τ in the absence of a magnetic field.

Thus, the method of a photon echo provides an opportunity to independently determine $\gamma_{ab}^{(1)}$ and $\gamma_{ab}^{(2)}$. Note that, for $1/2 \rightleftarrows 3/2$ transitions, formulas similar to (4.2.1) and (4.4.1)–(4.4.4) have been presented in [30], where the way of extracting the spectroscopic information concerning the relaxation parameter $\gamma_{ab}^{(2)}$ of the quadrupole moment of resonant $1/2 \rightleftarrows 3/2$ transitions has been also proposed.

4.5 Requirements to the Parameters of Pump Pulses Used for the Investigation of the Relaxation Parameter of the Dipole Moment of a Resonant Transition as Functions of the Modulus of the Velocity of Resonant Atoms (Molecules)

In this section, we discuss how the dependence of the relaxation parameter $\gamma_{ab}^{(1)}$ of the dipole moment of an optically allowed transition involving resonant levels with arbitrary angular momenta on the modulus of the velocity v of resonant atoms (molecules) influences the photon echo signal. In contrast to Sections 4.2–4.4 of this chapter, we will employ the collision integral calculated in the model of elastic depolarizing collisions and averaged only in the direction of the velocity of resonant atoms (molecules). Such a collision integral is described by the terms involved in the right-hand sides of equations (1.5.5)–(1.5.7).

Suppose that a photon echo signal is produced under the action of pump pulses (2.3.1)–(2.3.4). The area of the first pump pulse will be assumed to be small, while the area of the second pump pulses will be assumed to be arbitrary. Then, solving the set of equations (2.3.8)–(2.3.11) with initial conditions (2.2.11)–(2.2.13), we derive the following expression for the electric field strength in the photon echo signal [10, 32]:

$$\boldsymbol{E}^{e} = -\pi\hbar^{-3}\omega L c^{-1}|d|^{4} e^{(1)}\,\widetilde{T}_1\, e^{(2)\,2}\,\widetilde{T}_2^{\,2} N_0 \boldsymbol{\varepsilon}^{e} \exp[i(\omega t - ky + 2\Phi_2 - \Phi_1)] + \text{c.c.} \tag{4.5.1}$$

Here, the nonvanishing components of the vector $\boldsymbol{\varepsilon}^{e}$, which characterizes polarization properties, the waveform, and the decay of the photon echo signal, are written as

$$\varepsilon_x^{e} = (1/2)\sin\psi \sum_{m=-J}^{J} B_m(J_a, J_b)\, T_{m,m+1}, \tag{4.5.2}$$

$$\varepsilon_z^{e} = (1/3)\cos\psi \sum_{m=-J}^{J} A_m(J_a, J_b)\, T_{m,m}. \tag{4.5.3}$$

The quantity ω in formula (4.5.1) stands for the carrier frequency of the pump pulses; L is the length of the gas medium; $k = \omega/c$, $e^{(n)}$, Φ_n, and $\widetilde{T}_n$ are the amplitude, the phase, and the duration of the n-th pump pulse ($n = 1, 2$) respectively; N_0 is defined by formula (2.3.18); and ψ is the angle between the polarization vectors of the pump pulses. The quantity $T_{\mu,m}$ involved in expressions (4.5.2) and (4.5.3) is given by

$$T_{\mu,m} = 12 \int d\boldsymbol{v}\, f(\boldsymbol{v})(V_2)_{\mu,m}[(kv_y - \omega + \omega_0)\widetilde{T}_1]^{-1}\{\sin[(kv_y - \omega + \omega_0)\widetilde{T}_1] - i[1 - \cos((kv_y - \omega + \omega_0)\widetilde{T}_1)]\}\exp\{i[kv_y - \Delta\omega(\boldsymbol{v})](t' - \tau) - \gamma_{ab}^{(1)}(\boldsymbol{v})(t' + \tau)\}, \tag{4.5.4}$$

where $\Delta\omega(v)$ and t' are defined by formulas (2.3.13) and (2.3.33), respectively, and the function $(V_2)_{\mu,m}$ is given by formula (4.2.7), where the quantity $\Omega_{\eta}^{(2)}$ $(\eta = \mu,m)$ can be found from the equation

$$(\Omega_{\eta}^{(2)}\,\widetilde{T}_2)^2 = [(kv_y - \omega + \omega_0)\,\widetilde{T}_2\,]^2 + \theta_{2\eta}^2\,, \tag{4.5.5}$$

and $\theta_{2\eta}$ can be obtained from (4.2.10) with $n = 2$.

Finally, the quantities $A_m(J_a,J_b) = A_m(J_b,J_a)$ and $B_m(J_a,J_b) = B_m(J_b,J_a)$ are given by

$$A_m(J,J) = a^2(J)m^4,$$

$$B_m(J,J) = (2/3)a^2(J)m^2[J(J+1) - m^2 - 1],$$

$$A_m(J,J+1) = -\,(2/3)b^2(J)[(J+1)^4 - (J+1)^2(2m^2+1) + m^2(m^2+1)],$$

$$B_m(J,J+1) = b^2(J)[(J+1)^4 - 2(J+1)^2m^2 + m^4],$$

where the coefficients $a(J)$ and $b(J)$ are defined by formulas (4.3.3) and (4.3.4), respectively.

Generally, the photon echo signal (4.5.1)–(4.5.5) is elliptically polarized and propagates with a carrier frequency ω through a gas medium in the same direction as the pump pulses (2.3.1)–(2.3.4). As it follows from (4.5.1)–(4.5.5), the electric field strength in a photon echo signal produced in the case when the area of the first pump pulse is small depends only on the relaxation parameters $\gamma_{ab}^{(1)}(v)$ and $\Delta_{ab}^{(1)}(v)$ of the dipole moment of an optically allowed transition involving resonant levels with arbitrary angular momenta.

Recall that, in Section 2.4, we performed a detailed analysis of the quantity (2.3.38), which governs the decay and the waveform of the photon echo signal produced through $J_b = 1 \rightarrow J_a = 0$ transition. We should note that the analysis of the quantity $T_{\mu,m}$, which is defined by (4.5.4) and which characterizes the decay and the waveform of the photon echo signal produced through optically allowed transitions involving resonant levels with arbitrary angular momenta, is similar to the analysis of the quantity (2.3.38). Suppose, similar to Section 2.4, that w is the characteristic scale of variation of the functions $\gamma_{ab}^{(1)}(v)$ and $\Delta_{ab}^{(1)}(v)$. Then, if we assume that the field of the second pump pulse is weak, i.e., inequality (2.4.5) is satisfied for this pulse, and $kw >> ku$, the intensity of the photon echo signal (4.5.1)–(4.5.5) is given by formula (2.4.13), where u is the root-mean-square velocity of resonant atoms (molecules) in the gas.

As mentioned above, formula (2.4.13) has been already employed by Vasilenko *et. al.* [33] for processing the experimental data. Figure 4.3 taken from [33] presents the results of these measurements. The ordinate axis in Fig. 4.3 shows the ratio $\gamma_{ab}^{(1)}(v)/\gamma_{ab}^{(1)}(0)$, while the abscissa axis shows the dimensionless velocity v/u. As can be seen from this figure, the relaxation parameter $\gamma_{ab}^{(1)}$ is virtually independent of v in a pure SF_6 gas, while the presence of a buffer gas (krypton) makes this dependence easily detectable.

Note that theoretical analysis performed in this section is based on the results of [32, 34]. Closing this section, we will consider yet another possible manifestation of the dependence of the relaxation parameter of the dipole moment of a resonant transition on the modulus of the velocity of resonant atoms (molecules). As mentioned in Chapter 1, if the effective angle of scattering is sufficiently small and small-angle scattering plays a dominant role, then the part of the collision integral that governs velocity-changing scattering involves a small parameter and can be neglected in the analysis of certain problems of coherent optics of gas media. Obviously, this part of the collision integral can be neglected only when it does not give rise to qualitatively new effects.

The authors of [35] assumed in this context that the deviation of the decay of the photon echo intensity with the growth in the time interval τ between the pump pulses from the dependence predicted by (2.4.9), which was detected in experiments with atomic gases [36, 37], is one of such qualitatively new effects. However, as demonstrated in [38], this phenomenon can be also explained within the framework of the model of elastic depolarizing collisions.

Recall that the parameter $\mathrm{Re}\,\Gamma_{ab}^{(1)}(v)$, which governs collisional relaxation of the polarization vector of a medium related to a group of atoms moving with a velocity **v**, depends on v in the model of elastic depolarizing collisions and is determined by the amplitudes of scattering of a resonant atom in the states a and b from nonresonant atoms of the resonant or buffer gases. The calculation of these scattering amplitudes, in its turn, requires the knowledge of interaction potentials of colliding atoms. Thus, it is rather difficult to find the function $\mathrm{Re}\,\Gamma_{ab}^{(1)}(v)$ in the explicit form. Therefore, we will have to restrict our consideration to some limiting cases.

Note that Berman *et al.* [35] employed the model of absolutely solid

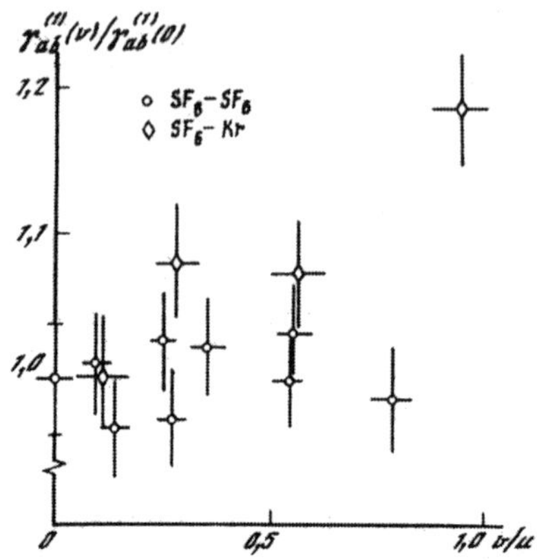

Fig. 4.3. Dependence of the relaxation parameter of the dipole moment of a resonant transition on the velocity modulus of SF_6 resonant molecules.

spheres to explain the deviation of the decay of the intensity of the photon echo produced through resonant levels belonging to different electronic states from the dependence predicted by (2.4.9). Using this model, the authors of [35] took into account the influence of the change in the velocity of resonant atoms in the process of collisions on the decay of the photon echo intensity.

In connection with the results of [35], we should note that the dependence of $\mathrm{Re}\,\Gamma_{ab}^{(1)}(v)$ on v gives rise to a deviation of the decay of the photon echo signal from the dependence predicted by (2.4.9). We will first find $\mathrm{Re}\,\Gamma_{ab}^{(1)}(v)$ using the model of impenetrable spheres, which was employed in [35]. In the case when the mass of a buffer-gas atom is much larger than the mass of a resonant atom, we find that

$$\mathrm{Re}\,\Gamma_{ab}^{(1)}(v) = \Gamma v u^{-1}. \tag{4.5.6}$$

Here, Γ is proportional to the density of buffer-gas atoms and is independent of v, u is the root-mean-square velocity of resonant atoms.

As it follows from [38], the intensity of the photon echo signal produced by small-area rectangular pump pulses in the case of a narrow inhomogeneously broadened spectral line of a resonant transition as a function of the time interval τ is proportional to the factor

$$I_e(\tau) \sim \exp[-2(\gamma_a^{(0)} + \gamma_b^{(0)})\tau]K^2(\Gamma\tau). \tag{4.5.7}$$

Here, as before, $1/\gamma_a^{(0)}$ and $1/\gamma_b^{(0)}$ are the relaxation times of the resonant levels a and b due to radiative decay and inelastic gas-kinetic collisions,

$$K(y) = -2\pi^{-1/2}y + (1 + 2y^2)\exp(y^2)[1 - \Phi(y)], \tag{4.5.8}$$

and $\Phi(y)$ is the error integral.

As can be seen from expression (4.5.7), the intensity of the photon echo signal is represented by a product of an exponential and a function depending on the parameter $\Gamma\tau$. Consequently, the deviation of formula (4.5.7) from the dependence predicted by (2.4.9) may be only due to the behavior of the function (4.5.8). The degree of this deviation can be seen from the plot of the function

$$\pi^{1/2}(4\Gamma\tau)^{-1}\ln[K(\Gamma\tau)], \tag{4.5.9}$$

which is shown in Fig. 4.4 (curve 1). Note that the choice of the function (4.5.9) for characterizing the deviation of the intensity decay of the photon echo signal with the growth in τ from the dependence predicted by (2.4.9) is associated with the following circumstance. If the quantity $K(\Gamma\tau)$ were an exponential function of the parameter $\Gamma\tau$, then the function (4.5.9) would be a constant, which could be set equal to -1 with an appropriate normalization. Therefore, any deviation of the function (4.5.9) from -1 implies that the intensity decay of the photon echo signal with the growth in τ deviates to some extent from the exponential behavior.

To show that this effect may be also manifested for other potentials (including potentials more realistic than the potential of absolutely solid spheres), we will calculate [38] $\mathrm{Re}\,\Gamma_{ab}^{(1)}(v)$ for a van der Waals-type potential, assuming that the ratio of the mass of a buffer-gas atom to the mass of a resonant atom is large. Such a calculation yields

$$\mathrm{Re}\,\Gamma_{ab}^{(1)}(v) = \Gamma_B(v/u)^{3/5}, \tag{4.5.10}$$

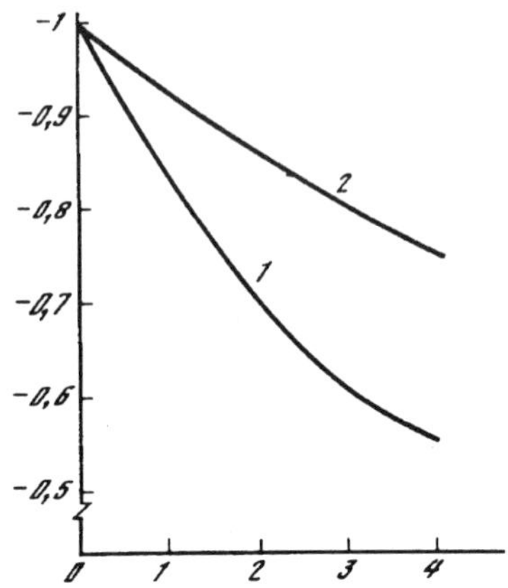

Fig. 4.4. The plots of the functions defined by (4.5.9) (curve 1) and (4.5.12) (curve 2). The abscissa axis corresponds to the parameter $\Gamma\tau$ for curve 1 and parameter $\Gamma_B\tau$ for curve 2.

where Γ_B is proportional to the density of buffer-gas atoms and is independent of v.

If the quantity $\text{Re}\,\Gamma_{ab}^{(1)}(v)$ is described by formula (4.5.10), then we again arrive at expression (4.5.7) [38] for the intensity of the photon echo signal produced by small-area rectangular pump pulses through a narrow spectral line. However, in the case under study, we have

$$K(y) = 4\pi^{-1/2} \int_0^\infty dx\, x^2 \exp(-x^2 - 2yx^{3/5}). \tag{4.5.11}$$

The function

$$\pi^{1/2}[4\Gamma_B\tau\, \Gamma(1.8)]^{-1} \ln[K(\Gamma_B\tau)], \tag{4.5.12}$$

which characterizes the deviation of the decay of the photon echo intensity from the exponential behavior, is presented by curve 2 in Fig. 4.4. Here,

$K(\Gamma_B\tau)$ is defined by formula (4.5.11), and $\Gamma(1.8)$ is the Γ-function. As can be seen from the comparison of curves 1 and 2, the degree of the deviation of the intensity decay of the photon echo signal with the increase in τ from the exponential behavior depends on the interaction potential of colliding atoms. Therefore, comparing the results of theoretical calculations with the experimental data, one has to be especially careful in choosing an adequate model of the interaction between atoms.

Thus, as can be seen from formulas (4.5.6)–(4.5.12), additional experiments are required to identify the mechanism responsible for the deviation of the intensity decay of the photon echo signal in gas media from the dependence predicted by (2.4.9).

4.6 The Possibility of Studying the Dependence of Relaxation Matrices on the Direction of the Velocity of Resonant Atoms (Molecules)

Let us consider how the dependence of the collision integral calculated in the model of elastic depolarizing collisions not only on the modulus but also on the direction of the velocity $\mathbf{v}$ of resonant atoms (molecules) may influence the intensity and polarization of a photon echo signal. We will assume that the photon echo signal is produced under the action of linearly polarized pump pulses with electric field strengths described by formulas (2.3.1)–(2.3.4). To solve this problem, we should employ equations of the form (2.3.8)–(2.3.11), where the collision integral is determined by the right-hand sides of (1.3.2)–(1.3.4) rather than by the right-hand sides of equations (1.5.5)–(1.5.7).

As an example, we will consider the case [34, 39] when a photon echo signal is produced through an optically allowed transition involving resonant levels with angular momenta $J_b = 1$ and $J_a = 0$. In Chapter 1, the collision integral not averaged in the direction of the velocity $\mathbf{v}$ of resonant atoms (molecules) for this case was analyzed in terms of the model of elastic depolarizing collisions. Formula (1.4.2)–(1.4.4) describe the explicit dependence of the relaxation matrix $\Gamma_{ab}(\mathbf{v})^{kk'}_{qq'}$ on the modulus and the direction of the velocity $\mathbf{v}$ of resonant atoms (molecules) for transitions of the considered type.

The procedure employed to find the electric field strength $\boldsymbol{E}^e$ of the photon echo signal in this section is similar to the technique described in Section 2.3. Therefore, we will omit intermediate calculations and present the final expression for $\boldsymbol{E}^e$ in the case when the photon echo signal is produced through $J_b = 1 \to J_a = 0$ transition. The electric field strength $\boldsymbol{E}^e$ in this case is given [34, 39] by formula (4.5.1), where the nonvanishing components of the vector ε^e, characterizing polarization properties, the decay, and the waveform of the echo signal, are written as

$$\varepsilon_x^e = (1/9)\sin\psi \int d\mathbf{v}\, T(\mathbf{v}) g(\tau)^* g(t'), \tag{4.6.1}$$

$$\varepsilon_z^e = (1/9)\cos\psi \int d\mathbf{v}\, T(\mathbf{v})[2h(\tau)^* h(t') + g(\tau)^* g(t')]. \tag{4.6.2}$$

Here, τ is the time interval between the pump pulses, ψ is the angle between the polarization vectors of these pulses, t' is defined by formula (2.3.33), and the other quantities are given by

$$T(\mathbf{v}) = 2\exp[-\tfrac{1}{2}(\gamma_a^{(0)} + \gamma_b^{(0)})(t - y/c)] f(\mathbf{v})(\Omega_1 \widetilde{T}_1 \Omega_2^2 \widetilde{T}_2^{\,2})^{-1} \times$$
$$\times [1 - \cos(\Omega_2 \widetilde{T}_2)\}][\sin(\Omega_1 \widetilde{T}_1) - i\Omega_1^{-1}(kv_y - \omega + \omega_0)(1 - \cos(\Omega_1 \widetilde{T}_1))] \times$$
$$\times \exp[i(kv_y - \omega + \omega_0)(t' - \tau)], \tag{4.6.3}$$

$$h(t) = \tfrac{1}{2}(1 + \cos^2\theta)\exp\{-[\Gamma_1(v) - i\Delta_1(v)]t\} +$$
$$+ \tfrac{1}{2}\sin^2\theta \exp\{-[\Gamma_0(v) - i\Delta_0(v)]t\}, \tag{4.6.4}$$

$$g(t) = \tfrac{1}{2}\sin^2\theta \exp\{-[\Gamma_1(v) - i\Delta_1(v)]t\} -$$
$$- \tfrac{1}{2}\sin^2\theta \exp\{-[\Gamma_0(v) - i\Delta_0(v)]t\}. \tag{4.6.5}$$

The quantity Ω_n in formulas (4.6.3)–(4.6.5) can be found from equation (2.3.20), $f(\mathbf{v})$ describes the Maxwell distribution of resonant atoms (molecules) in velocities $\mathbf{v}$, v_y is the projection of the velocity $\mathbf{v}$ on the Y axis, ω is the carrier frequency of the pump pulses, ω_0 is the frequency of the resonant transition $b \to a$, θ is the polar angle of the vector $\mathbf{v}$ with respect to an arbitrarily chosen axis, $\Gamma_1(v)$ and $\Delta_1(v)$ are the relaxation characteristics of the transverse (with respect to $\mathbf{v}$) component of the polarization of the medium, and $\Gamma_0(v)$ and $\Delta_0(v)$ are the relaxation characteristics of the longitudinal (with respect to $\mathbf{v}$) component of the polarization of the medium.

The photon echo signal described by (4.5.1) and (4.6.1)–(4.6.5) is elliptically polarized and propagates through a gas medium with a carrier frequency ω in the same direction as the pump pulses (2.3.1)–(2.3.4). Elastic depolarizing collisions, included by $\Gamma_1(v)$, $\Delta_1(v)$, $\Gamma_0(v)$, and $\Delta_0(v)$, influence the rotation of the axes of the polarization ellipse and the distortion of this ellipse. We should note an important feature of our finding. With $\psi = \pi/2$, i.e., in the case when polarization vectors of the pump pulses are orthogonal to each other, the polarization of the photon echo signal described by (4.5.1) and (4.6.1)–(4.6.5) becomes linear and coincides with the polarization vector of the first pump pulse. At the same time, as it follows from (2.3.36)–(2.3.38), when the collision integral calculated in the model of elastic depolarizing collisions and averaged over the direction of the velocity $\mathbf{v}$ of resonant atoms (molecules) is used to determine the electric field strength in the photon echo signal, the polarization vector of the photon echo signal produced through $J_b = 1 \rightarrow J_a = 0$ transition coincides with the polarization vector of the second pump pulse, while for $\psi = \pi/2$, the echo signal is not observed in this case. Thus, the dependence of the relaxation matrix $\Gamma_{ab}(\mathbf{v})_{qq'}^{kk'}$ on the direction of the velocity $\mathbf{v}$ of resonant atoms (molecules) gives rise to a qualitatively new effect in the case of $1 \rightarrow 0$ transition. The experimental observation of this effect would provide the information concerning the dependence of relaxation matrices on the direction of the velocity of resonant atoms (molecules). We emphasize that a similar effect is observed also in the case of $J_b = 0 \rightarrow J_a = 1$ transition.

The effect considered above should be observed (with given τ) when the pressure of a buffer gas is increased in such a way that the influence of elastic depolarizing collisions becomes significant. In addition, as it follows from (4.6.1) and (4.6.5), the magnitude of this effect is determined by the differences $\Gamma_1(v) - \Gamma_0(v)$ and $\Delta_1(v) - \Delta_0(v)$ of the transverse and longitudinal (with respect to $\mathbf{v}$) relaxation characteristics of the polarization of a medium. In Chapter 1 of this monograph, these differences were analyzed for the case of van der Waals interaction between resonant atoms and buffer-gas atoms. It was demonstrated that for the values of v on the order of the thermal velocity u, the quantities $\Gamma_1(v) - \Gamma_0(v)$ and $\Delta_1(v) - \Delta_0(v)$ noticeably differ from zero only in the case when the mass of buffer-gas atoms is much greater than the mass of resonant atoms. This circumstance should be taken into account in experiments on the photon echo in gases aimed at the observation of the effect predicted in [34, 39].

Closing this section, we should note that, in what follows, we will employ the collision integral calculated in the model of elastic depolarizing collisions and averaged in both the direction and the modulus of the velocity $\mathbf{v}$ of resonant atoms (molecules).

4.7 The Possibility of Measuring the Relaxation Parameters of Multipole Moments for Optically Forbidden Transitions

In this section, we present the theory of polarization properties of various modifications of the photon echo, including the three-level photon echo (TLPE) and modified three-level photon echo (MTLPE). This theory, which was developed in [40–45], allows us to propose an experimental technique for measuring the relaxation parameters of multipole moments for optically forbidden transitions.

The three-level photon echo was observed for the first time by Hartmann and his colleagues [46]. In contrast to the conventional photon echo, the TLPE is produced in a three-level system under the action of three pump pulses (Fig. 4.5).

Let us consider the physical scenario of TLPE formation in a gas medium. Suppose that the first pump pulse with a duration $\widetilde{T}_1$ and carrier frequency ω_1, which is resonant to the frequency ω_0 of an optically allowed transition $b \to a$, is incident on the $y = 0$ boundary of a gas medium at the moment of time $t = 0$ and propagates along the positive direction of the Y axis. The strength of the electric field in this pulse can be written as

$$\boldsymbol{E}_1 = e^{(1)}\boldsymbol{l}_1 g_1(\xi)\exp[i(\omega_1 t - k_1 y + \Phi_1)] + \text{c.c.}, \quad 0 \le \xi \le \widetilde{T}_1, \tag{4.7.1}$$

where $e^{(1)}$ is the constant amplitude, Φ_1 is the constant phase shift, $\boldsymbol{l}_1 = (\sin\psi_1, 0, \cos\psi_1)$ is the polarization vector, $\xi = t - y/c$, $k_1 = \omega_1/c$, and the function g_1 describing the pulse waveform is normalized in accordance with (4.1.1). This pump pulse excites multipole polarization moments in the medium, which are involved in fast oscillations with frequency ω_1. As the first pump pulse leaves the point y of the gas medium, these multipole moments of the polarization of the medium decay with characteristic times $1/\gamma_{ab}^{(k)}$ ($|J_a - J_b| \le k \le J_a + J_b$), where $\gamma_{ab}^{(k)}$ is the relaxation characteristic of the optical coherence matrix that is related to a dipole-allowed transition $b \to a$ and that includes radiative decay, as well as inelastic gas-kinetic and elastic depolarizing collisions. In addition, the multipole moments of the polarization of the medium, involved in fast oscillations with frequency ω_1, oscillate because of Doppler dephasing, acquiring a phase factor $(ik_1 v_y \tau_1)$ by the moment of time when the second pump pulse reaches the point y of the gas medium. Here, τ_1 is the time interval between the first and second pump pulses, and v_y is the projection of the velocity $\mathbf{v}$ of a resonant atom

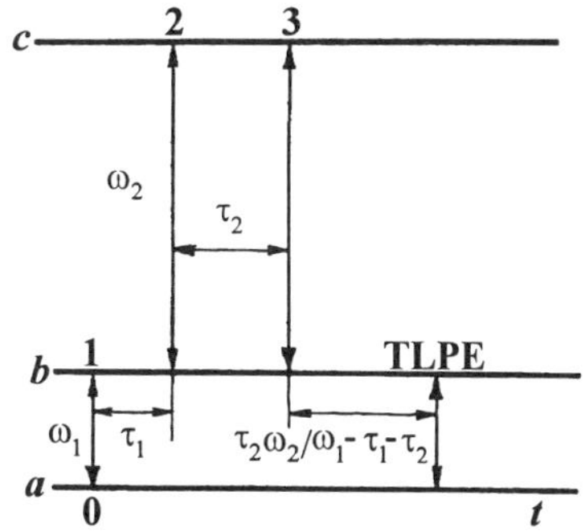

Fig. 4.5. Diagram of the formation of a three-level photon echo signal: 1–3, pump pulses; TLPE, three-level photon echo signal.

(molecule) on the Y axis.

Suppose that the second pump pulse with a duration $\widetilde{T}_2$ and carrier frequency ω_2, which is resonant to the frequency ϖ_0 of an optically allowed transition $c \to b$ ($E_c > E_b > E_a$), is incident on the $y = L$ boundary of the gas medium at the moment of time $t = \tau_1 + \widetilde{T}_1$ and propagates along the negative direction of the Y axis. The strength of the electric field in this pulse can be written as

$$\boldsymbol{E}_2 = e^{(2)}\boldsymbol{l}_2 g_2(\eta - L/c - \tau_1 - \widetilde{T}_1)\exp[i(\omega_2 t + k_2 y + \Phi_2)] + \text{c.c.},$$
$$0 \le \eta - L/c - \tau_1 - \widetilde{T}_1 \le \widetilde{T}_2, \tag{4.7.2}$$

where $e^{(2)}$ is the constant amplitude, Φ_2 is the constant phase shift, $\boldsymbol{l}_2 = (\sin\psi_2, 0, \cos\psi_2)$ is the polarization vector, $\eta = t + y/c$, $k_2 = \omega_2/c$, and the function g_2 describing the pulse waveform is normalized in accordance with (4.1.1). This pulse transfers the coherence from multipole polarization moments of the medium involved in fast oscillations with frequency ω_1 to multipole polarization moments of the medium involved in fast oscillations with frequency $\omega_1 + \omega_2$. As the second pump pulse leaves the point y of the gas medium, the multipole moments of the polarization of the medium involved in fast oscillations with frequency $\omega_1 + \omega_2$ decay with characteristic times $1/\gamma_{ac}^{(k)}$ ($|J_a - J_c| \le k \le J_a + J_c$), where $\gamma_{ac}^{(k)}$ are the relaxation characteristics of the optical coherence matrix that is related to a dipole-forbidden transition $c \to a$ and that includes radiative decay, as well as inelastic gas-kinetic and elastic depolarizing collisions. The multipole polarization moments of the medium involved in fast oscillations with frequency $\omega_1 + \omega_2$ acquire a phase factor $\exp[-i(k_2 - k_1)v_y\tau_2 + ik_1v_y\tau_1]$ by the moment of time when the third pump pulse reaches the point y of the gas medium. Here, τ_2 is the time interval between the second and third pump pulses.

Suppose that the third pump pulse with a duration $\widetilde{T}_3$ and carrier frequency ω_2, which is resonant to the frequency ϖ_0 of an optically allowed transition $c \to b$ ($E_c > E_b > E_a$), is incident on the $y = L$ boundary of the gas medium at the moment of time $t = \tau_1 + \tau_2 + \widetilde{T}_1 + \widetilde{T}_2$ and propagates along the negative direction of the Y axis. The strength of the electric field in this pulse can be written as

$$\boldsymbol{E}_3 = e^{(3)}\boldsymbol{l}_3 g_3(\eta - L/c - \tau_1 - \tau_2 - \widetilde{T}_1 - \widetilde{T}_2)\exp[i(\omega_2 t + k_2 y + \Phi_3)] + \text{c.c.},$$
$$0 \le \eta - L/c - \tau_1 - \tau_2 - \widetilde{T}_1 - \widetilde{T}_2 \le \widetilde{T}_3, \quad (4.7.3)$$

where $e^{(3)}$ is the constant amplitude, Φ_3 is the constant phase shift, $\boldsymbol{l}_3 = (0, 0, 1)$ is the polarization vector, and the function g_3 describing the pulse waveform is normalized in accordance with (4.1.1). This pulse transfers the coherence from multipole polarization moments of the medium involved in fast oscillations with frequency $\omega_1 + \omega_2$ to the dipole moment of the polarization of the medium involved in fast oscillations with frequency ω_1. As the third pump pulse leaves the point y of the gas medium, the part of the dipole moment of the polarization of the medium that contributes to the TLPE signal decays with a characteristic time $1/\gamma_{ab}^{(1)}$ and oscillates due to Doppler dephasing. This part of the dipole moment of the polarization of the medium is proportional to the phase factor $\exp[-ik_1v_y(t - k_2\tau_2/k_1 - \widetilde{T}_1 - \widetilde{T}_2 - \widetilde{T}_3 - y/c]$. Therefore, radiation-emitting particles in three-level photon echo are in phase at the point y of a gas medium at the moment of time $t \approx t_{te}$, where

$$t_{te} = \omega_2\tau_2/\omega_1 + \widetilde{T}_1 + \widetilde{T}_2 + \widetilde{T}_3 + y/c. \quad (4.7.4)$$

The TLPE thus produced propagates in the same direction as the first pump pulse with a carrier frequency ω_1. These two circumstances considerably simplify the separation of the TLPE signal from the second and third pump pulses. When the relaxation times $1/\gamma_{ab}^{(k)}$ of the optical coherence matrix related to the dipole-allowed transitions $b \to a$ are known, the investigation of the TLPE signal makes it possible to determine the relaxation times $1/\gamma_{ac}^{(k)}$ ($k = 0, 1, \ldots$) of the optical coherence matrix related to the dipole-forbidden transition $c \to a$. This possibility was highlighted for the first time by the authors of [40]. Note that the relaxation parameters $\gamma_{ab}^{(k)}$ can be determined, for example, from independent photon-echo experiments for the $b \to a$ transition.

We should emphasize that the three-level photon echo can be produced only when the inequality

$$\tau_1/\tau_2 < \omega_2/\omega_1 - 1 \tag{4.7.5}$$

is satisfied. Here, we assume, as usually, that the time intervals between the pump pulses are large as compared with pump pulse durations.

To find the electric field strength in the TLPE signal produced by pump pulses (4.7.1)–(4.7.3), we will employ equations (2.2.1)–(2.2.3) with initial condition (2.2.4) and use the collision integral calculated in the model of elastic depolarizing collisions and averaged in both the direction and the modulus of the velocity **v** of resonant atoms (molecules). Since the method of calculation of the electric field strength $\boldsymbol{E}^{te}$ in the TLPE signal has been described in detail in [41, 42], we present only the final expression for $\boldsymbol{E}^{te}$. As demonstrated in [41, 42], in the case when the areas of all the three pump pulses are small, we have

$$\begin{aligned}\boldsymbol{E}^{te} &= \pi\hbar^{-3}\omega_1 L c^{-1} |d|^2 |\bar{d}|^2 e^{(1)} \tilde{T}_1 e^{(2)} \tilde{T}_2 e^{(3)} \tilde{T}_3 N_0 \bar{S}\, \boldsymbol{e}^{te} \times \\ &\times \exp[-\gamma_{ab}^{(1)}(t - \tau_2 - \tilde{T}_1 - \tilde{T}_2 - \tilde{T}_3 - y/c)] \times \\ &\times \exp[i(\omega_1 t - k_1 y + \Phi_1 + \Phi_2 - \Phi_3)] + \text{c.c.},\end{aligned} \tag{4.7.6}$$

where d is the reduced matrix element of the dipole moment operator related to the resonant transition $b \to a$, and $\bar{d}$ is the analogous quantity for the resonant transition $c \to b$. The nonvanishing components of the vector $\boldsymbol{e}^{te}$, characterizing polarization properties of the TLPE signal, are given by

$$e_x^{te} = -[\bar{a}_1 \bar{b}_1(\tau_2)\sin(\psi_1 - \psi_2) + \bar{a}_2 \bar{b}_2(\tau_2)\sin(\psi_1 + \psi_2)], \tag{4.7.7}$$

$$e_z^{te} = -\frac{2}{3}[\bar{a}_0\, \bar{b}_0(\tau_2)\cos(\psi_1 - \psi_2) + 2\bar{a}_2\, \bar{b}_2(\tau_2)(\cos\psi_1\cos\psi_2 -$$

$$-\frac{1}{2}\sin\psi_1\sin\psi_2)], \tag{4.7.8}$$

where

$$\bar{a}_k = \begin{Bmatrix} k & 1 & 1 \\ J_b & J_a & J_c \end{Bmatrix}^2, \tag{4.7.9}$$

$$\bar{b}_2(\tau_2) = \exp[-\gamma_{ac}^{(k)}\tau_2 + i(\Delta_{ac}^{(k)} - \Delta_{ac}^{(1)})\tau_2]. \tag{4.7.10}$$

The quantity N_0 is defined by formula (2.3.18), the relaxation parameter $\gamma_{ab}^{(1)}$ is given by expression (2.3.15), and

$$\gamma_{ac}^{(k)} = (\gamma_a^{(0)} + \gamma_c^{(0)})/2 + \operatorname{Re}\Gamma_{ac}^{(k)}, \tag{4.7.11}$$

where $1/\gamma_a^{(0)}$ and $1/\gamma_c^{(0)}$ are the population relaxation times for the levels a and c, respectively, due to radiative decay and inelastic gas-kinetic collisions, and $\Gamma_{ac}^{(k)}$ and $\Delta_{ac}^{(k)}$ are the relaxation parameters of the optically forbidden transition $c \rightarrow a$ including elastic depolarizing collisions. Finally, the quantity $\bar{S}$, which characterizes the waveform of the TLPE signal, is written as

$$\bar{S} = \int c_1^*(v_y)\bar{c}_2(v_y)c_3^*(v_y)f(v_y)\times$$

$$\exp\{ik_1v_y[t - t_{te} + \frac{1}{2}\tilde{T}_1 + \frac{1}{2}(\tilde{T}_2 + \tilde{T}_3)(2 - k_2/k_1)]\}dv_y, \tag{4.7.12}$$

where t_{te} is defined by formula (4.7.4), and the quantities $c_1(v_y)$, $\bar{c}_2(v_y)$, and $c_3(v_y)$, are given by

$$c_1(v_y) = \frac{1}{\widetilde{T}_1} \int\limits_{-\widetilde{T}_1/2}^{\widetilde{T}_1/2} g_1(s + \widetilde{T}_1/2) \exp(ik_1 v_y s) ds , \tag{4.7.13}$$

$$\bar{c}_2(v_y) = \frac{1}{\widetilde{T}_2} \int\limits_{-\widetilde{T}_2/2}^{\widetilde{T}_2/2} g_2(s + \widetilde{T}_2/2) \exp(ik_2 v_y s) ds , \tag{4.7.14}$$

$$c_3(v_y) = \frac{1}{\widetilde{T}_3} \int\limits_{-\widetilde{T}_3/2}^{\widetilde{T}_3/2} g_3(s + \widetilde{T}_3/2) \exp(ik_2 v_y s) ds . \tag{4.7.15}$$

For the sake of simplicity, formulas (4.7.6)–(4.7.15) were written for the case of an exact resonance,

$$\omega_1 = \omega_0 + \Delta_{ab}^{(1)}, \qquad \omega_2 = \varpi_0 + \Delta_{ac}^{(1)} - \Delta_{ab}^{(1)}. \tag{4.7.16}$$

In the general case, the signal of three-level photon echo (4.7.6)–(4.7.15) is elliptically polarized and propagates with a carrier frequency ω_1 along the wave vector of the first pump pulse. Note that, in the case when $|\Delta_{ac}^{(k)} - \Delta_{ac}^{(1)}|\tau_2 \ll 1$, the TLPE signal (4.7.6)–(4.7.15) is linearly polarized. As it follows from (4.7.6)–(4.7.15), polarization properties of the TLPE signal produced by small-area pump pulses are independent of the waveforms of the pump pulses. This circumstance makes small-area pump pulses very attractive for the application in experiments on the three-level photon echo in gas media. In the case of small-area rectangular pump pulses with the same duration, formulas (4.7.6)–(4.7.15) are reduced to the corresponding expressions from [40], while in the case when the pump pulses have arbitrary waveforms, the relevant formulas were derived for the first time by Yevseyev *et al.* [41]. In the particular case when the TLPE signal is produced by small-area pump pulses and the polarization plane of the third pulse is orthogonal to the polarization plane of the first two pulses, i.e., $\psi_1 = \psi_2 = \pi/2$, the polarization properties of the photon echo signal are provided in the appendices to the experimental paper by Hartmann and his colleagues [47]. We should emphasize that these expressions were derived without allowance for elastic depolarizing collisions. The waveforms of the pump pulses in this study were assumed to be arbitrary. As demonstrated in [47], the polarization vector of the TLPE signal in the above-specified particular case is orthogonal to the polarization vector of the first pump pulse. Setting $\psi_1 = \psi_2 = \pi/2$, $\gamma_{ac}^{(0)} = \gamma_{ac}^{(2)} = \gamma_{ac}^{(1)}$, and $\Delta_{ac}^{(0)} = \Delta_{ac}^{(1)} = \Delta_{ac}^{(2)} = 0$ in formulas (4.7.6)–(4.7.15), we find that the polarization vector of the TLPE signal is also orthogonal to the polarization vector of the first pump pulse. This is one of the facts that indicate the reliability of formulas (4.7.6)–(4.7.15). Another indication of the reliability of these formulas is associated with the fact that, if we set $\gamma_{ac}^{(0)} = \gamma_{ac}^{(2)} = \gamma_{ac}^{(1)} = 0$, and $\Delta_{ac}^{(0)} = \Delta_{ac}^{(1)} = \Delta_{ac}^{(2)} = 0$ and assume that the pump pulses have rectangular waveforms, then formulas (4.7.6)–(4.7.15) are reduced to the expressions presented in paper [48], which was published later than [40].

Note that the waveform of the TLPE signal (4.7.6)–(4.7.15) has been analyzed in detail in [40–42], and these results will be discussed in Section 5.1. Here, we will only consider the possibility of extracting the experimental data on relaxation parameters $\gamma_{ac}^{(k)}$ of an optically forbidden transition $c \rightarrow a$. We assume that the quantity $\gamma_{ab}^{(1)}$ is known, since it can be determined, for example, from independent experiments on the decay of a conventional photon echo signal as a function of the time interval between the pump pulses. Suppose that $|\Delta_{ac}^{(k)} - \Delta_{ac}^{(1)}|\tau_2 \ll 1$ and recall that the TLPE signal (4.7.6)–(4.7.15) is linearly polarized in this case. Then, if the angles ψ_1 and ψ_2 are chosen in an appropriate way, then the terms decaying with a single relaxation parameter can be separated in $\boldsymbol{E}^{te}$. In other words, we can implement polarization echo spectroscopy using the signal of the three-level photon echo.

If, for example, $\psi_1 = -\psi_2$, then the X-component of the electric field strength in the TLPE signal decays with the increase in the time interval τ_2 between the second and third pump pulses proportionally to the factor

$$E_x^{te} \sim \bar{a}_1 \exp\{-[\gamma_{ab}^{(1)}(\omega_2/\omega_1 - 1) + \gamma_{ac}^{(1)}]\tau_2\}. \tag{4.7.17}$$

Provided that $\psi_1 = \psi_2$, this component of the electric field strength in the TLPE signal decays with the growth in τ_2 proportionally to the factor

$$E_x^{te} \sim \bar{a}_2 \exp\{-[\gamma_{ab}^{(1)}(\omega_2/\omega_1 - 1) + \gamma_{ac}^{(2)}]\tau_2\}. \tag{4.7.18}$$

Finally, if $\tan\psi_1 \tan\psi_2 = 2$, then the Z- component of the electric field strength in the TLPE signal decays with the growth in the time interval τ_2 proportionally to the factor

$$E_z^{te} \sim \bar{a}_0 \exp\{-[\gamma_{ab}^{(1)}(\omega_2/\omega_1 - 1) + \gamma_{ac}^{(0)}]\tau_2\}. \tag{4.7.19}$$

Expressions (4.7.17)–(4.7.19) were derived for the first time by Yevseyev *et al.* [40]. We should note also that the coefficient $\bar{a}_0$ differs from zero only in the case when $J_a = J_c$. We emphasize that, choosing parameters of the problem in accordance with formulas (4.7.17)–(4.7.19), one can extract the information concerning the differences in the relaxation parameters $\gamma_{ac}^{(k)}$ ($k = 0, 1, 2$).

As it follows from (4.7.6)–(4.7.15), there are also other relations between ψ_1 and ψ_2 that allow a decay process with a single relaxation time $1/\gamma_{ac}^{(0)}$, $1/\gamma_{ac}^{(1)}$, or $1/\gamma_{ac}^{(2)}$ to be separated in (4.7.7) or (4.7.8).

Note that expressions for the electric field strength in the TLPE signal produced by linearly polarized pump pulses with arbitrary areas whose polarization vectors lie in different planes were derived in [42]. Analysis of these formulas shows [10, 42] that, for elementary transitions involving resonant levels with low angular momenta, one can still extract in this case the spectroscopic information on the relaxation times of multipole moments of optically forbidden transitions.

Now, let us consider experiments on the three-level photon echo [47], where the resonant levels had the following angular momenta: $J_a = J_b = 1/2$ and $J_c = 1/2$ or $J_c = 3/2$. In the case when $J_a = J_b = J_c = 1/2$, the equalities $\gamma_{ac}^{(0)} = \gamma_{ac}^{(1)}$ and $\Delta_{ac}^{(0)} = \Delta_{ac}^{(1)}$ hold true [49]. Therefore, formulas (4.7.17) and (4.7.19) allow one to determine only one relaxation parameter.

In the case when $J_a = J_b = 1/2$ and $J_c = 3/2$, formulas (4.7.17)–(4.7.19) involve two relaxation parameters, $\gamma_{ac}^{(1)}$ and $\gamma_{ac}^{(2)}$. Therefore, properly arranged measurements and processing of the experimental data with the use of formulas (4.7.17) and (4.7.18) provide an opportunity to extract the spectroscopic information concerning these relaxation parameters.

In experiments [47], the angles ψ_1 and ψ_2 were equal to $\pi/2$. As can be seen from (4.7.7) and (4.7.8), the TLPE signal in this case is linearly polarized along the polarization vector of the third pump pulse. Such a polarization state of the echo signal was observed in experiments [47]. Relations

$$e_z^{te} = (1/18)\exp(-\gamma_{ac}^{(2)}\tau_2), \qquad e_x^{te} = 0 \tag{4.7.20}$$

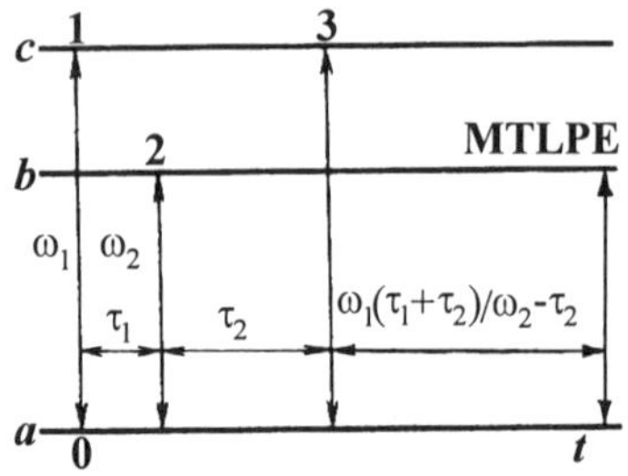

Fig. 4.6. Diagram of the formation of a modified three-level photon echo signal: 1–3, pump pulses; MTLPE, modified three-level photon echo signal.

should also hold true in this case. Thus, the decay due to $\gamma_{ac}^{(2)}$ was observed in experiments [47], where E_z^{te} was investigated as a function of τ_2 for $J_a = J_b = 1/2$ and $J_{c'} = 3/2$.

Note that, in the case of van der Waals interaction between atoms of resonant and buffer gases, the equality $\gamma_{ac}^{(2)} = \gamma_{ac}^{(1)}$ is satisfied (e.g., see [49]). It would be of interest to check this relation experimentally. For this purpose, we should measure the decay of the X-component of the electric field strength in the TLPE signal as a function of the time interval τ_2 for $\psi_1 = \psi_2$ and $\psi_1 = -\psi_2$. The difference in these dependences is due to the difference in the quantities $\gamma_{ac}^{(1)}$ and $\gamma_{ac}^{(2)}$.

As mentioned above, the three-level photon echo can be produced only when condition (4.7.5) is met, which imposes certain restrictions on the frequencies of resonant transitions.

A modified three-level photon echo (Fig. 4.6) is yet another scheme of the photon echo in a three-level system. The possibility of implementing such a scheme of the photon echo was discussed for the first time by Hartmann and his colleagues [47]. This scheme proved to be free of the above-specified limitation. Similar to the TLPE, the modified three-level photon echo is produced in a gas medium excited with three pulses whose frequencies are resonant to transitions between three energy levels a, b, and c. However, the modified three-level photon echo requires another system of resonant transitions. Transitions $c \to a$ and $b \to a$ with frequencies ω_0 and ϖ_0, respectively, should be optically allowed in this case. All the three pump pulses propagate in the same direction. The carrier frequency ω_1 of the first and third pump pulses is resonant to the frequency ω_0, and the carrier frequency ω_2 of the second pump pulse is resonant to the frequency $\overline{\omega}_0$.

The time required for the generation of the MTLPE signal is approximately equal to t_{mte}, where

$$t_{mte} = \tau_1 + \omega_1(\tau_1 + \tau_2)/\omega_2 + y/c. \tag{4.7.21}$$

In contrast to the three-level photon echo, the frequencies ω_1 and ω_2 in such a scheme are not subject to any additional restrictions other than the restriction related to the inequality $\omega_1 > \omega_2$, which follows from the diagram of energy transitions.

The MTLPE signal propagates with a carrier frequency ω_2 in the direction of incidence of the pump pulses. As was demonstrated for the first time by the authors of [43–45], this signal contains the spectroscopic information concerning the relaxation parameters $\gamma_{bc}^{(0)}$, $\gamma_{bc}^{(1)}$, and $\gamma_{bc}^{(2)}$ of an optically forbidden transition $c \rightarrow b$ if the photon echo signal is produced by small-area pump pulses. The procedure of polarization echo spectroscopy based on the MTLPE and the method of the extraction of the spectroscopic information on the relaxation parameters $\gamma_{bc}^{(0)}$, $\gamma_{bc}^{(1)}$, and $\gamma_{bc}^{(2)}$ were described in [43–45].

Let us consider the physical scenario of MTLPE formation in a gas medium. Suppose that the first pump pulse with a duration $\widetilde{T}_1$ and carrier frequency ω_1, which is resonant to the frequency ω_0 of an optically allowed transition $c \rightarrow a$, is incident on the $y = 0$ boundary of a gas medium at the moment of time $t = 0$ and propagates along the positive direction of the Y axis. The strength of the electric field in this pulse is given by formula (4.7.1). This pulse excites multipole polarization moments in the medium, which are involved in fast oscillations with frequency ω_1.

As the first pump pulse leaves the point y of the gas medium, these multipole moments of the polarization decay with characteristic times $1/\gamma_{ac}^{(k)}$ ($|J_a - J_c| \leq k \leq J_a + J_c$), where $\gamma_{ac}^{(k)}$ is the relaxation characteristic of the optical coherence matrix that is related to a dipole-allowed transition $c \rightarrow a$ and that includes radiative decay, as well as inelastic gas-kinetic and elastic depolarizing collisions. In addition, the multipole moments of the polarization of the medium, involved in fast oscillations with frequency ω_1, oscillate because of Doppler dephasing, acquiring a phase factor $\exp(ik_1v_y\tau_1)$ by the moment of time when the second pump pulse reaches the point y of the gas medium. Here, $k_1 = \omega_1/c$, τ_1 is the time interval between the first and second pump pulses, and v_y is the projection of the velocity $\mathbf{v}$ of resonant atoms (molecules) on the Y axis.

Suppose that the second pump pulse with a duration $\widetilde{T}_2$ and carrier frequency ω_2, which is resonant to the frequency ϖ_0 of an optically allowed transition $b \to a$ ($E_c > E_b > E_a$), is incident on the $y = 0$ boundary of the gas medium at the moment of time $t = \tau_1 + \widetilde{T}_1$ and propagates along the positive direction of the Y axis. The strength of the electric field in this pulse can be written as

$$\boldsymbol{E}_2 = e^{(2)}\boldsymbol{l}_2 g_2(\xi - \tau_1 - \widetilde{T}_1)\exp[i(\omega_2 t - k_2 y + \Phi_2)] + \text{c.c.},$$
$$0 \le \xi - \tau_1 - \widetilde{T}_1 \le \widetilde{T}_2, \qquad (4.7.22)$$

where $e^{(2)}$ is the constant amplitude, Φ_2 is the constant phase shift, $\boldsymbol{l}_2 = (\sin\psi_2, 0, \cos\psi_2)$ is the polarization vector, $\xi = t - y/c$, $k_2 = \omega_2/c$, and the function g_2 describing the pulse waveform is normalized in accordance with (4.1.1). This pulse transfers the coherence from multipole polarization moments of the medium involved in fast oscillations with frequency ω_1 to multipole moments related to an optically forbidden transition $c \to b$.

As the second pump pulse leaves the point y of the gas medium, the multipole moments related to the optically forbidden transition $c \to b$ decay with characteristic times $1/\gamma_{bc}^{(k)}$ ($|J_b - J_c| \le k \le J_b + J_c$), where $\gamma_{bc}^{(k)}$ are the relaxation characteristics of the optical coherence matrix that is related to the dipole-forbidden transition $c \to b$ and that includes radiative decay, as well as inelastic gas-kinetic and elastic depolarizing collisions. The multipole polarization moments related to the optically forbidden transition $c \to b$ acquire a phase factor $\exp\{iv_y[(k_1 - k_2)\tau_2 + k_1\tau_1]\}$ by the moment of time when the third pump pulse reaches the point y of the gas medium. Here, τ_2 is the time interval between the second and third pump pulses.

Suppose that the third pump pulse with a duration $\widetilde{T}_3$ and carrier frequency ω_1, which is resonant to the frequency ω_0 of an optically allowed transition $c \to a$, is incident on the $y = 0$ boundary of the gas medium at the moment of time $t = \tau_1 + \tau_2 + \widetilde{T}_1 + \widetilde{T}_2$ and propagates along the positive direction of the Y axis. The strength of the electric field in this pulse can be written as

$$\boldsymbol{E}_3 = e^{(3)}\boldsymbol{l}_3 g_3(\xi - \tau_1 - \tau_2 - \widetilde{T}_1 - \widetilde{T}_2)\exp[i(\omega_1 t - k_1 y + \Phi_3)] + \text{c.c.},$$

$$0 \le \xi - \tau_1 - \tau_2 - \widetilde{T}_1 - \widetilde{T}_2 \le \widetilde{T}_3, \qquad (4.7.23)$$

where $e^{(3)}$ is the constant amplitude, Φ_3 is the constant phase shift, $\boldsymbol{l}_3 = (0, 0, 1)$ is the polarization vector, and the function g_3 describing the pulse waveform is normalized according to (4.1.1) with $n = 3$. This pulse transfers the coherence from multipole polarization moments related to the optically forbidden transition $c \to b$ to the dipole moment of the polarization of the medium involved in fast oscillations with frequency ω_2.

As the third pump pulse leaves the point y of the gas medium, the part of this dipole moment that contributes to the MTLPE signal decays with a characteristic time $1/\gamma_{ab}^{(1)}$ and oscillates due to Doppler dephasing. This part of the dipole moment of the polarization of the medium is proportional to the phase factor $\exp[ik_2 v_y(t - t_{mte} - \widetilde{T}_1 - \widetilde{T}_2 - \widetilde{T}_3)]$. Therefore, radiation-emitting particles are in phase at the point y of a gas medium at the moment of time approximately equal to t_{mte} (4.7.21).

The MTLPE thus produced propagates in the same direction as the pump pulses with a carrier frequency ω_2. When the relaxation times $1/\gamma_{ab}^{(k)}$ of the optical coherence matrix related to the dipole-allowed transitions $b \to a$ are known, the investigation of the MTLPE signal makes it possible to determine the relaxation times $1/\gamma_{bc}^{(k)}$ (k = 0, 1, 2, ...) of the optical coherence matrix related to the dipole-forbidden transition $c \to b$. This possibility was highlighted for the first time by the authors of [43]. Note that the relaxation parameters $\gamma_{ab}^{(k)}$ can be determined, for example, from independent photon-echo experiments for the $b \to a$ transition.

To find the electric field strength in the MTLPE signal produced by pump pulses (4.7.1), (4.7.22), and (4.7.23), we will employ equations (2.2.1)–(2.2.3) with initial condition (2.2.4) and use the collision integral calculated in the model of elastic depolarizing collisions and averaged in both the direction and the modulus of the velocity **v** of resonant atoms (molecules). Since the method of calculation of the electric field strength $\boldsymbol{E}^{mte}$ in the MTLPE signal has been described in detail in [44, 45], we present only the final expression for $\boldsymbol{E}^{mte}$. As demonstrated in [43], in the case when the areas of all the three pump pulses are small, we have

$$\boldsymbol{E}^{mte} = -\pi\hbar^{-3}\omega_2 L c^{-1} |d|^2 |\bar{d}|^2 e^{(1)} \widetilde{T}_1 e^{(2)} \widetilde{T}_2 e^{(3)} \widetilde{T}_3 N_0 S_{mte}\, \boldsymbol{e}^{mte} \times$$

$$\times \exp(-\gamma_{ac}^{(1)}\tau_1 - \gamma_{ab}^{(1)} t')\exp[i(\omega_2 t - k_2 y - \Phi_1 + \Phi_2 + \Phi_3)] + \text{c.c.}, \qquad (4.7.24)$$

where L is the length of the gas medium; d and $\bar{d}$ are the reduced matrix elements of the dipole moment operators related to the resonant transitions $c \to a$ and $b \to a$, respectively; N_0 is the difference of population densities in the Zeeman sublevels of the resonant levels c and a before the irradiation of the medium with the first pump pulse; and $t' = t - \tau_1 - \tau_2 - \widetilde{T}_1 - \widetilde{T}_2 - \widetilde{T}_3 - y/c$. The nonvanishing components of the vector $\boldsymbol{e}^{mte}$, characterizing polarization properties of the MTLPE signal, are given by

$$e_x^{mte} = a_1 b_1{}^*\sin(\psi_2 - \psi_1) + a_2 b_2{}^*\sin(\psi_2 + \psi_1), \tag{4.7.25}$$

$$e_z^{mte} = (2/3)a_0 b_0{}^*\cos(\psi_2 - \psi_1) + (1/3)a_2 b_2{}^*[3\cos(\psi_2 + \psi_1) + \\ + \cos(\psi_2 - \psi_1)], \tag{4.7.26}$$

where

$$b_k = \exp\{-[\gamma_{bc}^{(k)} - i(\Delta_{bc}^{(k)} - \Delta_{bc}^{(1)})]\tau_2\}, \tag{4.7.27}$$

$$a_k = \begin{Bmatrix} k & 1 & 1 \\ J_a & J_b & J_c \end{Bmatrix}^2, \tag{4.7.28}$$

J_a, J_b, and J_c are the angular momenta of the resonant levels, and ψ_1 and ψ_2 are the angles between the polarization vectors of the first and second pump pulses and the polarization vector of the third pump pulse. The quantities $\gamma_{ac}^{(1)}$ and $\gamma_{ab}^{(1)}$ involved in (4.7.24)–(4.7.28) stand for the homogeneous half-widths of the spectral lines corresponding to optically allowed resonant transitions $c \to a$ and $b \to a$,

$$\gamma_{bc}^{(k)} = (\gamma_b^{(0)} + \gamma_c^{(0)})/2 + \mathrm{Re}\,\Gamma_{bc}^{(k)},$$

where $1/\gamma_b^{(0)}$ and $1/\gamma_c^{(0)}$ are the population relaxation times for the levels b and c, respectively, due to radiative decay and inelastic gas-kinetic collisions and $\mathrm{Re}\,\Gamma_{bc}^{(k)}$, and $\Delta_{bc}^{(k)}$ are the relaxation parameters of the optically forbidden transition $c \to b$ including elastic depolarizing collisions. Finally, the quantity S_{mte}, which characterizes the waveform of the MTLPE signal, is written as

$$S_{mte'} = \int c_1(v_y)\bar{c}_2^*(v_y)\bar{c}_3^*(v_y)f(v_y)\times$$

$$\exp\{iv_y[k_2(t - t_{mte} - \tilde{T}_1 - ½\tilde{T}_2) - k_1(½\tilde{T}_1 + \tilde{T}_2 + ½\tilde{T}_3)]\}dv_y, \qquad (4.7.29)$$

where the quantities $c_1(v_y)$ and $\bar{c}_2(v_y)$ are defined by formulas (4.7.13) and (4.7.14), respectively, and

$$\bar{c}_3(v_y) = \frac{1}{\tilde{T}_3}\int_{-\tilde{T}_3/2}^{\tilde{T}_3/2} g_3(s + \tilde{T}_3/2)\exp(ik_1 v_y s)ds\,. \qquad (4.7.30)$$

For the sake of simplicity, formulas (4.7.24)–(4.7.30) were written for the case of an exact resonance,

$$\omega_1 = \omega_0 + \Delta_{ac}^{(1)}, \qquad \omega_2 = \varpi_0 + \Delta_{ab}^{(1)}. \qquad (4.7.31)$$

In the general case, the MTLPE signal (4.7.24)–(4.7.30) is elliptically polarized and propagates with a carrier frequency ω_2 in the same direction as the pump pulses. Note that, in the case when $|\Delta_{bc}^{(k)} - \Delta_{bc}^{(1)}|\tau_2 \ll 1$, the MTLPE signal becomes linearly polarized. As it follows from (4.7.24)–(4.7.30), polarization properties of the MTLPE signal produced by small-area pump pulses are independent of the waveforms of the pump pulses. This circumstance makes small-area pump pulses very attractive for the application in experiments on the modified three-level photon echo in gas media. Note that the waveform of the MTLPE signal (4.7.24)–(4.7.30) has been analyzed in detail in [43–45], and these results will be discussed in Section 5.1. Therefore, we will proceed with the discussion of the possibility of applying formulas (4.7.24)–(4.7.30) to extract the spectroscopic information on the relaxation parameters $\gamma_{bc}^{(k)}$ $(k = 0, 1, 2)$ for an optically forbidden transition $c \to b$.

If the relaxation parameter $1/\gamma_{ab}^{(1)}$ is known, formulas (4.7.24)–(4.7.30) make it possible to extract the experimental information concerning the relaxation parameters $\gamma_{bc}^{(k)}$ and $\Delta_{bc}^{(k)}$ of an optically forbidden transition $c \to b$. The quantity $\gamma_{ab}^{(1)}$ involved in formulas (4.7.24)–(4.7.30) can be determined, for example, from independent experiments on the decay of the primary photon echo signal produced through an optically allowed transition $b \to a$ with the growth in the time interval between pump pulses. With an assumption that $|\Delta_{bc}^{(k)} - \Delta_{bc}^{(1)}|\tau_2 << 1$, the MTLPE signal (4.7.24)–(4.7.30) is linearly polarized. With properly chosen angles ψ_1 and ψ_2, we can separate the terms in $\boldsymbol{E}^{mte}$ that decay with a single relaxation parameter $\gamma_{bc}^{(k)}$. Thus, we finally arrive in the case of the MTLPE at formulas similar to expressions (4.7.17)–(4.7.19), derived for the three-level photon echo. Such formulas, which were presented for the first time in [43], can be employed to extract the spectroscopic information on the relaxation parameters $\gamma_{bc}^{(0)}$, $\gamma_{bc}^{(1)}$, and $\gamma_{bc}^{(2)}$ for an optically forbidden transition $c \to b$. However, relaxation parameters $\gamma_{bc}^{(0)}$, $\gamma_{bc}^{(1)}$, and $\gamma_{bc}^{(2)}$ are usually close to each other in MTLPE experiments. Therefore, the use of such an experimental approach requires a sufficiently high accuracy of experimental measurements.

To loosen the requirements to the accuracy of measurements, Yevseyev *et al.* [43] have proposed another experimental technique. In the case when $\psi_1 = \pi/2$ and $\psi_2 = 0$, we have $E_z^{mte} = 0$, and the MTLPE signal (4.7.24)–(4.7.30) is linearly polarized along the X axis. Let us denote the intensity of the MTLPE signal in this case as I_1. In the case when $\psi_1 = 0$ and $\psi_2 = \pi/2$, we have $E_z^{mte} = 0$, and the MTLPE signal is linearly polarized along the X axis too. Let us denote the intensity of the MTLPE signal in this case as I_2. The ratio of MTLPE intensities corresponding to these two cases is

$$I_1/I_2 = \{a_1 - a_2\exp[(\gamma_{bc}^{(1)} - \gamma_{bc}^{(2)})\tau_2]\}^2 \times$$
$$\times\{a_1 + a_2\exp[(\gamma_{bc}^{(1)} - \gamma_{bc}^{(2)})\tau_2]\}^{-2}. \qquad (4.7.32)$$

Thus, we can find the quantity $|\gamma_{bc}^{(1)} - \gamma_{bc}^{(2)}|$ by experimentally studying the ratio I_1/I_2 as a function of the time interval τ_2 between the second and third pump pulses.

For $\psi_1 = \psi_2 = \pi/2$ and $\psi_1 = \psi_2 = 0$, we have $E_x^{mte} = 0$, and the MTLPE signal (4.7.24)–(4.7.30) is linearly polarized along the Z axis. Let us denote the intensities of the MTLPE signal in this case as I_3 and I_4. Then, we can write

$$I_3/I_4 = \{a_0 - a_2\exp[(\gamma_{bc}^{(0)} - \gamma_{bc}^{(2)})\tau_2]\}^2 \times$$
$$\times\{a_1 + 2a_2\exp[(\gamma_{bc}^{(0)} - \gamma_{bc}^{(2)})\tau_2]\}^{-2}, \qquad (4.7.33)$$

which allows us to determine the quantity $|\gamma_{bc}^{(0)} - \gamma_{bc}^{(2)}|$ by experimentally studying the ratio I_3/I_4 as a function of the time interval τ_2.

We should emphasize that the dependence of the ratio of the MTLPE intensities on τ_2 in formulas (4.7.32) and (4.7.33) is due to the difference in the relevant relaxation parameters, which permits these differences to be determined with a high accuracy. Furthermore, such an approach does not require the knowledge of the relaxation parameter $\gamma_{ab}^{(1)}$. These circumstances make formulas (4.7.32) and (4.7.33) a convenient tool for processing the experimental data and for extracting the spectroscopic information on the forbidden transition $c \to b$ by means of the modified three-level photon echo.

4.8 Measurement of Population, Orientation, and Alignment Relaxation Times for Levels Involved in Resonant Transitions

As demonstrated in [16, 32, 42, 50–62], population, orientation, and alignment relaxation times of resonant levels can be measured by two modifications of the photon echo mentioned in Chapter 2 – stimulated photon echo (SPE) and modified stimulated photon echo (MSPE). The theory of polarization properties of the stimulated photon echo was developed in [16, 32, 50–55], while the theory of polarization properties of the modified stimulated photon echo was developed in [42, 56–62].

Let us first consider polarization properties of the SPE produced by small-area pump pulses with arbitrary waveforms through an optically allowed transition involving resonant levels with arbitrary angular momenta. Recall that, in Section 2.5, we considered the physical scenario of the formation of the SPE signal produced by linearly polarized rectangular pump pulses (2.5.1)–(2.5.6) through optically allowed transition changing the angular momentum $J_b = 1 \to J_a = 0$. We have also derived the expressions for the electric field strength in this signal. In this section, we assume that the SPE signal is produced under the action of linearly polarized pump pulses (2.5.1)–(2.5.3) whose polarization vectors make an angle ψ. However, now, we will consider pump pulses of arbitrary waveforms. In other words, we assume that the functions g_n involved in (2.5.1)–(2.5.3) and normalized in accordance with (4.1.1) are not necessarily described by (2.5.4)–(2.5.6).

To determine the electric field strength in the SPE signal, we should solve the set of equations (2.3.8)–(2.3.11) using the method described in Section 2.5. Performing this procedure, we derive the following expression for the electric field strength $\boldsymbol{E}^{se}$ of the SPE signal [16, 50]:

$$\boldsymbol{E}^{se} = -\pi\hbar^{-3}\omega L c^{-1} |d|^4 e^{(1)} \widetilde{T}_1 e^{(2)} \widetilde{T}_2 e^{(3)} \widetilde{T}_3 N_0 S_{se} \boldsymbol{e}^{se} \times$$

$$\times \exp[i(\omega t - ky - \Phi_1 + \Phi_2 + \Phi_3) - \gamma_{ab}^{(1)}(t - \tau_2 - y/c)] + \text{c.c.} \qquad (4.8.1)$$

Here ω is the carrier frequency of the pump pulses, $e^{(n)}$ and Φ_n are the constant amplitude and phase of the n-th pump pulse (n = 1, 2, 3), $\widetilde{T}_n$ is the effective duration of the pump pulse, $\gamma_{ab}^{(1)}$ is the homogeneous half-width of the spectral line corresponding to an optically allowed resonant transition $b \to a$, d is the reduced matrix element of the dipole moment operator related to this transition, N_0 is defined by formula (2.3.18), and τ_2 is the time interval between the second and third pump pulses. The nonvanishing components of the vector $\boldsymbol{e}^{se}$, which characterizes the polarization properties of the SPE signal, are given by

$$e_x^{se} = -V\sin(\psi_1 - \psi_2) + W\sin(\psi_1 + \psi_2), \qquad (4.8.2)$$

$$e_z^{se} = (2/3)[U\cos(\psi_1 - \psi_2) + W(2\cos\psi_1\cos\psi_2 - \sin\psi_1\sin\psi_2)], \qquad (4.8.3)$$

where

$$U = M(J_b, J_a)\exp(-\gamma_b^{(0)}\tau_2) + M(J_a, J_b)\exp(-\gamma_a^{(0)}\tau_2), \qquad (4.8.4)$$

$$V = N(J_b, J_a)\exp(-\gamma_b^{(1)}\tau_2) + N(J_a, J_b)\exp(-\gamma_a^{(1)}\tau_2), \qquad (4.8.5)$$

$$W = L(J_b, J_a)\exp(-\gamma_b^{(2)}\tau_2) + L(J_a, J_b)\exp(-\gamma_a^{(2)}\tau_2). \qquad (4.8.6)$$

The quantities $M(J_b, J_a)$, $N(J_b, J_a)$, and $L(J_b, J_a)$ involved in (4.8.4)–(4.8.6) are independent of τ_2, being the functions of the angular moments of the resonant levels:

$$M(J, J) = M(J, J+1) = [3(2J+1)]^{-1}, \quad N(J, J) = [6J(J+1)(2J+1)]^{-1},$$

$$M(J+1, J) = [3(2J+3)]^{-1}, \quad L(J, J) = (2J-1)(2J+3)[30J(J+1)(2J+1)]^{-1},$$

$$L(J, J+1) = J(2J-1)[30(J+1)(2J+1)(2J+3)]^{-1},$$

$$L(J+1, J) = (J+2)(2J+5)[30(J+1)(2J+1)(2J+3)]^{-1},$$

$$N(J, J+1) = J[6(J+1)(2J+1)]^{-1}, \quad N(J+1, J) = (J+2)[6(J+1)(2J+3)]^{-1}.$$

Recall that the relaxation parameters $\gamma_{a,b}^{(k)}$ ($k = 0, 1, 2$) of the levels of a resonant transition involved in (4.8.4)–(4.8.6) are given by formula (2.3.15) and are assumed to be independent of the modulus of the velocity **v** of resonant atoms (molecules). We also set $\Gamma_a^{(0)} = \Gamma_b^{(0)} = 0$, because elastic collisions do not change the populations of resonant levels. Finally, the quantity S_{se}, which characterizes the waveform of the SPE signal, is given by

$$S_{se} = \int c_1(v_y) c_2^*(v_y) c_3^*(v_y) f(v_y) \times$$
$$\times \exp\{ikv_y[t - t_{se} + ½(\widetilde{T}_3 - \widetilde{T}_1 - \widetilde{T}_2)]\} dv_y, \tag{4.8.7}$$

where

$$c_n(v_y) = \frac{1}{\widetilde{T}_n} \int_{-\widetilde{T}_n/2}^{\widetilde{T}_n/2} g_n(s + \widetilde{T}_n/2) \exp(ikv_y s) ds, \tag{4.8.8}$$

$n = 1, 2, 3$; t_{se} is determined by formula (2.5.7), and $f(v_y)$ describes the Maxwell distribution of resonant atoms (molecules) in projections v_y of their velocities **v** on the Y axis.

For simplicity, formulas (4.8.1)–(4.8.8) were written for the case of an exact resonance,

$$\omega = \omega_0 + \Delta_{ab}^{(1)}, \tag{4.8.9}$$

where $\Delta_{ab}^{(1)}$ is the shift of the spectral line corresponding to the resonant transition $b \to a$ due to elastic depolarizing collisions. In addition, we assumed, as usually, that the effective durations of the pump pulses are small as compared with the time intervals between these pulses and the relevant irreversible relaxation times. Finally, in writing (4.8.1)–(4.8.8), we omitted the terms including the radiative population of the lower resonant level a due to the spontaneous emission from the upper level b.

The SPE signal (4.8.1)–(4.8.8) propagates in the same direction as the pump pulses, having a carrier frequency ω and linear polarization. As it follows from (4.8.1)–(4.8.8), polarization properties of the SPE signal produced by small-area pump pulses are independent of the waveform of the pump pulses. This fact, which was pointed out for the first time by the authors of [16], is an important advantage of using small-area pump pulses in polarization echo spectroscopy based on the stimulated photon echo.

Let us consider first expression (4.8.7). We will analyze the case when the inhomogeneously broadened spectral line involved in the formation of the SPE signal is narrow (2.3.24) for all the pump pulses. In this case, formula (4.8.7) yields

$$S_{se} = \exp\{-\tfrac{1}{4}[ku(t - 2\tau_1 - \tau_2 - y/c)]^2\}. \tag{4.8.10}$$

Thus, in the case when the spectral line involved in the formation of the SPE signal is narrow for all the pump pulses, the intensity of the SPE signal reaches its maximum at the moment of time $t = 2\tau_1 + \tau_2 + y/c$, while the duration of the SPE signal is $(ku)^{-1}$. This result was obtained for the first time by Samartsev *et al.* [63].

It is of interest also to consider the cases when the spectral line of the resonant transition involved in the formation of the SPE signal is broad for one of the pump pulses and narrow for the remaining two pump pulses. These situations are discussed in Section 5.1.

Integration in (4.8.7) can be also performed in the analytic form in the case when the spectral line involved in the formation of the SPE signal is broad (2.3.25) for all the pump pulses and the pump pulses have a rectangular shape. For example, with $\widetilde{T}_1 = \widetilde{T}_2 = \widetilde{T}_3 = \widetilde{T}$, expression (4.8.7) again gives [50] formula (2.5.18), where the replacement $M(p_2, p_3^*) \to S_{se}$ should be made.

As it follows from (4.8.1)–(4.8.8), the electric field strength in the SPE signal as a function of the time interval τ_2 between the second and third pump pulses is determined by the dependence of the components of the vector $\boldsymbol{e}^{se}$ on τ_2. Therefore, the projection of the electric field strength of the SPE signal on the direction of the polarization vector of the third pump pulse depends on τ_2 through e_z^{se}, while the projection of this electric field on the perpendicular direction depends on τ_2 through e_x^{se}. Keeping this circumstance in mind, we will demonstrate how the information on the population, orientation, and alignment relaxation times of resonant levels can be extracted from the experimental data.

Formula (4.8.2) with $\psi_1 = \psi_2$ yields

$$E_x^{se}(\tau_2) \sim L(J_b, J_a)\exp(-\gamma_b^{(2)}\tau_2) + L(J_a, J_b)\exp(-\gamma_a^{(2)}\tau_2). \tag{4.8.11}$$

This expression makes it possible to measure the alignment relaxation times $1/\gamma_b^{(2)}$ and $1/\gamma_a^{(2)}$ of resonant levels in the case when these times are either close to each other or considerably differ from each other. To determine these parameters, one has to investigate the decay of E_x^{se} with the growth in τ_2 for $\psi_1 = \psi_2$. Formula (4.8.11) allows also the spectroscopic information on the quantities $d\gamma_b^{(2)}/dp$ and $d\gamma_a^{(2)}/dp$ to be obtained by changing the buffer-gas pressure p for a constant time interval τ_2.

If $\psi_2 = -\psi_1$, then formula (4.8.2) gives

$$E_x^{se}(\tau_2) \sim N(J_b, J_a)\exp(-\gamma_b^{(1)}\tau_2) + N(J_a, J_b)\exp(-\gamma_a^{(1)}\tau_2). \tag{4.8.12}$$

Therefore, increasing τ_2 and observing the decay of the projection of the electric field strength in the SPE signal on the direction perpendicular to the polarization vector of the third pump pulse, one can measure the orientation relaxation times $1/\gamma_b^{(1)}$ and $1/\gamma_a^{(1)}$ of resonant levels if these times are either close to each other or considerably differ from one another. One can also extract the spectroscopic information on the quantities $d\gamma_b^{(1)}/dp$ and $d\gamma_a^{(1)}/dp$ by studying the decay of E_x^{se} with the growth in the buffer-gas pressure p for a constant time interval τ_2.

When ψ_1 and ψ_2 are chosen in such a way that $\tan\psi_1\tan\psi_2 = 2$ formula (4.8.3) yields

$$E_z^{se}(\tau_2) \sim M(J_b, J_a)\exp(-\gamma_b^{(0)}\tau_2) + M(J_a, J_b)\exp(-\gamma_a^{(0)}\tau_2). \tag{4.8.13}$$

This expression makes it possible to measure the population relaxation times $1/\gamma_b^{(0)}$ and $1/\gamma_a^{(0)}$ of resonant levels if these times are either close to each other or considerably differ from one another. This formula also allows one to extract the spectroscopic information on the quantities $d\gamma_b^{(0)}/dp$ and $d\gamma_a^{(0)}/dp$ by changing the buffer-gas pressure p for a constant time interval τ_2.

Analysis of formulas (4.8.2) and (4.8.3) shows that there are also other relations between ψ_1 and ψ_2 that allow decay processes with population, orientation, or alignment relaxation times of resonant levels to be separated in (4.8.2) or (4.8.3).

Thus, the method of stimulated photon echo makes it possible to obtain the spectroscopic information concerning the population (orientation or alignment) relaxation times when these times are either close to each other or considerably differ from one another for resonant levels. First, we will analyze the case when these times are close for resonant levels, i.e., $\gamma_a^{(k)} \approx \approx \gamma_b^{(k)}$ ($k = 0, 1, 2$). One can encounter such a situation when the SPE is produced in molecular gases through vibrational-rotational transitions that belong to the same electronic state. We should note also that the radiative population of the lower resonant level due to the spontaneous emission from the upper level does not play a significant role for such transitions. In this case, expressions (4.8.11)–(4.8.13) yield

$$E_x^{se}(\tau_2) \sim [L(J_b, J_a) + L(J_a, J_b)]\exp(-\gamma_b^{(2)}\tau_2) \tag{4.8.14}$$

for $\psi_2 = \psi_1$,

$$E_x^{se}(\tau_2) \sim [N(J_b, J_a) + N(J_a, J_b)]\exp(-\gamma_b^{(1)}\tau_2) \tag{4.8.15}$$

for $\psi_2 = -\psi_1$, and

$$E_z^{se}(\tau_2) \sim [M(J_b, J_a) + M(J_a, J_b)]\exp(-\gamma_b^{(0)}\tau_2) \tag{4.8.16}$$

for $\tan\psi_1\tan\psi_2 = 2$. Formulas (4.8.14)–(4.8.16) provide an opportunity to measure the relaxation times $1/\gamma_b^{(0)}$, $1/\gamma_b^{(1)}$, and $1/\gamma_b^{(2)}$ through the experimental investigation of the decay of the components of the electric field strength in the SPE signal with the growth in τ_2. However, when SPE experiments are performed in molecular gases, the relaxation characteristics $\gamma_b^{(0)}$, $\gamma_b^{(1)}$, and $\gamma_b^{(2)}$ are close to each other. Therefore, the use of formulas (4.8.14)–(4.8.16) for the measurement of these parameters requires a sufficiently high accuracy of experimental studies. To loosen the requirements to the accuracy of measurements, Yevseyev *et al.* [64] have proposed another experimental technique allowing the differences $\gamma_b^{(2)}-\gamma_b^{(0)}$ and $\gamma_b^{(1)}-\gamma_b^{(2)}$ of relaxation parameters to be measured directly. The former difference can be determined by measuring the ratio η of the intensities I_{se} of SPE signals as a function of τ_2 for the following two cases. In the first case, the polarization plane of the third pump pulse should be orthogonal to the polarization plane of the first two pump pulses ($\psi_2 = \psi_1 = \pi/2$). In the second case, the polarization planes of all the three pump pulses should coincide with each other ($\psi_2 = \psi_1 = 0$). As it follows from (4.8.1)–(4.8.3), the ratio $\eta = I_{se}(\psi_1 = \pi/2, \psi_2 = \pi/2)/I_{se}(\psi_1 = 0, \psi_2 = 0)$ with $\gamma_a^{(k)} \approx \gamma_b^{(k)}$ can be represented as

$$\eta = \{1-(2J-1)(2J+3)[10J(J+1)]^{-1}\exp[-(\gamma_b^{(2)}-\gamma_b^{(0)})\tau_2]\}^2\times$$
$$\times\{1+2(2J-1)(2J+3)[10J(J+1)]^{-1}\exp[-(\gamma_b^{(2)}-\gamma_b^{(0)})\tau_2]\}^{-2} \qquad (4.8.17)$$

for $J\rightarrow J$ transitions (*Q*-branch) and

$$\eta = \{1-(2J^2+4J+5)[20(J+1)^2]^{-1}\exp[-(\gamma_b^{(2)}-\gamma_b^{(0)})\tau_2]\}^2\times$$
$$\times\{1+2(2J^2+4J+5)[20(J+1)^2]^{-1}\exp[-(\gamma_b^{(2)}-\gamma_b^{(0)})\tau_2]\}^{-2} \qquad (4.8.18)$$

for $J\rightleftarrows J+1$ transitions (*P* or *R* branch). Formulas (4.8.17) and (4.8.18) allow the difference $\gamma_b^{(2)}-\gamma_b^{(0)}$ to be measured directly through the experimental investigation of the ratio η as a function of τ_2.

To find the difference $\gamma_b^{(1)}-\gamma_b^{(2)}$, the authors of [64] have proposed to measure the ratio ξ of the intensities I_{se} of SPE signals as a function of the time interval τ_2 for other two cases. In the first case, the polarization plane of the first pump pulse should be orthogonal to the polarization plane of the other two pump pulses ($\psi_1 = \pi/2$, $\psi_2 = 0$). In the second case, the polarization plane of the second pump pulse should be orthogonal to the polarization plane of the other two pump pulses ($\psi_1 = 0$, $\psi_2 = \pi/2$). As it follows from (4.8.1)–(4.8.3), the ratio $\xi = I_{se}(\psi_1 = \pi/2, \psi_2 = 0)/I_{se}(\psi_1 = 0, \psi_2 = \pi/2)$ with $\gamma_a^{(k)} \approx \gamma_b^{(k)}$ can be represented as

$$\xi = \{(2J-1)(2J+3) - 5\exp[-(\gamma_b^{(1)} - \gamma_b^{(2)})\tau_2]\}^2 \times$$
$$\times \{(2J-1)(2J+3) + 5\exp[-(\gamma_b^{(1)} - \gamma_b^{(2)})\tau_2]\}^{-2} \qquad (4.8.19)$$

for $J \to J$ transitions (Q-branch) and

$$\xi = \{1 - 5(2J^2 + 4J + 1)(2J^2 + 4J + 5)^{-1}\exp[-(\gamma_b^{(1)} - \gamma_b^{(2)})\tau_2]\}^2 \times$$
$$\times \{1 + 5(2J^2 + 4J + 1)(2J^2 + 4J + 5)^{-1}\exp[-(\gamma_b^{(1)} - \gamma_b^{(2)})\tau_2]\}^{-2} \qquad (4.8.20)$$

for $J \rightleftarrows J + 1$ transitions (P or R branch). Formulas (4.8.19) and (4.8.20) allow the difference $\gamma_b^{(1)} - \gamma_b^{(2)}$ to be measured directly through the experimental investigation of the ratio ξ as a function of the time interval τ_2.

In experiments on the stimulated photon echo in SF_6 molecules excited with CO_2-laser pulses, the rotational quantum numbers J of resonant levels are high ($J >> 1$). In the limiting case when $J >> 1$, formulas (4.8.17)–(4.8.20) give

$$\eta = \{1 - (2/5)\exp[-(\gamma_b^{(2)} - \gamma_b^{(0)})\tau_2]\}^2 \times$$
$$\times \{1 + (4/5)\exp[-(\gamma_b^{(2)} - \gamma_b^{(0)})\tau_2]\}^{-2}, \qquad (4.8.21)$$

$$\xi = 1 \qquad (4.8.22)$$

for the Q-branch and

$$\eta = \{1 - (1/10)\exp[-(\gamma_b^{(2)} - \gamma_b^{(0)})\tau_2]\}^2 \times$$

$$\times \{1 + (1/5)\exp[-(\gamma_b^{(2)} - \gamma_b^{(0)})\tau_2]\}^{-2}, \qquad (4.8.23)$$

$$\xi = \{1 - 5\exp[-(\gamma_b^{(1)} - \gamma_b^{(2)})\tau_2]\}^2\{1 + 5\exp[-(\gamma_b^{(1)} - \gamma_b^{(2)})\tau_2]\}^{-2} \qquad (4.8.24)$$

for the P or R branch.

We should note that formulas (4.8.21) and (4.8.23) have been already employed by the authors of [65] to process the results of photon-echo experiments performed for the $Q(38)$ transition of SF_6 molecules, where the ratio η was measured as a function of the time interval τ_2 (Fig. 4.7).

Processing the results of these measurements with the use of formula (4.8.21), the authors of [65] were able to obtain the spectroscopic information on the difference $\gamma_b^{(2)} - \gamma_b^{(0)}$ $((\gamma_b^{(2)} - \gamma_b^{(0)})p^{-1} = (20 \pm 5)\times 10^3\ c^{-1} \times$ mTorr^{-1}). Finally, using the population relaxation time, which was known for the studied resonant levels from other experiments ($p/\gamma_b^{(0)} = 36\ ns \times$ Torr), the authors of [65] determined the relaxation parameter $\gamma_b^{(2)}$ of the alignment of resonant levels ($\gamma_b^{(2)}/p = (50 \pm 5)\ )\times 10^3\ c^{-1} \times$ mTorr^{-1}). We emphasize that the authors of [65] were the first to measure the alignment relaxation time for the resonant levels involved in the $Q(38)$ transition of SF_6 molecules. The authors of [65] also carried out similar measurements for other resonant transitions and performed experiments in the presence of buffer gases. Thus, the results of [65] demonstrate that the SPE is an efficient technique for measuring the relaxation parameters of resonant molecular levels.

Now, let us consider the case when the relaxation parameters $\gamma_a^{(k)}$ and $\gamma_b^{(k)}$ appreciably differ from each other. In this case, formulas (4.8.11)–(4.8.13) allow one to find these parameters separately of each other, because, for small and large τ_2, the decay in (4.8.11)–(4.8.13) can be described with a single exponential with different exponents.

Recall that all the formulas presented in this section were derived with an assumption that the radiative population of the lower resonant level due to the spontaneous emission from the upper level does not play a significant role. Note that this process was taken into account in [10, 53, 54], where the theory of polarization properties of the stimulated photon echo produced by pump pulses with arbitrary areas through transitions involving resonant levels with low angular momenta was developed. The necessity of including the radiative population of the lower level in this case was associated with a fact that the formulas derived in these papers were applied to atomic systems, where the relevant terms may sometimes play an important role.

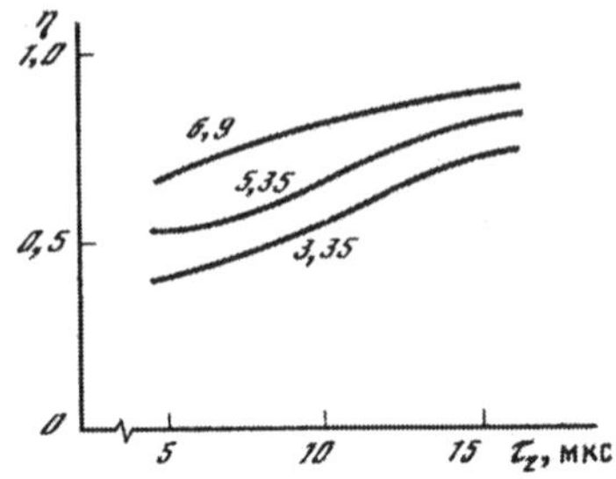

Fig. 4.7. Dependence of the ratio η on the time interval τ_2 between the second and third pump pulses for the Q(38) transition in SF_6. The SF_6 pressure in millitorrs is given near the curves.

The influence of this radiative population process for optically allowed resonant transitions involving levels with arbitrary angular momenta was analyzed by the authors of [55], whose aim was to determine the characteristic time parameter limiting the time interval τ_2 between the second and third pump pulses. The answer to this question is important for the application of the photon echo in atomic systems for data storage. Note that the results obtained in [55] are thoroughly discussed in Section 5.2.

Now, let us consider the possibilities of using the modified stimulated photon echo for extracting the spectroscopic information on the relaxation parameters of resonant levels. As mentioned in Chapter 2, the MSPE effect was predicted by the authors of [66]. This modification of the photon echo was observed for the first time by Hartmann and his group [67]. The theory of polarization properties of the modified stimulated photon echo was developed in [42, 56–62].

Let us consider the physical scenario of the formation of the modified stimulated photon echo [10, 42]. Similar to the case of the stimulated photon echo, a gas medium is excited with three pump pulses. The first two pump pulses with durations $\widetilde{T}_1$ and $\widetilde{T}_2$ and carrier frequency ω_1, which is resonant to the frequency ω_0 of an optically allowed transition $b \to a$, are incident on the $y = 0$ boundary of the gas medium at the moments of time $t = 0$ and $t = \tau_1 + \widetilde{T}_1$. Suppose that the electric field strength $\boldsymbol{E}_1$ in the first pump pulse is defined by formula (4.7.1) and the strength of the electric field in the second pump pulse can be written as

$$\boldsymbol{E}_2 = e^{(2)}\boldsymbol{l}_2 g_2(\xi - \tau_1 - \widetilde{T}_1)\exp[i(\omega_1 t - k_1 y + \Phi_2)] + \text{c.c.},$$
$$0 \le \xi - \tau_1 - \widetilde{T}_1 \le \widetilde{T}_2, \qquad (4.8.25)$$

where $e^{(2)}$ is the constant amplitude, Φ_2 is the constant phase shift, $k_1 = \omega_1/c$, $\boldsymbol{l}_2 = (\sin\psi_2, 0, \cos\psi_2)$ is the polarization vector, $\xi = t - y/c$, and the function g_2 characterizing the waveform of the second pulse is normalized in accordance with (4.1.1). Similar to the case of the stimulated photon echo (see Section 2.5), the first two pump pulses (4.7.1) and (4.8.25) induce a coherence in multipole moments of the resonant levels b and a. These multipole moments of the levels b and a acquire a phase factor $\exp(-ik_1 v_y \tau_1)$ by the moment of time when the third pump pulse reaches the point y of the gas medium. Here, τ_1 is the time interval between the first and second pump pulses and v_y is the projection of the velocity **v** of a resonant atom (molecule) on the Y axis.

The third pump pulse with a duration $\widetilde{T}_3$ in modified stimulated photon echo is incident on the $y = 0$ boundary of the gas medium at the moment of time $t = \tau_1 + \tau_2 + \widetilde{T}_1 + \widetilde{T}_2$. In contrast to the stimulated photon echo, this pulse has a carrier frequency ω_2, which is resonant to the frequency ϖ_0 of an optically allowed transition $c \to b$. The strength of the electric field in this pulse can be written as

$$\boldsymbol{E}_3 = e^{(3)}\boldsymbol{l}_3 g_3(\xi - \tau_1 - \tau_2 - \widetilde{T}_1 - \widetilde{T}_2)\exp[i(\omega_2 t - k_2 y + \Phi_3)] + \text{c.c.},$$
$$0 \le \xi - \tau_1 - \tau_2 - \widetilde{T}_1 - \widetilde{T}_2 \le \widetilde{T}_3, \qquad (4.8.26)$$

where $e^{(3)}$ is the constant amplitude, Φ_3 is the constant phase shift, $k_2 = \omega_2/c$, $\boldsymbol{l}_3 = (0, 0, 1)$ is the polarization vector, and the function g_3 characterizing the pulse waveform is normalized according to (4.1.1) with $n = 3$. The third pump pulse transfers the coherence induced by the first two pump pulses in the multipole moments of the common level b to the dipole moment of the polarization of the medium involved in fast oscillations with frequency ω_2. As the system evolves in time, the part of the dipole moment of the polarization of the medium that contributes to the MSPE signal acquires the phase factor $\exp[ik_2 v_y(t - \tau_1 - k_1\tau_1/k_2 - \tau_2 - \widetilde{T}_1 - \widetilde{T}_2 - \widetilde{T}_3 - y/c)]$, where τ_2 is the time interval between the second and third pump pulses. Thus, radiation-emitting particles at the point y of a gas medium are in phase at the moment of time $t \approx t_{mse}$, where

$$t_{mse} = (1 + \omega_1/\omega_2)\tau_1 + \tau_2 + \widetilde{T}_1 + \widetilde{T}_2 + \widetilde{T}_3 + y/c. \tag{4.8.27}$$

This part of the dipole moment of the polarization of the medium serves as a source for the MSPE signal (see Fig. 2.1). The MSPE signal has the carrier frequency ω_2 and propagates in the same direction as the pump pulses.

Note that the third pump pulse in the MSPE scheme in a gas medium may irradiate the $y = L$ boundary of the medium, propagating in the direction opposite of the first two pulses. In this case, the MSPE signal with the carrier frequency ω_2 propagates in the same direction as the third pump pulse.

We emphasize that the MSPE signal contains the spectroscopic information concerning the relaxation times $1/\gamma_b^{(k)}$ ($k = 0, 1, 2$) of the common level b. This circumstance was pointed out for the first time by the authors of [56, 57].

We should note that the modified stimulated photon echo has certain advantages over the stimulated photon echo. The MSPE scheme makes it possible to measure the relaxation times $1/\gamma_b^{(k)}$ of a common resonant level b (with different k), whereas the SPE signal depends on the relaxation times of both resonant levels a and b.

Let us consider polarization properties of the MSPE signal produced by small-area pump pulses through optically allowed transitions involving resonant levels with arbitrary angular momenta. The method of calculation of the electric field strength $\boldsymbol{E}^{mse}$ in the MSPE signal has been described in detail in [42, 61]. Therefore, we present only the final expression for $\boldsymbol{E}^{mse}$. Applying expressions presented in [41, 42], we have

$$\begin{aligned} \boldsymbol{E}^{mse} &= \pi\hbar^{-3}\omega_2 L c^{-1} |d|^2 |\bar{d}|^2 e^{(1)} \widetilde{T}_1 e^{(2)} \widetilde{T}_2 e^{(3)} \widetilde{T}_3 N_0 S_{mse} \boldsymbol{e}^{mse} \times \\ &\times \exp[i(\omega_2 t - k_2 y - \Phi_1 + \Phi_2 + \Phi_3)] \times \\ &\times \exp[-\gamma_{ab}^{(1)}\tau_1 - \gamma_{bc}^{(1)}(t - \tau_1 - \tau_2 - y/c)] + \text{c.c.}, \end{aligned} \tag{4.8.28}$$

where N_0 is defined by formula (2.3.18); L is the length of the gas medium; $\gamma_{ab}^{(1)}$ and $\gamma_{bc}^{(1)}$ are the homogeneous half-widths of the spectral lines; and d

and $\overline{d}$ are the reduced matrix element of the dipole moment operators related to the optically allowed resonant transitions $b \rightarrow a$ and $c \rightarrow b$, respectively. The quantity S_{mse}, which characterizes the waveform of the MSPE signal, is written as

$$S_{mse} = \int c_1 (v_y) c_2^*(v_y) c_3^*(v_y) f(v_y) \times$$
$$\exp\{ik_2 v_y[t - t_{mse} + ½(\widetilde{T}_3 - k_1 \widetilde{T}_1 / k_2 - k_1 \widetilde{T}_2 / k_2)]\} dv_y, \quad (4.8.29)$$

where t_{mse} is given by formula (4.8.27), the quantities $c_1(v_y)$ and $c_3(v_y)$ are defined by formulas (4.7.13) and (4.7.15), respectively, $f(v_y)$ describes the Maxwell distribution of resonant atoms (molecules) in projections v_y of their velocities **v** on the Y axis, and

$$c_2(v_y) = \frac{1}{\widetilde{T}_2} \int_{-\widetilde{T}_2/2}^{\widetilde{T}_2/2} g_2(s + \widetilde{T}_2/2) \exp(ik_1 v_y s) ds\,. \quad (4.8.30)$$

Finally, the nonvanishing components of the vector e^{mse}, characterizing polarization properties of the MSPE signal, are given by

$$e_x^{mse} = gb_1(\tau_2)\sin(\psi_1 - \psi_2) + ¼\, rb_2(\tau_2)\sin(\psi_1 + \psi_2)], \quad (4.8.31)$$

$$e_z^{msc} = sb_0(\tau_2)\cos(\psi_1 - \psi_2) + (1/3) rb_2(\tau_2)(\cos\psi_1 \cos\psi_2 -$$
$$- ½\sin\psi_1 \sin\psi_2), \quad (4.8.32)$$

where

$$b_k(\tau_2) = \exp(-\gamma_b^{(k)} \tau_2), \quad (4.8.33)$$

the relaxation parameter $\gamma_b^{(0)}$ includes only radiative decay and inelastic gas-kinetic collisions, and the relaxation parameters $\gamma_b^{(k)}$ with $k \neq 0$ depend also on elastic depolarizing collisions through the quantities $\Gamma_b^{(k)}$ [see formula (2.3.15)]. The quantities s, r, and g involved in (4.8.31) and (4.8.32) are independent of τ_2, being the functions of the angular moments of the resonant levels:

$$s = 2[9(2J_b + 1)]^{-1}, \qquad g = S(J_c)S(J_a)[24(2J_b + 1)X(J_b)]^{-1},$$

$$r = T(J_c)T(J_a)[30(4J_b^2 - 1)(2J_b + 3)X(J_b)]^{-1}.$$

Here,

$$X(J) = J(J + 1), \quad S(J) = X(J) - X(J_b) - 2, \quad T(J) = 3S(J)[S(J) + 1] - 8X(J_b).$$

For the sake of simplicity, formulas (4.8.28)–(4.8.33) were written in the case of an exact resonance,

$$\omega_1 = \omega_0 + \Delta_{ab}^{(1)}, \qquad \omega_2 = \varpi_0 + \Delta_{bc}^{(1)}, \tag{4.8.34}$$

where $\Delta_{ab}^{(1)}$ and $\Delta_{bc}^{(1)}$ are the shifts of the spectral lines corresponding to the transitions $b \to a$ and $c \to b$ due to elastic depolarizing collisions. Furthermore, we assumed that the durations of the pump pulses are small as compared with the time intervals between these pulses and the relevant irreversible relaxation times.

The MSPE signal (4.8.28)–(4.8.33) is linearly polarized and propagates with a carrier frequency ω_2 in the same direction as the pump pulses. As it follows from (4.8.28)–(4.8.33), polarization properties of the MSPE signal produced by small-area pump pulses are independent of the waveforms of the pump pulses. This circumstance, which was pointed out for the first time in [42], is an important advantage of using small-area pump pulses in the polarization echo spectroscopy of gas media. Note that, in the case of rectangular pump pulses with the same duration, formulas (4.8.28)–(4.8.33) for the electric field strength in the MSPE signal were derived for the first time by Yevseyev *et al.* [56].

Let us consider expression (4.8.29). We will first analyze the case when resonant spectral lines involved in the formation of the MSPE signal are narrow for all the three pump pulses [$k_1 u \widetilde{T}_1 \ll 1$, $k_1 u \widetilde{T}_2 \ll 1$, and $k_2 u \widetilde{T}_3 \ll 1$, where, as before, u is the root-mean-square thermal velocity of resonant atoms (molecules)]. In this case, formula (4.8.29) yields

$$S_{mse} = \exp\{-\tfrac{1}{4}[k_2 u(t - (1 + \omega_1/\omega_2)\tau_1 - \tau_2 - y/c)]^2\}. \tag{4.8.35}$$

Thus, in the case when the spectral lines involved in the formation of the MSPE signal are narrow for all the pump pulses, the maximum of the MSPE signal reaches the observation point y at the moment of time $t = (1 + \omega_1/\omega_2)\tau_1 + \tau_2 + y/c$, while the duration of the MSPE signal is $\sim (k_2 u)^{-1}$.

It is of interest also to consider the cases when the resonant spectral lines involved in the formation of the MSPE signal are broad for one of the pump pulses and narrow for the remaining two pump pulses (see Section 5.1).

Integration in (4.8.29) can be also performed in the analytic form in the case when the spectral lines involved in the formation of the MSPE signal are broad for all the three pump pulses, and the pump pulses have a rectangular shape. For example, with $\widetilde{T}_1 = \widetilde{T}_2 = \widetilde{T}_3 = \widetilde{T}$, expression (4.8.29) with $k_2/k_1 \geq 2$ gives

$$\begin{aligned} S_{mse} &= \pi^{1/2}(k_2 u \widetilde{T})^{-1}(a + b)^2 [\theta(a + b) - \theta(a + b - 1)] - (a^2 - 4b - 4a + 2 + \\ &+ b^2 + 2ab)[\theta(a + b - 1) - \theta(a + b - 2)] + 2[\theta(a + b - 2) - \theta(a)] - \\ &- (a^2 - 2)[\theta(a) - \theta(a - 1)] + (a - 2)^2[\theta(a - 1) - \theta(a - 2)], \end{aligned} \tag{4.8.36}$$

where

$$a = b(t - t_{mse})\widetilde{T}^{-1}, \qquad b = k_2/k_1,$$

t_{mse} can be determined from (4.8.27) with $\widetilde{T}_1 = \widetilde{T}_2 = \widetilde{T}_3 = \widetilde{T}$, and the function $\theta(x)$ is given by (2.3.5). Thus, the duration of the MSPE signal produced in the case of broad spectral lines is on the order of $\omega_2 \widetilde{T}/\omega_1$. Note that formula (4.8.36) was derived for the first time by Yevseyev *et al.* [42].

As it follows from (4.8.28)–(4.8.33), the electric field strength in the MSPE signal as a function of the time interval τ_2 between the second and third pump pulses is completely determined by the dependence of the component of the vector $\boldsymbol{e}^{mse}$ on τ_2. Therefore, the projection of the electric field strength of the MSPE signal on the direction of the polarization vector of the third pump pulse depends on τ_2 through e_z^{mse}, while the projection of this electric field on the perpendicular direction depends on τ_2 through e_x^{mse}. Keeping this circumstance in mind, we will demonstrate how the information on the population, orientation, and alignment relaxation times of the common level b can be extracted from the experimental data.

Formula (4.8.31) with $\psi_1 = \psi_2$ yields

$$E_x^{mse}(\tau_2) \sim \exp(-\gamma_b^{(2)}\tau_2). \tag{4.8.37}$$

This expression makes it possible to measure the alignment relaxation time $1/\gamma_b^{(2)}$ of the level b. For this purpose, one has to investigate the decay of E_x^{mse} with the growth in τ_2 for $\psi_1 = \psi_2$.

If $\psi_2 = -\psi_1$, then formula (4.8.31) gives

$$E_x^{mse}(\tau_2) \sim \exp(-\gamma_b^{(1)}\tau_2). \tag{4.8.38}$$

Therefore, increasing τ_2 and observing the decay of the projection of the electric field strength in the MSPE signal on the direction perpendicular to the polarization vector of the third pump pulse with $\psi_2 = -\psi_1$, one can directly measure the orientation relaxation time $1/\gamma_b^{(1)}$ of the level b.

When ψ_1 and ψ_2 are chosen in such a way that $\tan\psi_1\tan\psi_2 = 2$, formula (4.8.32) yields

$$E_z^{mse}(\tau_2) \sim \exp(-\gamma_b^{(0)}\tau_2). \tag{4.8.39}$$

This expression makes it possible to directly measure the population relaxation time $1/\gamma_b^{(0)}$ of the level b.

Analysis of formulas (4.8.31) and (4.8.32) shows that there are also other relations between ψ_1 and ψ_2 that allow decay processes with a single relaxation time related to population, orientation, or alignment of the common level b to be separated in (4.8.31) or (4.8.32).

We should note that formulas (4.8.37)–(4.8.39) allow also the spectroscopic information on the quantities $d\gamma_b^{(0)}/dp$, $d\gamma_b^{(1)}/dp$, and $d\gamma_b^{(2)}/dp$ to be obtained. For this purpose, one has to change the buffer-gas pressure p for a constant time interval τ_2, and process the experimental data with the use of formulas (4.8.37)–(4.8.39).

It should be mentioned that these expressions were derived for the first time by Yevseyev *et al.* in paper [56], which was published in 1981. In 1984, paper [68] by Keller and Le Gouet was published, where the MSPE signal was observed for ${}^3S_1 \to {}^3P_1 \to {}^1S_0$ $(c \to b \to a)$ transitions in a ^{174}Yb vapor. The authors of [68] detected a strong dependence of the MSPE signal intensity on the polarizations of the pump pulses. The authors of this paper investigated the following two cases. In the first case, all the three pump pulses were linearly polarized and their polarization vectors coincided with each other ($\psi_1 = \psi_2 = 0$). In the second case, the polarization vectors of the first and second pump pulses were orthogonal to the polarization vector of the third pump pulse ($\psi_1 = \psi_2 = \pi/2$). The MSPE signal intensity $I_{mse}(\psi_1 = 0, \psi_2 = 0)$ in the former case was three orders of magnitude lower than the intensity of the MSPE signal $I_{mse}(\psi_1 = \pi/2, \psi_2 = \pi/2)$ in the latter case. The authors of [68] did not specify the values of the areas θ_1, θ_2, and θ_3 of the pump pulses meeting the relation

$$\mu = I_{mse}(\psi_1 = 0, \psi_2 = 0)/I_{mse}(\psi_1 = \pi/2, \psi_2 = \pi/2) \approx 2\times10^{-3}.$$

Yevseyev *et al.* [60] employed formulas (4.8.28)–(4.8.33) to find the ratio μ in the case when $J_a = 0$ and $J_b = J_c = 1$, which corresponds to the experimental conditions implemented in [68]. Indeed, as it follows from formulas (4.8.31) and (4.8.32), the nonvanishing components of the vector e^{mse}, which characterizes the polarization properties of the MSPE signal for transitions involving resonant levels with angular momenta $J_a = 0$ and $J_b = J_c = 1$, are given by

$$e_x^{mse} = (1/18)[\exp(-\gamma_b^{(1)}\tau_2)\sin(\psi_1 - \psi_2) - \\ - \exp(-\gamma_b^{(2)}\tau_2)\sin(\psi_1 + \psi_2)], \tag{4.8.40}$$

$$e_z^{mse} = (2/27)[\exp(-\gamma_b^{(0)}\tau_2)\cos(\psi_1 - \psi_2) - \exp(-\gamma_b^{(2)}\tau_2)(\cos\psi_1\cos\psi_2 - \\ - ½\sin\psi_1\sin\psi_2). \tag{4.8.41}$$

As can be seen from (4.8.40) and (4.8.41), the MSPE signal is linearly polarized in both cases (when $\psi_1 = \psi_2 = 0$ and when $\psi_1 = \psi_2 = \pi/2$), and its polarization vector coincides with the polarization vector of the third pump pulse, which is consistent with the results of experiments [68]. In the former case (when $\psi_1 = \psi_2 = 0$), formula (4.8.41) gives the following expression for the only nonvanishing component of the vector $\boldsymbol{e}^{mse}$:

$$e_z^{mse} = (2/27)[\exp(-\gamma_b^{(0)}\tau_2) - \exp(-\gamma_b^{(2)}\tau_2)]. \tag{4.8.42}$$

In the latter case (when $\psi_1 = \psi_2 = \pi/2$), formula (4.8.41) yields

$$e_z^{mse} = (2/27)[\exp(-\gamma_b^{(0)}\tau_2) + \tfrac{1}{2}\exp(-\gamma_b^{(2)}\tau_2)]. \tag{4.8.43}$$

The results of experiments [68] can be understood if we assume that the quantities $\gamma_b^{(0)}$ and $\gamma_b^{(2)}$ are sufficiently close to each other, because, in this case, the ratio

$$\mu = \{1 - \exp[-(\gamma_b^{(2)} - \gamma_b^{(0)})\tau_2]\}^2\{1 + \tfrac{1}{2}\exp[-(\gamma_b^{(2)} - \gamma_b^{(0)})\tau_2]\}^{-2} \tag{4.8.44}$$

becomes much less than unity. Unfortunately, the authors of [68] do not specify the buffer-gas pressure and the time interval τ_2 corresponding to $\mu \approx 2\times10^{-3}$. Since the areas of the pump pulses employed in [68] are unknown, it is important to mention that, as demonstrated in [60], formula (4.8.44) holds true for $J_a = 0$ and $J_b = J_c = 1$ with arbitrary areas of the pump pulses. The relevant expressions describing the polarization properties of the MSPE signal for $J_a = 0$ and $J_b = J_c = 1$ were presented in [61].

Applying formula (4.8.44) with the ratio of intensities equal to 2×10^{-3} and τ_2 equal to 75 ns (in [68], the time interval τ_2 was varied from 60 to 90 ns), we find that $\Gamma_b^{(2)} = \gamma_b^{(2)} - \gamma_b^{(0)}$ is equal to $0.9\times10^6\ c^{-1}$. Using also the lifetime of the level b given in [68] $1/\gamma_b^{(0)} = 875\ ns$, we have $\gamma_b^{(0)} \approx 1.15\times10^6\ c^{-1}$. Thus, the relaxation parameter, $\gamma_b^{(2)} \approx 2.05\times10^6\ c^{-1}$, is of the same order of magnitude as the relaxation parameter $\gamma_b^{(0)}$.

Unfortunately, the authors of [68] did not investigate μ for different pressures p. Therefore, the data of these experiments do not allow one to extract the information concerning $d\Gamma_b^{(2)}/dp$.

Thus, the authors of [60] were the first to employ the photon echo to obtain the spectroscopic information on the alignment relaxation time of a resonant level.

The authors of [60] have also pointed to the fact that, apparently, the experimental scheme implemented in [68] makes it possible to determine also the relaxation parameter $\gamma_b^{(1)}$. Indeed, as it follows from (4.8.40) and (4.8.41), the only nonvanishing component of the vector $\boldsymbol{e}^{mse}$ for $\psi_1 = \pi/2$ and $\psi_2 = 0$ is written as

$$e_x^{mse} = (1/18)[\exp(-\gamma_b^{(1)}\tau_2) - \exp(-\gamma_b^{(2)}\tau_2)]. \tag{4.8.45}$$

Similar to the case considered above, we can apply formulas (4.8.40) and (4.8.41) to find that, with $\psi_1 = 0$ and $\psi_2 = \pi/2$,

$$e_x^{mse} = -(1/18)[\exp(-\gamma_b^{(1)}\tau_2) + \exp(-\gamma_b^{(2)}\tau_2)]. \tag{4.8.46}$$

Therefore, measuring the ratio of the intensities

$$\xi = I_{mse}(\psi_1 = \pi/2, \psi_2 = 0)/I_{mse}(\psi_1 = 0, \psi_2 = \pi/2) =$$

$$\{1 - \exp[-(\gamma_b^{(1)} - \gamma_b^{(2)})\tau_2]\}^2\{1 + \exp[-(\gamma_b^{(1)} - \gamma_b^{(2)})\tau_2]\}^{-2}, \tag{4.8.47}$$

one can extract the spectroscopic information concerning the difference $\Gamma_b^{(1)} - \Gamma_b^{(2)}$. In combination with the information on $\Gamma_b^{(2)}$, these spectroscopic data would provide an opportunity to find the quantity $\Gamma_b^{(1)}$. Furthermore, if the accuracy of measurements is sufficiently high, such an experimental scheme would allow the results of theoretical calculations for the ratio $\Gamma_b^{(1)}/\Gamma_b^{(2)}$ to be checked. Note that calculations of the ratio $\Gamma_b^{(1)}/\Gamma_b^{(2)}$ with an assumption of van der Waals interaction between resonant and buffer-gas atoms yield a value of 1.13 (see Chapter 1).

We emphasize that formulas analogous to (4.8.44) and (4.8.47) can be also derived for resonant levels with other angular momenta. Expressions (4.8.31) and (4.8.32) should be employed for this purpose. Such an approach would make it possible to obtain the spectroscopic information on the differences $\gamma_b^{(2)} - \gamma_b^{(0)}$ and $\gamma_b^{(1)} - \gamma_b^{(2)}$ by measuring the dependences of the ratios μ and ξ on the time interval τ_2.

Note that, in the case when the third pump pulse propagates in the direction opposite of the first two beams, polarization properties of the MSPE signal were considered in [57]. Population, orientation, and alignment relaxation times of a common resonant level *b* can be also measured in this case with the use of the methods considered above when we studied the scheme where all the three pump pulses propagate through a gas medium in the same direction. However, in contrast to the case when all the three pump pulses propagate through a gas medium in the same direction, a delay effect, predicted in [57], when different points *y* of a gas medium are characterized by different time intervals between the second and third pump pulses, is observed in the scheme with a backward third pulse.

Polarization properties of the MSPE signal produced by rectangular pump pulses with arbitrary areas through optically allowed transitions involving resonant levels with low angular momenta have been investigated in [42, 61, 62]. These studies have shown how the expressions for the electric field strength $\boldsymbol{E}^{mse}$ in the MSPE signal produced through such transitions can be employed to extract the spectroscopic information concerning the relaxation times of a common level *b* in the case when the areas of the pump pulses are not small.

Thus, we have demonstrated in this section that the signals of stimulated and modified stimulated photon echo have been already employed and can be employed in the future for polarization echo spectroscopy of gas media, allowing the spectroscopic information on the population, orientation, and alignment relaxation times of levels involved in inhomogeneously broadened resonant transitions to be extracted.

4.9 The Possibility of Measuring the Lifetime of the Upper Resonant State with Respect to Spontaneous Decay to the Lower Resonant State

In this section, we will present the results obtained in [51]. Initially, this study was made in order to predict the non-Faraday rotation of the polarization vector of the stimulated photon echo signal in the presence of a permanent uniform longitudinal magnetic field. However, this study not only allowed the prediction of this effect, but also demonstrated how this effect can be employed to measure the lifetime of the upper resonant state with respect to spontaneous decay to the lower state for $J_b = ½ \rightarrow J_a = ½$ transition changing the angular momentum.

The choice of $J_b = ½ \rightarrow J_a = ½$ transition was associated with the following circumstance. As mentioned in Chapter 2, the non-Faraday rotation of the polarization vector of the primary photon echo signal in a permanent uniform longitudinal magnetic field was predicted in [69] for $0 \rightleftarrows 1$ transitions changing the angular momentum. As shown in [70], this effect is not observed for $J_b = ½ \rightarrow J_a = ½$ transition. Finally, the study [26] has demonstrated that the non-Faraday rotation of the polarization vector of the primary photon echo signal occurs for all the optically allowed resonant transitions, except for $J_b = ½ \rightarrow J_a = ½$ transition, which motivated the investigation of the non-Faraday rotation of the polarization vector of the stimulated photon echo for such transition in [51]. (Recall that the results of [26] were discussed in Section 4.1.)

Let us consider first the case when all the three pump pulses involved in the formation of the stimulated photon echo signal propagate through a gas medium in the same direction. The case when the third pump pulse propagates through a gas medium in the direction opposite of the first two pulses will be considered later.

Suppose that the pump pulses are linearly polarized in the same plane, irradiating the $y = 0$ boundary of a gas medium at the moments of time $t = 0$, $t = \tau_1 + \widetilde{T}_1$, and $t = \tau_1 + \tau_2 + \widetilde{T}_1 + \widetilde{T}_2$, where τ_1 (τ_2) is the time interval between the first and second (second and third) pump pulses, and $\widetilde{T}_1$ and $\widetilde{T}_2$ are the effective durations of the first and second pulses. We will assume that the pump pulses propagate along the strength vector $\boldsymbol{H}$ of a permanent uniform magnetic field and the electric field strengths of the pump pulses are described by formulas (2.5.1)–(2.5.3) with polarization vectors $\boldsymbol{l}_1 = (0, 0, 1)$, $\boldsymbol{l}_2 = (0, 0, 1)$, and $\boldsymbol{l}_3 = (0, 0, 1)$. Let us explain the mechanism behind the non-Faraday rotation of the polarization vector of the stimulated photon echo signal produced by pump pulses propagating through a gas medium in the same direction in the case of $J_b = ½ \rightarrow J_a = ½$ transition. In our analysis, we will assume, as usual, that inequalities (4.1.35) with $n = 1, 2, 3$ are satisfied, allowing us to neglect the influence of the magnetic field during the periods of time when pump pulses propagate through the gas medium. We emphasize also that, in both cases under consideration, the first two pump pulses have the same effect on the medium in the formation of the stimulated photon echo.

The first pump pulse with a carrier frequency ω, which is resonant to the frequency ω_0 of an atomic (molecular) transition $b \rightarrow a$ propagates in the direction of the vector $\boldsymbol{H}$ in a nonexcited gas medium, where the Zeeman sublevels of resonant levels are characterized by a negative difference of population densities $N_0 = n_b - n_a$. Here, n_b and n_a are the population densities of the Zeeman sublevels of the resonant levels b and a before the moment of time when the first pump pulse reaches the point y of the gas medium. The first pump pulse induces polarization in the medium in such a way that, when this pulse leaves the point y, the polarization vector $\boldsymbol{P}$ of the medium related to the group of atoms (molecules) moving with a velocity v differs from zero. In the constant-field approximation, the vector $\boldsymbol{P}$ is directed along the polarization vector of the first pump pulse. Within the time interval between the first and second pump pulses, the Doppler dephasing of radiation-emitting particles is accompanied by a clockwise precession of the vector $\boldsymbol{P}$ (if we look at this vector along the vector $\boldsymbol{H}$) with the frequency

$$\varepsilon = (\varepsilon_a + \varepsilon_b)/2. \tag{4.9.1}$$

Here, ε_a and ε_b are defined by (4.1.30). Such a precession has been already considered in Section 4.1. Thus, by the moment of time when the second pump pulse arrives at the point y of the gas medium, the angle of the clockwise rotation of the vector $\boldsymbol{P}$ related to $J_b = ½ \rightarrow J_a = ½$ transition reaches the value of $\varepsilon\tau_1$.

As mentioned in Section 2.5, the action of the second pump pulse on a medium in the formation of the stimulated photon echo signal is reduced to the transfer of coherence from multipole moments of the resonant transition to multipole moments of resonant levels. Since, in the case when $J_{a,b} = ½$, resonant levels are completely characterized by the population ($k = 0$) and orientation ($k = 1$), the second pump pulse transfers coherence to the populations and orientations of resonant levels. The populations of resonant levels under these conditions are modulated with a factor $\cos(\varepsilon\tau_1)$, while the orientations of resonant levels are modulated with a factor $\sin(\varepsilon\tau_1)$. In what follows, we will define the amplitude of the population difference of resonant levels as the difference of the populations in these levels without the factor $\cos(\varepsilon\tau_1)$, while the amplitude of the orientation difference of resonant levels will be defined as the difference of the orientations without the factor $\sin(\varepsilon\tau_1)$. For our further analysis, it is important to note that the amplitude of the population difference for resonant levels in the case under study is equal to the amplitude of the orientation difference of these levels.

If the radiative population of the lower resonant level due to spontaneous emission from the upper level is insignificant, then the population difference and orientation difference of resonant levels relax in the same way within the time interval τ_2, because the equalities $\gamma_a^{(1)} = \gamma_a^{(0)}$ and $\gamma_b^{(1)} = \gamma_b^{(0)}$ are satisfied for $J_b = J_a = ½$ [49], where $1/\gamma_{a,b}^{(1)}$ is the orientation relaxation time of the resonant level a (b) and $1/\gamma_{a,b}^{(0)}$ is the population relaxation time of the level a (b). Therefore, in the case under consideration, the amplitude of the population difference for resonant levels is equal to the amplitude of the orientation difference for these levels by the moment of time when the third pump pulse reaches the point y of a gas medium. In the opposite case, when the radiative population of the lower resonant level due to spontaneous decay from the upper level plays an important role, the amplitude of the population difference for resonant levels is not equal to the amplitude of the orientation difference for these levels by the moment of time when the third pump pulse reaches the point y of a gas medium, which eventually gives rise to the non-Faraday rotation of the polarization vector of the stimulated photon echo signal produced by copropagating pump pulses through $J_b = ½ \rightarrow J_a = ½$ transition.

In contrast to the first pump pulse, the third pump pulse irradiates a medium where not only the population difference, but also the orientation difference of resonant levels differs from zero. Therefore, even in the constant-field approximation, as the third pump pulse leaves the point y of a gas medium, the orientation of the vector $\boldsymbol{P}$ differs from the orientation of the polarization vector of the third pump pulse. Note that the component P_z, which is parallel to the polarization vector of the third pump pulse, is related to the population difference of the resonant levels and is modulated with a factor $\cos(\varepsilon\tau_1)$, while the component P_x, which is orthogonal to the polarization vector of the third pump pulse, arises due to the fact that the orientation difference of resonant levels differs from zero and is modulated with a factor $\sin(\varepsilon\tau_1)$. In what follows, the amplitude of the component P_z of the polarization of a medium will be defined as a premultiplier of $\cos(\varepsilon\tau_1)$, while the amplitude of the component P_x will be defined as a premultiplier of $\sin(\varepsilon\tau_1)$. Recall (see Section 2.5) that only some part of the vector $\boldsymbol{P}$, produced in a medium under the action of three pump pulses, denoted by $\boldsymbol{P}^{se}$, contributes to the electric field strength of the stimulated photon echo signal. This component of $\boldsymbol{P}$ differs for the cases when all the three pump pulses propagate through the medium in the same direction and when the third pump pulse propagates in the direction opposite of the first two beams. If the radiative population of the lower resonant level due to spontaneous emission from the upper level is insignificant, then the amplitudes of the components P_z^{se} and P_x^{se} are equal in their absolute values and opposite in sign. The angle of counterclockwise rotation of the vector $\boldsymbol{P}^{se}$, which contributes to the electric field strength in the stimulated photon echo, is equal to $\varepsilon\tau_1$ with respect to the polarization vector of the third pump pulse under these conditions if we look at this vector along $\boldsymbol{H}$. If the amplitudes of the components P_z^{se} and P_x^{se} differ from each other in their absolute values, which is the case when the radiative population of the lower resonant level due to spontaneous emission from the upper level plays a significant role, then the angle of counterclockwise rotation of the vector $\boldsymbol{P}^{se}$ with respect to the polarization vector of the third pump pulse exceeds $\varepsilon\tau_1$ (we assume that $\gamma_b^{(0)} > \gamma_a^{(0)}$). The further evolution of the system in time, first, leads to the phasing of radiation-emitting particles within a time interval approximately equal to τ_1 after the moment of time when the third pump pulse leaves the point y of the gas medium and, second, gives rise to the clockwise rotation of the vector $\boldsymbol{P}^{se}$ at the moment of time when radiation-emitting particles are in phase by the angle $\varepsilon\tau_1$ if we look at this vector along $\boldsymbol{H}$. Consequently, if the radiative population of the lower resonant level due to spontaneous emission from the upper level is negligible, then the vector $\boldsymbol{P}^{se}$ is directed along the polarization vectors of the pump pulses at the moment of time when the radiation-emitting particles are in phase, and no rotation of the vector $\boldsymbol{E}^{se}$ is observed. Otherwise, we observe a counterclockwise rotation of the vector $\boldsymbol{E}^{se}$ if we look at this vector along $\boldsymbol{H}$. This rotation is exclusively due to the radiative population of the lower resonant level through spontaneous emission from the upper level.

The method of calculation of the electric field strength in the stimulated photon echo signal employed in [51] is similar to the approach employed in Section 2.5. In other words, we assume that the durations $\widetilde{T}_n$ ($n = 1, 2, 3$) of the pump pulses are small as compared with the time intervals τ_1 and τ_2 between these pulses and the relevant irreversible relaxation times. Furthermore, we assume that inequalities (4.1.35) are satisfied with $n = 1, 2, 3$, which make it possible to ignore the Zeeman splitting of resonant levels within periods of time when pump pulses propagate through the gas medium.

The electric field strength $\boldsymbol{E}^{se}$ of the stimulated photon echo signal in the case of rectangular pump pulses (2.5.1)–(2.5.6) with coinciding polarization vectors ($\boldsymbol{l}_n = (0, 0, 1)$, $n = 1, 2, 3$) is determined [51] by formula (4.8.1), where the vector $\boldsymbol{e}^{se}$ should be replaced by the vector $\boldsymbol{e}^{se}(t'')$. The quantity S_{se}, which characterizes the waveform of the stimulated photon echo signal, is written as

$$S_{se} = \int dv_y\, f(v_y)(\Omega_1 \widetilde{T}_1)^{-1}(\Omega_2 \widetilde{T}_2)^{-1}(\Omega_3 \widetilde{T}_3)^{-1}[\sin(\Omega_1 \widetilde{T}_1) - ikv_y\Omega_1^{-1}(1 -$$

$$- \cos\Omega_1 \widetilde{T}_1)][\sin(\Omega_2 \widetilde{T}_2) - ikv_y\Omega_2^{-1}(1 - \cos\Omega_2 \widetilde{T}_2)][\sin(\Omega_3 \widetilde{T}_3) + ikv_y\Omega_3^{-1}(1 -$$

$$- \cos\Omega_3 \widetilde{T}_3)]\exp[ikv_y (t - t_{se} + ½ (\widetilde{T}_3 - \widetilde{T}_1 - \widetilde{T}_2))]. \tag{4.9.2}$$

Here, t_{se} is given by formula (2.5.7), the quantity Ω_n is related to the area θ_n of the n-th pump pulse through expression (2.3.20), θ_n can be obtained from (2.4.1) with $J = 1/2$, and $f(v_y)$ as before, describes the Maxwell distribution of resonant atoms (molecules) in projections v_y of their velocities **v** on the Y axis. Finally, the vector $\boldsymbol{e}^{se}(t'')$, which characterizes the polarization properties of the stimulated photon echo signal, has the following nonvanishing components:

$$e_x^{se} (t'') = (1/9)[A_0(\tau_2)\cos(\varepsilon\tau_1)\sin(\varepsilon t'') - A_1(\tau_2)\sin(\varepsilon\tau_1)\cos(\varepsilon t'')], \tag{4.9.3}$$

$$e_z^{se} (t'') = (1/9)[A_0(\tau_2)\cos(\varepsilon\tau_1)\cos(\varepsilon t'') + A_1(\tau_2)\sin(\varepsilon\tau_1)\sin(\varepsilon t'')]. \tag{4.9.4}$$

Here,

$$A_0(t) = [1 + \gamma_{ab}(\gamma_b^{(0)} - \gamma_a^{(0)})^{-1}]\exp(-\gamma_b^{(0)} t) + \\ + [1 - \gamma_{ab}(\gamma_b^{(0)} - \gamma_a^{(0)})^{-1}]\exp(-\gamma_a^{(0)} t), \tag{4.9.5}$$

$$A_1(t) = [1 + (1/3)\gamma_{ab}(\gamma_b^{(0)} - \gamma_a^{(0)})^{-1}]\exp(-\gamma_b^{(0)} t) + \\ + [1 - (1/3)\gamma_{ab}(\gamma_b^{(0)} - \gamma_a^{(0)})^{-1}]\exp(-\gamma_a^{(0)} t), \tag{4.9.6}$$

$$t'' = t - \tau_1 - \tau_2 - \widetilde{T}_1 \quad \widetilde{T}_2 - \widetilde{T}_3 - y/c, \tag{4.9.7}$$

and $1/\gamma_{ab}$ is the lifetime of the state b with respect to the spontaneous decay of this state to the state a. ·

The stimulated photon echo signal described by (4.8.1) and (4.9.2)–(4.9.7) is linearly polarized and propagates through a gas medium with a carrier frequency ω in the same direction as the pump pulses.

Analysis of expression (4.9.2) performed in [51] for $\omega = \omega_0$ shows that the replacement of t'' by τ_1 in formulas (4.9.3) and (4.9.4) gives nonvanishing components of the vector $\boldsymbol{e}^{se}(\tau_1)$, characterizing polarization properties of the stimulated photon echo signal at the maximum of the echo signal intensity in the case of a narrow spectral line (2.3.24), and, by virtue of inequalities (4.1.35) with n = 1, 2, 3, nonvanishing components of the total photon echo signal produced through a broad (2.3.25) spectral line. Formulas (4.9.3) and (4.9.4) yield after such a replacement

$$e_x^{se}(\tau_1) = (1/18)[A_0(\tau_2) - A_1(\tau_2)]\sin(2\varepsilon\tau_1), \tag{4.9.8}$$

$$e_z^{se}(\tau_1) = (1/18)\{A_0(\tau_2) + A_1(\tau_2) + [A_0(\tau_2) - A_1(\tau_2)]\cos(2\varepsilon\tau_1)\}, \tag{4.9.9}$$

where the difference $A_0(\tau_2) - A_1(\tau_2)$, which characterizes the non-Faraday rotation of the polarization vector of the stimulated photon echo signal produced in a gas medium by copropagating pump pulses, can be written, with allowance for (4.9.5) and (4.9.6), as

$$A_0(\tau_2) - A_1(\tau_2) = (2/3)\gamma_{ab}(\gamma_b^{(0)} - \gamma_a^{(0)})^{-1}[\exp(-\gamma_b^{(0)}\tau_2) - \\ - \exp(-\gamma_a^{(0)}\tau_2)]. \tag{4.9.10}$$

If the radiative population of the lower resonant level due to spontaneous emission from the upper level does not play a significant role, then, as can be seen from (4.9.10), we have $A_0(\tau_2) = A_1(\tau_2)$, and we do not observe any non-Faraday rotation of the polarization vector of the stimulated photon echo signal produced by copropagating pump pulses in a gas medium through the considered transition. Consequently, the rotation of the polarization vector of the SPE signal produced by copropagating pump pulses through $J_b = ½ \to J_a = ½$ transition is exclusively due to the radiative population of the lower resonant level through spontaneous emission from the upper level. Yevseyev *et al.* [51] proposed to employ this effect for the experimental measurement of the time $1/\gamma_{ab}$.

Now, let us consider the case when the third pump pulse with a duration $\widetilde{T}_3$ propagates in the negative direction of the Y axis, irradiating the $y = L$ boundary of a gas medium at the moment of time $t = \tau_1 + \tau_2 + \widetilde{T}_1 + \widetilde{T}_2$. In this case, the electric field strengths of the pump pulses are described by formulas (2.5.1) and (2.5.2) with $\boldsymbol{l}_1 = (0, 0, 1)$ and $\boldsymbol{l}_2 = (0, 0, 1)$, and we have

$$\boldsymbol{E}_3 = e^{(3)}\boldsymbol{l}_3 g_3(\eta - L/c - \tau_1 - \tau_2 - \widetilde{T}_1 - \widetilde{T}_2)\exp[i(\omega t + ky + \Phi_3)] + \text{c.c.},$$
$$0 \leq \eta - L/c - \tau_1 - \tau_2 - \widetilde{T}_1 - \widetilde{T}_2 \leq \widetilde{T}_3, \tag{4.9.11}$$

where $e^{(3)}$ is the constant amplitude, Φ_3 is the constant phase shift, $\boldsymbol{l}_3 = (0, 0, 1)$ is the polarization vector, $\eta = t + y/c$, and the function g_3, which characterizes the waveform of the pump pulse, is normalized according to (4.1.1) with $n = 3$.

In the case under consideration [51], we deal with a delay effect, analogous to that observed for a modified stimulated photon echo signal (see Section 4.8), when different points y of a gas medium are characterized by different time intervals between the second and third pump pulses. In addition, in contrast to the case of copropagating pump pulses, the part of the polarization of the medium $\boldsymbol{P}$ denoted as $\boldsymbol{P}^{se}$, which contributes to the stimulated photon echo signal, displays clockwise rotation with respect to the vector $\boldsymbol{l}_3$ if we look at this vector along $\boldsymbol{H}$ at the moment of time when the third pump pulse leaves the gas medium at the point y of the gas medium. Recall that, in the case of copropagating pump pulses, $\boldsymbol{P}^{se}$ exhibits a counterclockwise rotation if we look at this vector along the vector $\boldsymbol{H}$.

Within the time interval between the third pump pulse and the stimulated photon echo signal, the vector $\boldsymbol{P}^{se}$ displays a clockwise precession around $\boldsymbol{H}$. As a result, the electric field strength in the stimulated photon echo signal produced through $J_b = ½ \rightarrow J_a = ½$ transition in the case when the third pump pulse propagates in the direction opposite of the first two pulses exhibits clockwise rotation relative to the vector $\boldsymbol{l}_3$ if we look at this vector along $\boldsymbol{H}$ regardless of the lifetime $1/\gamma_{ab}$ of the state b with respect to the spontaneous decay of this state to the state a.

The electric field strength $\boldsymbol{E}^{se}$ of the stimulated photon echo signal in the case when the third pump pulse propagates in the direction opposite of the first two pump pulses and when all the pump pulses have a rectangular shape can be derived from (4.8.1) with replacements $k \rightarrow -k$, $\Phi_2 \rightarrow -\Phi_2$, $\Phi_1 \rightarrow -\Phi_1$, $y/c \rightarrow (L - y)/c$, and $\boldsymbol{e}^{se} \rightarrow \boldsymbol{e}^{se}(\widetilde{t}\,'')$. Then, we can find the quantity S_{se} involved in (4.8.1) by introducing the notation $(L - y)/c$ instead of y/c in (4.9.2). In this case, the vector $\boldsymbol{e}^{se}(\widetilde{t}\,'')$, characterizing polarization properties of the stimulated photon echo signal, has the following nonvanishing components:

$$e_x^{se}(\tilde{t}\,'') = (1/9)[B_0(\tau_2)\cos(\varepsilon\tau_1)\sin(\varepsilon\tilde{t}\,'') +$$

$$+ B_1(\tau_2)\sin(\varepsilon\tau_1)\cos(\varepsilon\tilde{t}\,'')], \qquad (4.9.12)$$

$$e_z^{se}(\tilde{t}\,'') = (1/9)[B_0(\tau_2)\cos(\varepsilon\tau_1)\cos(\varepsilon\tilde{t}\,'') -$$

$$- B_1(\tau_2)\sin(\varepsilon\tau_1)\sin(\varepsilon\tilde{t}\,'')]. \qquad (4.9.13)$$

Here, $\tilde{t}\,'' = t'' - (L - 2y)/c$, and the quantities $B_k(t)$ can be obtained from the quantities $A_k(t)$, defined by formulas (4.9.5) and (4.9.6), with the replacement

$$\exp(-\gamma_{a,b}^{(0)}t) \to \exp(-\gamma_{a,b}^{(0)}t)(\gamma_{a,b}^{(0)}L/c)^{-1}\mathrm{sh}(\gamma_{a,b}^{(0)}L/c).$$

Note that the appearance of additional factors depending on $\gamma_{a,b}^{(0)}L/c$ is associated with the above-mentioned nonequivalence of the time intervals between the second and third pump pulses for different points y inside a gas medium.

As can be seen from formulas (4.8.1) and (4.9.2) after appropriate replacements and expressions (4.9.12) and (4.9.13), in the case when the third pump pulse propagates in the direction opposite of the first two pulses, the stimulated echo signal remains linearly polarized. This signal propagates with a carrier frequency ω in the same direction as the third pump pulse, and its polarization vector displays clockwise rotation with respect to the polarization vectors of the pump pulses if we look at these vectors along $\boldsymbol{H}$ regardless of whether the radiative population of the lower resonant level due to spontaneous emission from the upper level is significant or not.

To simplify our calculations, we restrict our analysis to the case when

$$\gamma_a^{(0)}L/c \ll 1, \qquad \gamma_b^{(0)}L/c \ll 1, \qquad (4.9.14)$$

and we can ignore the delay effects. Then, as it follows from formulas (4.9.12) and (4.9.13), the nonvanishing components of the vector $\boldsymbol{e}^{se}(\tau_1)$, characterizing polarization properties of the stimulated photon echo signal at the maximum of its intensity in the case of a narrow spectral line and polarization properties of the total photon echo signal in the case of a broad spectral line, are written as

$$e_x^{se}(\tau_1) = (1/18)[A_0(\tau_2) + A_1(\tau_2)]\sin(2\varepsilon\tau_1), \tag{4.9.15}$$

$$e_z^{se}(\tau_1) = (1/18)\{A_0(\tau_2) - A_1(\tau_2) + [A_0(\tau_2) + A_1(\tau_2)]\cos(2\varepsilon\tau_1)\}. \tag{4.9.16}$$

Hence, the non-Faraday rotation of the polarization vector of the stimulated photon echo signal produced through $J_b = ½ \rightarrow J_a = ½$ transition in the case when the third pump pulse propagates in the direction opposite of the first two pump pulses is observed independently of whether the radiative population of the lower resonant level due to spontaneous emission from the upper level is significant or not.

Closing this section, we should note that the rotation of the polarization vector of the SPE signal predicted in [51] was called the non-Faraday rotation because the angle of this rotation is independent of the length of a gas cell and is determined by the lifetime of the upper resonant state with respect to the spontaneous decay of this state to the lower level and the time intervals τ_1 and τ_2 between the pump pulses.

4.10 Polarization Echo Spectroscopy of Atoms with Nonzero Nuclear Spins

Let us discuss how the polarization echo spectroscopy method can be applied to atoms with a nonzero nuclear spin. To describe the behavior of such atoms in resonant external fields of pump pulses, the authors of [71–73] proposed to employ the d'Alembert equation (2.2.1) and the quantum-mechanical equation for the density matrix $\hat{\rho}$ of resonant atoms (2.2.2). Under these conditions, the vector $\boldsymbol{P}$ of polarization of a medium related to a group of atoms moving with a velocity $\mathbf{v}$ can be expressed in terms of the density matrix $\hat{\rho}$ through relationship (2.2.3). Such an approach is now widely accepted.

In the case of atoms with nonzero nuclear spins, the theory of the photon echo and its modifications becomes much more complicated than in the case of atoms with zero nuclear spins. Difficulties of the theory of the photon echo for atoms with nonzero nuclear spins stem from two circumstances. First, since, in experiments, the spectral width of pump pulses usually exceeds the hyperfine splitting of one or all resonant levels, signals of the photon echo and its modifications are produced through the entire set of hyperfine components of resonant levels. Second, the description of collisional relaxation processes becomes much more complicated.

Let us specify the initial conditions for equations (2.2.1)–(2.2.3) in the case of atoms with nonzero nuclear spins. Suppose that the carrier frequency of the first pump pulse ω is close to the frequency ω_0 of an optically allowed resonant electron transition $b \to a$ ($E_b > E_a$). We restrict our consideration to the elements of the density matrix related to two resonant levels b and a and the resonant transition $b \to a$ between these levels. The Zeeman sublevels of hyperfine components of the resonant levels b and a will be characterized, along with the total angular momenta F_b and F_a, by the projections M_b and M_a of the total angular momenta on the quantization axis. The Y axis will be chosen along the direction of propagation of the first pump pulse through the gas medium.

We assume, as usual, that the first pump pulse irradiates the $y = 0$ boundary of the gas medium at the moment of time $t = 0$ and propagates through the gas medium in the positive direction of the Y axis. The initial moment of time for each point y of the gas medium then corresponds to $t - y/c = 0$. We assume also that resonant atoms in the gas medium are characterized by the Maxwellian velocity distribution, by a uniform distribution in Zeeman sublevels of hyperfine components of resonant levels, and by a uniform distribution in space before the irradiation of the gas medium with the first pump pulse. The density matrix of resonant atoms at the initial moment of time is then written as

$$\hat{\rho}(0) = f(\mathbf{v})[n_a \sum |F_a M_a\rangle\langle F_a M_a| + n_b \sum |F_b M_b\rangle\langle F_b M_b|]. \qquad (4.10.1)$$

Here, summation is performed over all possible values of F_a, F_b, M_a, and M_b; n_a and n_b are the population densities of the Zeeman sublevels of hyperfine components of resonant levels a and b for $t - y/c \leq 0$; and $f(\mathbf{v})$ describes the Maxwell distribution of resonant atoms in their velocities $\mathbf{v}$.

Now, let us consider the relaxation under the action of elastic depolarizing collisions. Relaxation in systems with nonzero nuclear spins can be adequately described in terms of the model of hyperfine-coupling breaking during collisions [49]. This model implies that the electrostatic interaction of colliding atoms has virtually no influence on the nuclear spin, the energy of this interaction is much higher than the energy of the hyperfine structure, and the mean duration of collisions is much less than the characteristic time for the hyperfine structure. Therefore, we can assume that the interaction time is too short for the nuclear spin I to change its orientation, while the electron angular moment J behaves in the same way as in the absence of the hyperfine structure. The hyperfine structure is recovered after collisions, and the momenta J and I add up again to produce the total momentum F. The model of hyperfine-coupling breaking is an analog of the Franck–Condon principle, which is well known in molecular spectroscopy and which is applied in the case under study to nuclear spin states.

The collision integral, governing the relaxation of a hyperfine multiplet in the model of hyperfine-coupling breaking in the process of interaction, has been derived and investigated in detail by Rebane [49]. Generally, the form of this integral is rather complex. However, as demonstrated for the first time in [52, 72, 74], the form of this integral can be considerably simplified with some assumptions on the time interval τ between the pump pulses. For example, if the inequalities

$$|\gamma_{ab}^{(k)} - \gamma_{ab}^{(1)}|\tau << 1, \qquad |\Delta_{ab}^{(k)} - \Delta_{ab}^{(1)}|\tau << 1 \tag{4.10.2}$$

are satisfied, then the relaxation of the optical coherence matrix $\langle F_a M_a|\, \hat{\rho}\, |F_b M_b\rangle$ is described by the expression

$$\langle F_a M_a|\, \Gamma\hat{\rho}\, |F_b M_b\rangle = (\gamma_{ab}^{(1)} - i\Delta_{ab}^{(1)})\langle F_a M_a|\, \hat{\rho}\, |F_b M_b\rangle. \tag{4.10.3}$$

Here, $\gamma_{ab}^{(1)}$ is the homogeneous half-width of the spectral line of the resonant electron transition $b \to a$ with angular momenta of resonant levels equal to J_b and J_a, $\Delta_{ab}^{(1)}$ is the shift of the spectral line of this transition due to elastic depolarizing collisions, and $\gamma_{ab}^{(k)}$ and $\Delta_{ab}^{(k)}$ ($|J_a - J_b| \leq k \leq J_a + J_b$) are the relaxation parameters of the multipole moments of the resonant electron transition $b \to a$.

We should emphasize that, for $J_b = 1/2 \to J_a = 1/2$ and $J_b = 3/2 \to J_a = 1/2$ transitions changing the electron angular momentum, the relaxation parameters $\gamma_{ab}^{(k)}$ and $\Delta_{ab}^{(k)}$ coincide with $\gamma_{ab}^{(1)}$ and $\Delta_{ab}^{(1)}$, respectively, for any k (e.g., see [49]). Thus, inequalities (4.10.2) and expression (4.10.3) are satisfied for such transitions with any τ. It should be noted that all the experiments on the photon echo with atoms with nonzero nuclear spins have been performed so far for transitions of this type (see Chapter 3 and review papers [75, 76]).

We deal with a similar situation when we consider the collision integral for the components of the density matrix $\langle F_b M_b | \hat{\rho} | F_b' M_b' \rangle$ and $\langle F_a M_a | \hat{\rho} | F_a' M_a' \rangle$. Specifically, if the inequality

$$|\gamma_b^{(k)} - \gamma_b^{(0)}|\tau << 1, \tag{4.10.4}$$

is satisfied, then the relaxation of density matrix components $\langle F_b M_b | \hat{\rho} | F_b' M_b' \rangle$ is governed by the expression [52]

$$\langle F_b M_b | \Gamma \hat{\rho} | F_b' M_b' \rangle = \gamma_b^{(0)} \langle F_b M_b | \hat{\rho} | F_b' M_b' \rangle. \tag{4.10.5}$$

Here, $\gamma_b^{(k)}$ $(0 \le k \le 2J_b)$ are the relaxation parameters of the multipole moments of the level b involved in thc resonant electron transition $b \to a$. In particular, $\gamma_b^{(0)}$ is the relaxation parameter of the population in the level b characterizing the radiative decay of this level and including inelastic gas-kinetic collisions.

We should note that, for levels with an electron angular momentum equal to 1/2, the equality $\gamma_b^{(1)} = \gamma_b^{(0)}$ is satisfied [49]. Consequently, expression (4.10.5) for such levels holds true with any τ. This circumstance is very important, because most of the experiments on the photon echo and its modifications in atoms with nonzero nuclear spins were performed with resonant levels whose electron angular momenta were equal to 1/2 (see Chapter 3 and review papers [75, 76]).

In the first studies [52, 58, 59, 71–74, 77–80] devoted to the polarization properties of the photon echo and its modifications in the case of atoms with nonzero nuclear spins, calculations were performed with an assumption that inequalities (4.10.2) and (4.10.4) are satisfied. In these papers, equations (2.2.1)–(2.2.3) subject to the initial condition (4.10.1) were solved, and polarization properties of the primary photon echo [59, 71–74, 78], stimulated photon echo [52], modified stimulated photon echo [58], coherent emission in time-separated fields [77, 79], and backward photon echo [80] were analyzed. These studies made it possible to propose experiments [52, 58, 59, 71–74, 77–80] aimed at measuring the homogeneous half-width of the resonant spectral line of an electron transitions and the widths of resonant levels and at identifying the structure of a resonant transition. Here, the identification of the structure of resonant transitions is understood as the determination of hyperfine structure components of resonant levels involved in the formation of the photon echo or its modifications.

Subsequently, the theory of polarization properties of the photon echo and its modifications in the case of atoms with nonzero nuclear spins was developed in two directions [44, 45, 81–84]. First, the approach described above was employed to investigate the modifications of the photon echo that were not considered in [52, 58, 59, 71–74, 77–80]. Specifically, the authors of [81, 83] developed the theory of polarization properties of the three-level photon echo for such atoms, and the authors of [44, 45] developed the theory of polarization properties of the modified three-level photon echo. The second direction includes the studies aimed at describing the relaxation of matrix density components related to resonant transitions and resonant levels with the use of relationships (4.10.3) and (4.10.5), respectively. In particular, in addition to (4.10.3), the authors of [81, 83] employed the dependence obtained in the so-called secular approximation [49], and Yevseyev *et al.* [84] have shown how the orientation relaxation time of $^2P_{3/2}$ levels of thallium atoms with nonzero nuclear spins can be measured.

We should emphasize that both theoretical and experimental studies devoted to the polarization echo spectroscopy of atoms with nonzero nuclear spins have adopted an approach based on the use of small-area pump pulses [71, 72]. The main advantages of using the results of theoretical studies performed for small-area pump pulses are associated with the fact that the polarization and the intensity of the photon echo and its modifications in this case depend on a small number of parameters of a medium. Theoretical expressions derived with such an approach can be conveniently employed for processing the experimental data. Furthermore, polarization properties of the photon echo and its modifications in this case are independent of the waveforms and the areas of the pump pulses. The latter circumstance allows us to introduce the criterion of smallness for the areas of pump pulses in experiments on the photon echo and its modifications. To be able to apply formulas derived for small-area pump pulses to the results of photon-echo measurements, we should decrease the areas of the pump pulses starting with their optimal values, corresponding to the maximum intensity of the photon echo or its modifications, until the polarization properties of these signals become independent of the areas of the pump pulses.

Thus, a large number of theoretical studies [44, 45, 52, 58, 59, 71–74, 77–84] have been devoted to polarization properties of the photon echo and its modifications in the case of atoms with nonzero nuclear spins.

Unfortunately, the information concerning polarization properties of the photon echo or its modifications is usually missing from the existing literature on photon-echo experiments with such atoms (see Chapter 3). However, whenever such data are presented, the theory provides a reasonable explanation for all the polarization dependences observed in experiments.

4.11 Advantages of the Polarization Echo Spectroscopy of Gas Media

As shown in the previous sections of this chapter, a large number of theoretical studies [32, 42, 75, 76, 85–91] have been devoted to the polarization echo spectroscopy of gas media. In this section, we discuss the advantages of polarization echo spectroscopy and summarize theoretical predictions discussed above in order to stimulate further experimental investigations in this area.

Since polarization echo spectroscopy is a part of optical echo spectroscopy, we will first consider the main advantages of optical echo spectroscopy in general. First, the fact that spectral lines in optical echo spectroscopy are free of the influence of inhomogeneous broadening makes it possible to perform high-precision measurements within the contour of an inhomogeneously broadened spectral line. Second, the high resolution of optical echo spectroscopy in the time domain achieved in the case when ultrashort nano- and picosecond pump pulses are employed provides an opportunity to investigate fast relaxation processes. Third, in contrast to, for example, nonlinear laser spectroscopy, relaxation processes investigated in optical echo spectroscopy are not subject to perturbations due to the action of high-intensity laser radiation. We should also mention the flexibility of optical echo spectroscopy, which includes a broad variety of modifications aimed at extracting spectroscopic data of different kinds. Finally, since the intensity of the photon echo and its modifications is proportional to the square of the number of resonant atoms or molecules, optical echo spectroscopy offers many advantages over spectroscopic techniques based on incoherent phenomena, especially for low-pressure gas media.

Now, let us consider the characteristic features that distinguish polarization echo spectroscopy from its optical counterpart. First, polarization echo spectroscopy makes it possible to identify resonant transitions or the type of resonant transitions. Second, polarization echo spectroscopy can be employed to identify the structure of resonant transitions, i.e., to determine which hyperfine components of resonant levels are involved in the formation of the photon echo and its modifications. Finally, polarization echo spectroscopy makes it possible to separately measure the relaxation parameters of multipole moments of resonant levels and transitions. The knowledge of these parameters provides a deeper insight into the interaction potentials of atoms (molecules) in a gas medium.

Polarization echo spectroscopy provides an opportunity

to identify resonant transitions or their type;

to identify the structure of resonant transitions;

to measure the homogeneous half-width of an inhomogeneously broadened spectral line of a resonant transition;

to measure relaxation parameters of the quadrupole and octupole moments of a resonant transition;

to investigate the dependence of relaxation parameters of the dipole moment related to a resonant transition on the modulus of the velocity of resonant atoms (molecules);

to investigate the dependence of the collision integral on the direction of the velocity of resonant atoms (molecules);

to measure relaxation parameters of multipole moments for optically forbidden transitions;

to measure population, orientation, and alignment relaxation times of resonant levels;

to measure the lifetime of an upper resonant level with respect to the spontaneous decay of this level to the lower level.

Thus, the method of polarization echo spectroscopy is characterized by a high sensitivity and makes it possible to obtain a comprehensive spectroscopic information concerning the integral of elastic depolarizing collisions. Therefore, we expect that this technique will receive broad applications for spectroscopic investigations of gas media in the nearest future.

In conclusion, we should mention that Yevseyev and Reshetov [92] have predicted a modification of the photon echo referred to as collision-induced stimulated photon echo. The existence of such a modification of the photon echo is exclusively due to elastic depolarizing collisions. The observation of the photon echo in this modification would also provide an opportunity to extract the information concerning the collision integral in the model of elastic depolarizing collisions.

REFERENCES

1. N.M. Pomerantsev, *Usp. Fiz. Nauk* **65**: 87–110 (1958).
2. A. Löshe, *Kerninduktion* (Berlin: Wissenschaften, 1957) (in German).

3. A. Abraham, *The Principles of Nuclear Magnetism* (London: Oxford Univ. Press, 1961).
4. K.M. Salikhov, A.G. Semenov, Yu.D. Tsvetkov, *Elektronnoe spinovoe ekho i ego primenenie* (*Electron Spin Echo and Its Applications*) (Novosibirsk: Nauka, 1976) (in Russian).
5. L. Allen, J. Eberly, *Opticheskii rezonans i dvukhurovnevye atomy* (*Optical Resonance and Two-Level Atoms*) (Moscow: Mir, 1978) (in Russian).
6. E.A. Manykin, V.V. Samartsev, *Opticheskaya ekho-spektroskopiya* (*Optical Echo Spectroscopy*) (Moscow: Nauka, 1984) (in Russian).
7. V.A. Golenishchev-Kutuzov, V.V. Samartsev, B.M. Khabibullin, *Impul'snaya opticheskaya i akusticheskaya kogerentnaya spektroskopiya* (*Pulsed Optical and Acoustic Coherent Spectroscopy*) (Moscow: Nauka, 1988) (in Russian).
8. I.V. Yevseyev, V.M. Yermachenko, *Pis'ma Zh. Eksp. Teor. Fiz.* **28**: 689–692 (1978).
9. I.V. Yevseyev, V.M. Yermachenko, *Zh. Eksp. Teor. Fiz.* **76**: 1538–1546 (1979).
10. I.V. Yevseyev, *Teoriya polyarizatsionnoi ekho-spektroskopii atomov i molekul vzaimodeistvuyushchikh posredstvom uprugikh depolyarizuyushchikh stolknovenii* (*The Theory of Polarization Echo Spectroscopy of Atoms and Molecules Interacting through Elastic Depolarizing Collisions*) (Moscow: Doctor of Phys. and Math. Science Dissertation, 1987) (in Russian).
11. A.I. Alekseyev, I.V. Yevseyev, *Zh. Eksp. Teor. Fiz.* **56**: 2118–2128 (1969).
12. J.P. Gordon, C.H. Wang, C.K.N. Patel, *et al.*, *Phys. Rev.* **179**: 294–309 (1969).
13. A.I. Alekseyev, I.V. Yevseyev, *Zh. Eksp. Teor. Fiz.* **57**: 1735–1744 (1969).
14. A.I. Alekseyev, I.V. Yevseyev, *Zh. Eksp. Teor. Fiz.* **68**: 456–464 (1975).
15. A.I. Alekseyev, A.M. Basharov, *Izv. Akad. Nauk, Ser. Fiz.* **46**: 557–573 (1982).
16. I.V. Yevseyev, V.A. Reshetov, *Opt. Spektrosk.* **53**: 796–799 (1982).
17. C.H. Wang, *Phys. Rev.* **B 1**: 156–163 (1970).
18. A.I. Alekseyev, V.N. Beloborodov, *Photon Echo in Quasiclassical Description of the Rotational Motion of Atoms, in Nelineinye elektromagnitnye yavleniya v veshchestve* (*Nonlinear Electromagnetic Phenomena in Matter*) (Moscow: Energoatomizdat, 1984) (in Russian), pp. 54–69.
19. I.V. Yevseyev, V.A. Reshetov, *Opt. Acta* **29**: 119–130 (1982).
20. A.V. Yevseyev, I.V. Yevseyev, V.M. Yermachenko, *Opt. Spektrosk.* **50**: 77–84 (1981).
21. L.S. Vasilenko, N.N. Rubtsova, *Investigation of Relaxation Processes in a Gas with the Use of Coherent Transient Processes, in Lazernye sistemy* (*Laser Systems*) (Novosibirsk: Sib. Div. USSR Acad. Sci., 1982) (in Russian), pp. 143–154.
22. S.S. Alimpiev, N.V. Karlov, *Zh. Eksp. Teor. Fiz.* **63**: 482–490 (1972).
23. I.V. Yevseyev, V.M. Yermachenko, V.A. Reshetov, *Phys. Lett.* **A 77**: 126–128 (1980).
24. I.V. Yevseyev, V.M. Yermachenko, *Zh. Eksp. Teor. Fiz.* **77**: 2211–2219 (1979).
25. L.S. Vasilenko, N.N. Rubtsova, *Proc. IV All-Union Symp. Optical Echo and Methods of Its Practical Applications* (Kuibyshev: Kuibyshev State Univ., 1989) (in Russian), p. 46.
26. I.V. Yevseyev, V.M. Yermachenko, *Opt. Spektrosk.* **47**: 1139–1144 (1979).
27. I.V. Yevseyev, V.A. Reshetov, *Opt. Spektrosk.* **57**: 869–874 (1984).
28. I.I. Popov, I.S. Bikbov, I.V. Yevseyev, V.V. Samartsev, *Zh. Prikl. Spektrosk.* **52**: 794–798 (1990).
29. I.V. Yevseyev, V.M. Yermachenko, *Phys. Lett.* **A 60**: 187–189 (1977).

30. D.S. Bakaev, I.V. Yevseyev, V.M. Yermachenko, *Zh. Eksp. Teor. Fiz.* **76**: 1212–1225 (1979).
31. I.V. Yevseyev, V.M. Yermachenko, *Proc. VI Vavilov Conf. on Nonlinear Optics* (Novosibirsk: Sib. Div. USSR Acad. Sci., 1979), Part 2, pp. 155–158.
32. A.V. Yevseyev, I.V. Yevseyev, V.M. Yermachenko, *Fotonnoe ekho v gazakh: Vliyanie depolyarizuyushchikh stolknovenii* (*Photon Echo in Gases: The Influence of Depolarizing Collisions*) (Moscow: Inst. Atom. Energ., Preprint No. 3602/1, 1982) (in Russian).
33. L.S. Vasilenko, N.N. Rubtsova, V.P. Chebotayev, *Pis'ma Zh. Eksp. Teor. Fiz.* **38**: 39–393 (1983).
34. A.I. Alekseyev, I.V. Yevseyev, V.M. Yermachenko, *Zh. Eksp. Teor. Fiz.* **73**: 470–480 (1977).
35. P.R. Berman, T.W. Mossberg, S.R. Hartmann, *Phys. Rev.* **A 25**: 2550–2571 (1982).
36. T.W. Mossberg, R. Kachru, S.R. Hartmann, *Phys. Rev. Lett.* **44**: 73–77 (1980).
37. R. Kachru, T.J. Chen, S.R. Hartmann, *et al. Phys. Rev. Lett.* **47**: 902–905 (1981).
38. I.V. Yevseyev, V.M. Yermachenko, *Pis'ma Zh. Eksp. Teor. Fiz.* **38**: 388–391 (1983).
39. V.K. Matskevich, I.V. Yevseyev, V.M. Yermachenko, *Opt. Spektrosk.* **45**: 17–22 (1978).
40. I.V. Yevseyev, V.M. Yermachenko, *Phys. Lett.* **A 90**: 37–40 (1982).
41. I.V. Yevseyev, V.M. Yermachenko, V.A. Reshetov, *Opt. Acta* **30**: 817–829 (1983).
42. A.V. Yevseyev, I.V. Yevseyev, V.A. Reshetov, *Fotonnoe ekho v gazakh: Trekhurovnevye sistemy* (*Photon Echo in Gases: Three-Level Systems*) (Moscow: Inst. Atom. Energ., Preprint No. 3849/1, 1983) (in Russian).
43. I.V. Yevseyev, Yu.V. Men'shikova, V.N. Tsikunov, *Opt. Spektrosk.* **63**: 47–52 (1987).
44. I.V. Yevseyev, Yu.V. Men'shikova, V.N. Tsikunov, *Modifitsirovannoe trekhurovnevoe fotonnoe ekho, sformirovannoe na atomakh s otlichnym ot nulya spinom yadra* (*Modified Three-Level Photon Echo Produced on Atoms with a Nonzero Nuclear Spin*) (Moscow: Moscow Eng. Phys. Inst., Preprint No. 009, 1978) (in Russian).
45. I.V. Yevseyev, Yu.V. Men'shikova, V.N. Tsikunov, *J. Phys.* **B 22**: 1863–1883 (1989).
46. T. Mossberg, A. Flusberg, R. Kachru, S.R. Hartmann, *Phys. Rev. Lett.* **39**: 1523–1526 (1977).
47. A. Flusberg, R. Kachru, T. Mossberg, S.R. Hartmann, *Phys. Rev.* **A 19**: 1607–1621 (1979).
48. A.I. Alekseyev, A.M. Basharov, *Opt. Spektrosk.* **54**: 739–741 (1983).
49. V.N. Rebane, *Collisional Relaxation of Multipole Moments of the Density Matrix and Its Manifestation in Atomic Spectroscopy* (Leningrad: Doctor of Phys. and Math. Sci. Dissertation, 1980) (in Russian).
50. I.V. Yevseyev, V.M. Yermachenko, V.A. Reshetov, *Zh. Eksp. Teor. Fiz.* **78**: 2213–2221 (1980).
51. I.V. Yevseyev, V.M. Yermachenko, V.A. Reshetov, *Opt. Spektrosk.* **52**: 444–449 (1982).
52. A.V. Yevseyev, I.V. Yevseyev, V.A. Reshetov, *Opt. Spektrosk.* **58**: 518–523 (1985).
53. I.V. Yevseyev, V.N. Tsikunov, *Opt. Spektrosk.* **59**: 1372–1373 (1985).

54. I.V. Yevseyev, V.N. Tsikunov, *Phys. Lett.* **A 112**: 381–384 (1985).
55. I.V. Yevseyev, V.A. Reshetov, *Pis'ma Zh. Eksp. Teor. Fiz.* **44**: 160–162 (1986).
56. A.V. Yevseyev, I.V. Yevseyev, V.M. Yermachenko, *Dokl. Akad. Nauk SSSR* **256**: 57–60 (1981).
57. I.V. Yevseyev, V.M. Yermachenko, *Phys. Lett.* **A 80**: 253–255 (1980).
58. I.V. Yevseyev, P.V. Nesterov, V.A. Reshetov, *Opt. Commun.* **52**: 346–350 (1985).
59. I.V. Yevseyev, V.M. Yermachenko, V.A. Reshetov, *Novyi metod izmereniya vremen relaksatsii naselennosti, orientatsii i vystraivaniya* (Moscow: Moscow Eng. Phys. Inst., Preprint No. 011, 1984) (in Russian).
60. I.V. Yevseyev, V.M. Yermachenko, V.A. Reshetov, *Pis'ma Zh. Eksp. Teor. Fiz.* **41**: 132–133 (1985).
61. I.V. Yevseyev, V.M. Yermachenko, V.A. Reshetov, *J. Phys.* **B 19**: 185–198 (1986).
62. I.V. Yevseyev, V.N. Tsikunov, *Dokl. Akad. Nauk SSSR* **288**: 857–861 (1986).
63. V.V. Samartsev, R.G. Usmanov, G.M. Ershov, B.Sh. Khamidullin, *Zh. Eksp. Teor. Fiz.* **74**: 1979–1987 (1978).
64. I.V. Yevseyev, V.M. Yermachenko, V.N. Tsikunov, *Proc. VII All-Union Symp. on High- and Ultrahigh-Resolution Molecular Spectrosc.* (Tomsk: Sib. Div. USSR Acad. Sci., 1986), Part 3, pp. 266–270.
65. N.S. Belousov, L.S. Vasilenko, I.D. Matveenko, N.N. Rubtsova, *Opt. Spektrosk.* **63**: 34–38 (1987).
66. A.I. Sirasiev, V.V. Samartsev, *Opt. Spektrosk.* **39**: 730–734 (1975).
67. T. Mossberg, A. Flusberg, R. Kachru, S.R. Hartmann, *Phys. Rev. Lett.* **42**: 1665–1669 (1979).
68. J.-C. Keller, J.-L. Le Gouët, *Phys. Rev. Lett.* **52**: 2034–2037 (1984).
69. A.I. Alekseyev, *Pis'ma Zh. Eksp. Teor. Fiz.* **9**: 472–475 (1969).
70. A.I. Alekseyev, E.A. Manykin, *Phys. Lett.* **A 35**: 87–88 (1971).
71. A.V. Yevseyev, I.V. Yevseyev, V.A. Reshetov, *Dokl. Akad. Nauk SSSR* **275**: 64–67 (1984).
72. I.V. Yevseyev, V.M. Yermachenko, V.A. Reshetov, *Zh. Eksp. Teor. Fiz.* **87**: 1200–1210 (1984).
73. A.V. Yevseyev, I.V. Yevseyev, V.A. Reshetov, *Polyarizatsionnye svoistva fotonnogo ekha c uchetom sverkhtonkoi struktury rezonansnykh urovnei* (*Polarization Properties of the Photon Echo with Allowance for the Hyperfine Structure of Resonant Levels*) (Moscow: Inst. Atom. Energ., Preprint No. 3910/1, 1984) (in Russian).
74. I.V. Yevseyev, P.V. Nesterov, V.A. Reshetov, *Opt. Acta* **32**: 357–369 (1985).
75. M.A. Gubin, I.V. Yevseyev, V.A. Reshetov, *Fotonnoe ekho v gazakh: Eksperimental'nye metody formirovaniya i raznovidnosti* (*Photon Echo in Gases: Experimental Methods of Formation and Modifications*) (Moscow: P.N. Lebedev Phys. Inst., Preprint No. 214, 1984) (in Russian).
76. I.V. Yevseyev, V.M. Yermachenko, *Izv. Akad. Nauk SSSR, Ser. Fiz.* **50**: 1545–1550 (1986).
77. A.V. Yevseyev, I.V. Yevseyev, V.A. Reshetov, *Kvantovaya Elektron.* **12**: 494–500 (1985).
78. I.V. Yevseyev, V.A. Reshetov, *Phys. Lett.* **A 106**: 243–245 (1984).
79. I.V. Yevseyev, V.A. Reshetov, *Opt. Spektrosk.* **58**: 276–280 (1985).
80. I.V. Yevseyev, V.A. Reshetov, *Opt. Spektrosk.* **59**: 265–270 (1985).

81. A.I. Alekseyev, V.N. Beloborodov, O.V. Zhemerdeev, *Issledovanie uprugikh atomnykh stolknovenii po nestatsionarnomu kombinatsionnomu rasseyaniyu sveta v gaze* (*Investigation of Elastic Atomic Collisions Using Nonstationary Raman Scattering of Light in a Gas*) (Moscow: Moscow Eng. Phys. Inst., Preprint No. 022-85, 1985) (in Russian).
82. I.V. Yevseyev, V.A. Reshetov, *Opt. Spektrosk.* **60**: 1002–1007 (1986).
83. A.I. Alekseyev, O.V. Zhemerdeev, *Izv. Akad. Nauk SSSR, Ser. Fiz.* **50**: 1520–1529 (1986).
84. I.V. Yevseyev, V.A. Reshetov, *Phys. Lett.* **A 123**: 75–78 (1987).
85. A.V. Yevseyev, I.V. Yevseyev, *Fotonnoe ekho v gazakh: Polyarizatsionnye svoistva* (*Photon Echo in Gases: Polarization Properties*) (Moscow: Inst. Atom. Energ., Preprint No. 3328/1, 1980) (in Russian).
86. I.V. Yevseyev, *Izv. Akad. Nauk SSSR, Ser. Fiz.* **46**: 614–619 (1982).
87. M.A. Gubin, I.V. Yevseyev, V.M. Yermachenko, *Fotonnoe ekho v gazakh: Teoreticheskie rezul'taty, primeneniya i perspektivy dal'neishego ispol'zovaniya* (*Photon Echo in Gases: Theoretical Results and Applications*) (Moscow: P.N. Lebedev Phys. Inst., Preprint No. 7, 1985) (in Russian).
88. I.V. Yevseyev, V.M. Yermachenko, *Proc. IV Int. Symp. Selected Problems of Statistical Mechanics* (Dubna: Un. Inst. Nucl. Res., 1987) (in Russian), pp. 107–114.
89. I.V. Yevseyev, V.M. Yermachenko, *The State of Art in the Theory and Experiment in the Photon Echo and Its Modifications in Gas Media with Nonzero Nuclear Spin of Resonant Atoms*, in Problemy kvantovoi optiki (Problems of Quantum Optics) (Dubna: Un. Inst. Nucl. Res., 1988) (in Russian), pp. 29–38.
90. I.V. Yevseyev, V.M. Yermachenko, *Ekho yavleniya v kvantovoi optike* (*Echo Phenomena in Quantum Optics*) (Moscow: Moscow Eng. Phys. Inst., 1989) (in Russian).
91. I.V. Yevseyev, V.M. Yermachenko, *Polarization Echo Spectroscopy, in Interaction of Electromagnetic Field with Condensed Matter* (Singapore: World Sci., 1990), Vol. 7, pp. 162–209.
92. I.V. Yevseyev, V.A. Reshetov, *Opt. Spektrosk.* **65**: 376–380 (1988).

Chapter 5

APPLICATION OF THE PHOTON ECHO IN A GAS MEDIUM FOR DATA WRITING, STORAGE, AND PROCESSING

5.1 Correlation of Signal Shapes in Photon Echo and Its Modifications in Two-, Three-, and Four-Level Systems

As it was emphasized several times above, the phenomenon of the photon echo is an optical analog of the spin echo that has been discovered [1] about a decade earlier. That is why many ideas from the theory of the spin echo has been transferred to the theory of the photon echo and appeared to be fruitful there. One of these ideas put forward by Fernbach and Proctor [2] is the idea of the possibility of the use of the phenomenon of the spin echo for writing, storage, and processing of the data. The authors [2] determined the conditions under which the signals of the spin and stimulated spin echo can reproduce (direct or time-reversed) shape of one of the exciting pulses. The theoretical analysis was proved by the corresponding experiments.

Elyutin, Zakharov, and Manykin [3] demonstrated the possibility of correlation of the shape of signal of the primary photon echo with time shape of the first exciting pulse in the case of formation of the signal in the solid phase. This result was extended to the case of the gaseous media in [4]. It was demonstrated in [4] that the electric field intensity $\boldsymbol{E}^{e}$ of the signal of the photon echo formed at the inhomogeneously broadened spectral line of the optically allowed transition with arbitrary angular momenta of the resonance levels by the excitation pulses of small area is given by formula (2.3.36). In this case the quantity S that characterizes the decay and shape of the signal of the photon echo is given by (4.1.36). If one assumes that the signal of the photon echo is formed under the condition of exact resonance ($\omega = \omega_0 + \Delta^{(1)}_{ab}$) and the spectral line of the inhomogeneously broadened resonance transition $b \rightarrow a$ appears to be wide ($ku\widetilde{T}_1 >> 1$) for the first exciting pulse and narrow ($ku\widetilde{T}_2 << 1$) for the second one, then it follows from (4.1.36) that

$$S = \exp[-\gamma^{(1)}_{ab}(t - y/c)]2\pi^{1/2}(ku\widetilde{T}_1)^{-1}g_1(-t + t_e + \tfrac{1}{2}\widetilde{T}_1)[\theta(t - t_e + \tfrac{1}{2}\widetilde{T}_1) - \theta(t - t_e - \tfrac{1}{2}\widetilde{T}_1)]. \quad (5.1.1)$$

Here, $\gamma_{ab}^{(1)}$ is the homogeneous half-width of the spectral line of the resonance transition $b \rightarrow a$; $\Delta_{ab}^{(1)}$ is the shift of the spectral line of this transition due to elastic depolarizing collisions; $k = \omega/c$; ω is the carrier frequency of the exciting pulses; u is the root-mean square thermal speed of the resonance atoms (molecules); $\widetilde{T}_1$ ($\widetilde{T}_2$) is the effective duration of the first (second) exciting pulse; the value of t_e is determined by (2.3.35); the function g_1 describes the shape of the first exciting pulse the intensity of the electric field of which is given by (2.3.1); and the function $\theta(x)$ is determined by the expression (2.3.5).

It follows from (5.1.1) that in the case of formation of the signal of the photon echo by the exciting pulses of small area at the wide (for the first excitation pulse) and narrow (for the second excitation pulse) spectral line of the inhomogeneously broadened resonance transition, the echo signal reproduces the shape of the first exciting pulse reversed in time. Note that the mentioned effect takes place in gases also in the case when the area of the second exciting pulse is not small but the pulse complies with the strong field limit (2.4.4) [5].

The effect of correlation of the shape of the signal of the photon echo with the shape of the first exciting pulse was observed for the first time in the solid state [6] and then in gases [7–9].

It is demonstrated below that the stimulated photon echo in two-, three-, and four-level systems appears to be the most promising one among all the variants of the photon echo for writing, storage, and processing of the data. That is why the experimental work [6] in which the effect of correlation of the shape of the signal of the stimulated photon echo (in two-level systems) with the shape of one of the exciting pulses was observed for the first time appeared to be an important step in the development of the optical echo processors. The conditions of the effect of correlation of the shape of the signal of the stimulated photon echo with the shape of one of the exciting pulses were determined in theoretical works [4, 10]. Yevseyev and Reshetov [4] considered the degenerate resonance levels of gas atoms (molecules), whereas Elyutin, Zakharov, and Manykin [10] studied the nondegenerate levels of admixture paramagnetic ions in solid state.

Consider the results obtained in [4] for the stimulated photon echo. It was demonstrated in Section 4.8 that the value of S_{se} given by (4.8.7) characterizes the shape of the SPE signal formed in gas by exciting pulses of small area at the optically allowed transition with arbitrary angular momenta of the resonance levels. Analyze the expression (4.8.7) for the cases when the signal of the stimulated photon echo is formed at the spectral line of the resonance transition wide for one and narrow for another two exciting pulses.

If $ku\widetilde{T}_1 >> 1$, $ku\widetilde{T}_2 << 1$, and $ku\widetilde{T}_3 << 1$, then it follows from (4.8.7) that

$$S_{se} = 2\pi^{1/2}(ku\widetilde{T}_1)^{-1}g_1(-t + t_{se} + \widetilde{T}_1 + ½\widetilde{T}_2 - ½\widetilde{T}_3)[\theta(t - t_{se} - ½\widetilde{T}_2 + ½\widetilde{T}_3) - \theta(t - t_{se} - \widetilde{T}_1 - ½\widetilde{T}_2 + ½\widetilde{T}_3)], \tag{5.1.2}$$

where the quantity t_{se} is given by (2.5.7), $\widetilde{T}_n$ is the effective duration of the n-th (n = 1, 2, 3) exciting pulse and the other quantities were determined above. Hence, in the considered case the shape of SPE signal reproduces the shape of the first exciting pulse reversed in time.

Then, if $ku\widetilde{T}_1 << 1$, $ku\widetilde{T}_2 >> 1$, and $ku\widetilde{T}_3 << 1$, from (4.8.7) we have

$$S_{se} = 2\pi^{1/2}(ku\widetilde{T}_2)^{-1}g_2(t - t_{se} - ½\widetilde{T}_1 + ½\widetilde{T}_3)[\theta(t - t_{se} - ½\widetilde{T}_1 + ½\widetilde{T}_3) - \theta(t - t_{se} - ½\widetilde{T}_1 - \widetilde{T}_2 + ½\widetilde{T}_3)]. \tag{5.1.3}$$

Thus, in this case the signal of the stimulated photon echo reproduces the shape of the second exciting pulse.

And, finally, if $ku\widetilde{T}_1 << 1$, $ku\widetilde{T}_2 << 1$, and $ku\widetilde{T}_3 >> 1$ it follows from (4.8.7) that

$$S_{se} = 2\pi^{1/2}(ku\widetilde{T}_3)^{-1}g_3(t - t_{se} - ½\widetilde{T}_1 - ½\widetilde{T}_2 + \widetilde{T}_3)[\theta(t - t_{se} - ½\widetilde{T}_1 - ½\widetilde{T}_2 + \widetilde{T}_3) - \theta(t - t_{se} - ½\widetilde{T}_1 - ½\widetilde{T}_2)]. \tag{5.1.4}$$

Thus, in this case the signal of the stimulated photon echo reproduces the shape of the third exciting pulse. Note that three considered cases can be realized by means of variation of the duration of the exciting pulses.

One can demonstrate that the signal of the stimulated photon echo in gas reproduces the shape of the exciting pulse of small area. It is not necessary that the areas of the other two exciting pulses are small as well but the strong field limit (2.4.4) must be realized for them [5].

The effect of correlation of the time shape of the SPE signal with that of one of the excitation pulses has been reported for both solid state [6] and gases [7, 9, 11]. The latter is an indication of the possibility of the use of the gaseous media as working bodies for optical processors.

Note that the effect of correlation of the time shape of the signals of the photon and stimulated photon echo with the time shape of one of the exciting pulses is an analog of the corresponding effects in the spin echo. It was mentioned already that these effects for spin and stimulated spin echo were predicted and observed in [2]. New features of the effect of the correlation of shapes that have no analogues in the phenomenon of the spin echo are observed in three-level systems. In these systems one can observe not only the simple reproduction of the time shape of one of the excitation pulses with the help of the corresponding signals of some variant of the photon echo but also the reproduction with stretching or compression in time [12–17].

Such variants of the photon echo in three-level systems as the mentioned modified stimulated photon echo, three-level photon echo, and the modified three-level photon echo were realized in experiments only in gaseous media. That is why the use of the gaseous media as working substance of the optical echo processors allows one to read the recorded data with simultaneous stretching or compression in time.

The effect of correlation of the time shape of the variants of the photon echo in three-level systems with the shape of one of the excitation pulses with simultaneous stretching or compression in time was predicted theoretically for the modified stimulated photon echo in [12–14], for the three-level photon echo in [12, 14, 15], and for the modified three-level photon echo in [16, 17]. Consider, first, the modified stimulated photon echo.

Consider the case of formation of the MSPE signal by the exciting pulses of small area at the broad (for one) and narrow (for two others) spectral lines of the optically allowed resonance transitions. If $k_1 u\tilde{T}_1 \gg 1$, $k_1 u\tilde{T}_2 \ll 1$, and $k_2 u\tilde{T}_3 \ll 1$ then it follows from (4.8.29) that

$$S_{mse} = 2\pi^{1/2}(k_1 u\tilde{T}_1)^{-1} g_1(-\omega_2(t - t_{mse} + \tfrac{1}{2}\tilde{T}_3)/\omega_1 + \tilde{T}_1 + \tfrac{1}{2}\tilde{T}_2)[\theta(t - t_{mse} - \\ - \tfrac{1}{2}\omega_1\tilde{T}_2/\omega_2 + \tfrac{1}{2}\tilde{T}_3) - \theta(t - t_{mse} - \omega_1\tilde{T}_1/\omega_2 - \tfrac{1}{2}\omega_1\tilde{T}_2/\omega_2 + \tfrac{1}{2}\tilde{T}_3)], \quad (5.1.5)$$

where $k_1 = \omega_1/c$; ω_1 is the carrier frequency of the first and second exciting pulses; $k_2 = \omega_2/c$; ω_2 is the carrier frequency of the third exciting pulse; and the quantity t_{mse} is given by the expression (4.8.27). Hence, in the considered case the signal of the modified stimulated photon echo reproduces the time shape of the first exciting pulse. The duration of the MSPE signal equals $\omega_1 \widetilde{T}_1/\omega_2$. Then, if $k_1 u \widetilde{T}_1 \ll 1$, $k_1 u \widetilde{T}_2 \gg 1$, and $k_2 u \widetilde{T}_3 \ll 1$ for the quantity S_{mse} we have from (4.8.29):

$$S_{mse} = 2\pi^{1/2}(k_1 u \widetilde{T}_2)^{-1} g_2(\omega_2(t - t_{mse} + \tfrac{1}{2}\widetilde{T}_3)/\omega_1 - \tfrac{1}{2}\widetilde{T}_1)[\theta(t - t_{mse} - \\ - \tfrac{1}{2}\omega_1 \widetilde{T}_1/\omega_2 + \tfrac{1}{2}\widetilde{T}_3) - \theta(t - t_{mse} - \tfrac{1}{2}\omega_1 \widetilde{T}_1/\omega_2 - \omega_1 \widetilde{T}_2/\omega_2 + \tfrac{1}{2}\widetilde{T}_3)]. \quad (5.1.6)$$

Thus, in this case the MSPE signal reproduces the time shape of the second excitation pulse and its duration equals $\omega_1 \widetilde{T}_2/\omega_2$.

Consider in a more detailed way the results (5.1.5) and (5.1.6). As in the case of formation of the signal of the modified stimulated photon echo the ratio ω_1/ω_2 can be both larger and smaller than unity, it can reproduce (in contrast to SPE signal) the signals of the first and second exciting pulses with simultaneous compression and stretching. This effect was predicted in [12–14] and can be used in data processing with the application of the phenomenon of the photon echo.

Finally, if $k_1 u \widetilde{T}_1 \ll 1$, $k_1 u \widetilde{T}_2 \ll 1$, and $k_2 u \widetilde{T}_3 \gg 1$ then it follows from (4.8.29) that:

$$S_{mse} = 2\pi^{1/2}(k_2 u \widetilde{T}_3)^{-1} g_3(t - t_{mse} + \widetilde{T}_3 - \tfrac{1}{2}\omega_1(\widetilde{T}_1 + \widetilde{T}_2)/\omega_2)[\theta(t - t_{mse} - \\ - \tfrac{1}{2}\omega_1(\widetilde{T}_1 + \widetilde{T}_2)/\omega_2 + \widetilde{T}_3) - \theta(t - t_{mse} - \tfrac{1}{2}\omega_1(\widetilde{T}_1 + \widetilde{T}_2)/\omega_2)]. \quad (5.1.7)$$

Thus, in this case the MSPE signal reproduces the time shape of the third excitation pulse. We stress that the three considered cases can be realized by means of variation of the duration of the excitation pulses. Note that the influence of the shape of the excitation pulses on the shape of the MSPE signal has not been observed yet in experiments on modified stimulated photon echo. Therefore, it is worthwhile to perform such experiments in order to verify the theoretical relationships (5.1.5)–(5.1.7).

Recall also that the condition of smallness of areas of all three excitation pulses for reproduction of the time shape of one of them appears to be too strict. In fact it is necessary that only the area of the reproduced pulse is small whereas the areas of the other two excitation pulses can be arbitrary

[14]. However, the approximation of the strong field (2.4.4) must be realized for them.

Consider now the effect of correlation of the shape of the signal of three-level photon echo formed by excitation pulses of small area with time shape of one of the excitation pulses. Analyze the quantity $\bar{S}$ determined by the formula (4.7.12) that characterizes the shape of the TLPE signal. Let it be formed at the broad (for one of the excitation pulses) and narrow (for two others) spectral lines of the resonance transitions. If $k_1 u \tilde{T}_1 >> 1$, $k_2 u \tilde{T}_2 << 1$, and $k_2 u \tilde{T}_3 << 1$, then from (4.7.12) we have:

$$\bar{S} = 2\pi^{1/2}(k_1 u \tilde{T}_1)^{-1} g_1(t - t_{te} + \tilde{T}_1 + \tilde{T}_2 + \tilde{T}_3 - \tfrac{1}{2}\omega_2(\tilde{T}_2 + \tilde{T}_3)/\omega_1)[\theta(t - t_{te} + \tilde{T}_1 + \tilde{T}_2 + \tilde{T}_3 - \tfrac{1}{2}\omega_2(\tilde{T}_2 + \tilde{T}_3)/\omega_1) - \theta(t - t_{te} + \tilde{T}_2 + \tilde{T}_3 - \tfrac{1}{2}\omega_2(\tilde{T}_2 + \tilde{T}_3)/\omega_1)]. \tag{5.1.8}$$

Here t_{te} is given by the formula (4.7.4) and the other quantities are determined above. Therefore, in the considered case the TLPE signal reproduces the time shape of the first excitation pulse. Then, if $k_1 u \tilde{T}_1 << 1$, $k_2 u \tilde{T}_2 >> 1$, and $k_2 u \tilde{T}_3 << 1$ from (4.7.12) we have:

$$\bar{S} = 2\pi^{1/2}(k_2 u \tilde{T}_2)^{-1} g_2(\omega_1(-t + t_{te} + \tfrac{1}{2}\tilde{T}_1 - \tilde{T}_2 - \tilde{T}_3)/\omega_2 + \tilde{T}_2 + \tfrac{1}{2}\tilde{T}_3)[\theta(t - t_{te} + \tfrac{1}{2}\tilde{T}_1 + \tilde{T}_2 + \tilde{T}_3 - \tfrac{1}{2}\omega_2 \tilde{T}_3/\omega_1) - \theta(t - t_{te} + \tfrac{1}{2}\tilde{T}_1 + \tilde{T}_2 + \tilde{T}_3 - \tfrac{1}{2}\omega_2(\tilde{T}_3 + \tilde{T}_2)/\omega_1)]. \tag{5.1.9}$$

Thus, in this case the TLPE signal reproduces the time-reversed shape of the second excitation pulse. However, its duration in this case is determined as $\omega_2 \tilde{T}_2/\omega_1$. Recall that inequality in (4.7.5) is met if $\omega_2/\omega_1 > 1$. That is why in the considered case the duration of the TLPE signal is ω_2/ω_1 times larger than the duration of the second (reproduced) excitation pulse. Finally, if $k_1 u \tilde{T}_1 << 1$, $k_2 u \tilde{T}_2 << 1$, and $k_2 u \tilde{T}_3 >> 1$ then from (4.7.12) we have:

$$\bar{S} = 2\pi^{1/2}(k_2 u \tilde{T}_3)^{-1} g_3(\omega_1(t - t_{te} + \tfrac{1}{2}\tilde{T}_1 + \tilde{T}_2 + \tilde{T}_3)/\omega_2 - \tfrac{1}{2}\tilde{T}_2)[\theta(t -$$

$$- t_{te} + \tfrac{1}{2}\widetilde{T}_1 + \widetilde{T}_2 + \widetilde{T}_3 - \tfrac{1}{2}\omega_2\widetilde{T}_2/\omega_1) - \theta(t - t_{te} + \tfrac{1}{2}\widetilde{T}_1 + \widetilde{T}_2 + \widetilde{T}_3 -$$
$$- \tfrac{1}{2}\omega_2(\widetilde{T}_3 + \widetilde{T}_2)/\omega_1)], \qquad (5.1.10)$$

and TLPE signal reproduces the time shape of the third excitation pulse and its duration equals $\omega_2\widetilde{T}_3/\omega_1$. Note that the influence of the shape of the excitation pulses on the shape of the TLPE signal has not been observed yet in experiments on three-level photon echo. Therefore, it is worthwhile to perform such experiments in order to verify the theoretical relationships (5.1.8)–(5.1.10). Note also that the smallness of areas of all three excitation pulses is not compulsory for reproduction of the time shape of one of them by means of the signal of the three-level photon echo. In fact it is necessary that only the area of the reproduced pulse is small whereas the areas of the other two excitation pulses can be arbitrary [14]. However, the approximation of the strong field (2.4.4) must be realized for them.

Discuss finally the formation of the signal of the modified three-level photon echo at the broad (for one of the excitation pulses) and narrow (for the other two) spectral lines of the inhomogeneously broadened resonance transitions. If $k_1u\widetilde{T}_1 \gg 1$, $k_2u\widetilde{T}_2 \ll 1$, and $k_1u\widetilde{T}_3 \ll 1$ then the quantity S_{mte} that characterizes the shape of the signal is determined according to (4.7.29) as:

$$S_{mte} = 2\pi^{1/2}(k_1u\widetilde{T}_1)^{-1}g_1(-\omega_2(t - t_{mte} - \widetilde{T}_1 - \tfrac{1}{2}\widetilde{T}_2)/\omega_1 + \widetilde{T}_1 + \widetilde{T}_2 + \tfrac{1}{2}\widetilde{T}_3)\times$$
$$\times [\theta(t - t_{mte} - \widetilde{T}_1 - \tfrac{1}{2}\widetilde{T}_2 - \omega_1(\widetilde{T}_2 + \tfrac{1}{2}\widetilde{T}_3)/\omega_2) - \theta(t - t_{mte} - \widetilde{T}_1 - \tfrac{1}{2}\widetilde{T}_2 -$$
$$- \omega_1(\widetilde{T}_1 + \widetilde{T}_2 + \tfrac{1}{2}\widetilde{T}_3)/\omega_2)]. \qquad (5.1.11)$$

It follows from (5.1.11) that in this case the MTLPE signal reproduces the time shape of the first excitation pulse reversed in time and the duration of the signal equals $\omega_1\widetilde{T}_1/\omega_2$, i.e., it is ω_1/ω_2 times larger than the duration of the reproduced pulse. If $k_1u\widetilde{T}_1 \ll 1$, $k_2u\widetilde{T}_2 \gg 1$, and $k_1u\widetilde{T}_3 \ll 1$ from (4.7.29) we have:

$$S_{mte} = 2\pi^{1/2}(k_2u\widetilde{T}_2)^{-1}g_2(t - t_{mte} - \widetilde{T}_1 - \omega_1(\tfrac{1}{2}\widetilde{T}_1 + \widetilde{T}_2 + \tfrac{1}{2}\widetilde{T}_3)/\omega_2)\times$$
$$\times [\theta(t - t_{mte} - \widetilde{T}_1 - \omega_1(\tfrac{1}{2}\widetilde{T}_1 + \widetilde{T}_2 + \tfrac{1}{2}\widetilde{T}_3)/\omega_2) - \theta(t - t_{mte} - \widetilde{T}_1 - \widetilde{T}_2 -$$

$$- \omega_1(\tfrac{1}{2}\widetilde{T}_1 + \widetilde{T}_2 + \tfrac{1}{2}\widetilde{T}_3)/\omega_2)], \quad (5.1.12)$$

MTLPE signal reproduces the shape of the second excitation pulse, and the duration of the echo signal equals $\widetilde{T}_2$. If $k_1 u\widetilde{T}_1 \ll 1$, $k_2 u\widetilde{T}_2 \ll 1$, and $k_1 u\widetilde{T}_3 \gg 1$ then it follows from (4.7.29) that

$$S_{mte} = 2\pi^{1/2}(k_1 u\widetilde{T}_3)^{-1} g_3(\omega_2(t - t_{mte} - \widetilde{T}_1 - \tfrac{1}{2}\widetilde{T}_2)/\omega_1 - \tfrac{1}{2}\widetilde{T}_1 - \widetilde{T}_2)\times$$

$$\times [\theta(t - t_{mte} - \widetilde{T}_1 - \tfrac{1}{2}\widetilde{T}_2 - \omega_1(\tfrac{1}{2}\widetilde{T}_1 + \widetilde{T}_2)/\omega_2) - \theta(t - t_{mte} - \widetilde{T}_1 - \tfrac{1}{2}\widetilde{T}_2 -$$

$$- \omega_1(\tfrac{1}{2}\widetilde{T}_1 + \widetilde{T}_2 + \tfrac{1}{2}\widetilde{T}_3)/\omega_2)]. \quad (5.1.13)$$

In this case the signal of the modified three-level photon echo reproduces the time shape of the third excitation pulse, and the duration of the echo signal equals $\omega_1\widetilde{T}_3/\omega_2$, i.e., it is ω_1/ω_2 times larger than the duration of the third (reproduced) excitation pulse.

Note that the relationships (5.1.11)–(5.1.13) has not been verified yet. Therefore, it is worthwhile to carry out the corresponding experiments. Note also that the condition of smallness of areas of all three excitation pulses for reproduction of the time shape of one of them appears to be too strict. It is sufficient that only the area of the reproduced pulse is small whereas the areas of the other two excitation pulses can be arbitrary [17]. However, the approximation of the strong field (2.4.4) must be realized for them.

The variants of the photon echo in three-level systems (modified stimulated photon echo, three-level photon echo, and modified three-level photon echo) analyzed in this Section make it possible to compress and stretch the prerecorded data in course of its reproduction but do not allow its long-term storage. This is explained by the fact that the time interval τ_2 between the second and the third excitation pulses that determines the time of data storage cannot be long for this variants of the photon echo. For the mentioned variants of the photon echo τ_2 is limited by the times $1/\gamma_b^{(k)}$, $1/\gamma_{ac}^{(k)}$, and $1/\gamma_{bc}^{(k)}$, respectively. These times are determined by the radiation lifetime of the excited state (b, c, or b, respectively) that is coupled by the optically allowed transition with the state a (normally, the ground one). That is why it was proposed in [18] to form the stimulated photon echo in a four-level system (Fig. 5.1) in order to increase the time of the data storage with preserving the possibility of data stretching or compression in time during the

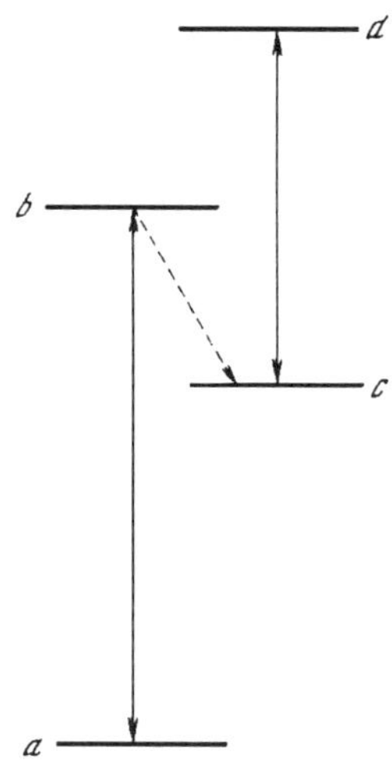

Fig. 5.1. The scheme of levels illustrating the second mechanism of formation of the long-lived stimulated photon echo.

reproduction. This variant was classified in [18] as a four-level stimulated photon echo.

In course of its formation the first two excitation pulses have the carrier frequency ω_1 that is resonant to the frequency ω_0 of the optically allowed transition $b \to a$. These pulses provide coherence in the multipole moments of the resonance levels b and a. In the considered system of levels the level b is coupled by the optically allowed transition with the metastable level c. The coherence is transferred from the resonance level b to the metastable level c due to radiation decay. The third excitation pulse with the carrier frequency ω_2 that is resonant to the frequency ϖ_0 of the optically allowed transition $d \to c$ (level d can coincide with level b) forms the signal of the stimulated photon echo. The latter can reproduce the time shape of one of the excitation pulses with simultaneous compression and stretching in time under the conditions defined in [18]. The coefficient of compression or stretching is determined by the ratio of the frequencies ω_1 and ω_2.

Note finally that the influence of the shapes of the excitation pulses on the shape and intensity of the signals of photon echo is studied extensively at present. In connection with this mention the works [19–23] where the new analytically solvable models of the excitation pulses were proposed for the theory of the photon echo.

5.2 Mechanisms of the Formation of the Long-Lived Stimulated Photon Echo

The increase of the time of storage of the processed data is one of the most important problems in the development of optical echo processors. Stimulated photon echo formed in two-, three-, and four-level systems appeared to be the most promising variant from this point of view. If the signal of the stimulated photon echo is formed at two-level systems with nondegenerate resonance levels, then the first two ("recording") excitation pulses separated by the time interval τ_1 generate nonequilibrium modulated populations of the upper and lower resonance levels. This nonequilibrium results in formation of the signal of the stimulated photon echo delayed by τ_1 ("the time of reproduction") relative to the third ("reading") pulse.

Thus, the time interval τ_2 between the second and the third excitation pulses during which the nonequilibrium modulated populations are maintained appears to be the time of data storage determined by the relaxation times of populations of the resonance levels at which the signal of the stimulated photon echo is formed. Therefore, the lower resonance level must correspond to the long-lived (ground or metastable) state in order to provide long times of data storage.

In the case of formation of the stimulated photon echo one often meets a situation when the upper resonance level b experiences radiation decay only into the ground state a that corresponds to the lower resonance level. The time of data storage in such a system in the case when both resonance levels are not degenerate does not exceed the lifetime $1/\gamma_{ab}$ of the excited state b (radiation decay to the ground state a). This is related with the fact that the sum of modulated nonequilibrium populations generated by the first two excitation pulses is not modulated (see, for example, [20]). During the time $1/\gamma_{ab}$ the particles go from the upper resonance level b to the lower level a in such a way that the nonequilibrium population of the lower resonance level becomes nonmodulated. That is why the third excitation pulse does not form the signal of the stimulated photon echo in such a system if $\tau_2 > 1/\gamma_{ab}$. At the same time in experiments on stimulated photon echo [24–26] the time interval between the second and the third excitation pulses was much longer than the lifetime of the excited state relative to its spontaneous decay to the lower level. Specifically, in [25] it was as long as 30 min.

Some mechanisms of formation of the long-lived stimulated photon echo were considered in theoretical works [18, 27–30]. The first mechanism was proposed in [27] where the theoretical study of formation of the signal of the stimulated photon echo at three nondegenerate resonance levels a, b, and c (Fig. 5.2) was carried out. The first two excitation pulses with the carrier frequency resonance to the frequency of the optically allowed transition $b \rightarrow a$ generate nonequilibrium modulated populations at the levels b and a. However, the modulated population of the level a is not compensated any more due to radiation decay from the level b during the lifetime $1/\gamma_{ab}$. It happens so because there is the second channel of radiation decay from the level b to the metastable level c. The third excitation pulse with the carrier frequency in resonance with the frequency of the optically allowed transition $b \rightarrow a$ arrives at the system after the time interval shorter than the

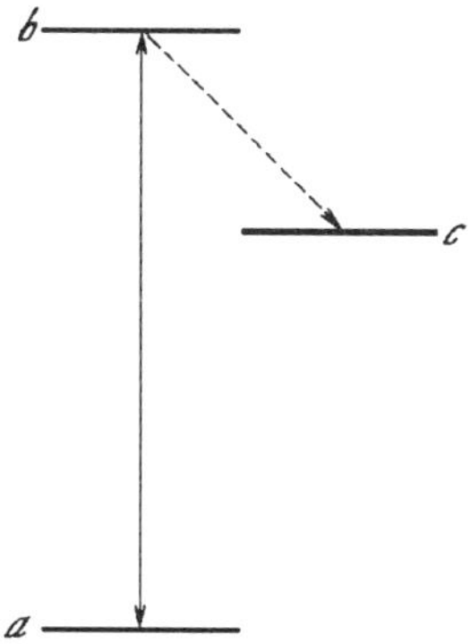

Fig. 5.2. The scheme of levels illustrating the first mechanism of formation of the long-lived stimulated photon echo.

lifetime of the metastable level c and forms the signal of the stimulated photon echo. Thus, in the case of the first mechanism the time of data storage is determined by the lifetime of the metastable level c.

The second mechanism of formation of the long-lived stimulated photon echo was considered in [18]. It was proposed to form the signal of the stimulated photon echo in a system of four levels a, b, c, and d (Fig. 5.1). The carrier frequency of the recording pulses is in resonance with the optically allowed transition $b \to a$ and the carrier frequency of the reading pulse is resonant to the frequency of the optically allowed transition $d \to c$. The recording pulses produce nonequilibrium modulated population at the level b that is transferred in course of the spontaneous radiation decay to both level a and metastable level c. The modulated population exists at the level c during its lifetime. The nonequilibrium modulated population of the level c results in formation of the signal of the stimulated photon echo after passing of the reading pulse through the medium. In this case the time of data storage is determined like before by the lifetime of the metastable level c. However, in contrast to the previous case the carrier frequency of the signal of the stimulated photon echo does not coincide with that of the recording pulses. As it was mentioned in the previous Section the signal of the four-level stimulated photon echo can experience stretching or compression in time which depends on the ratio of the frequencies of the optically allowed resonance transitions $b \to a$ and $d \to c$. This signal can be used not only for long-term data storage but also for fast and slow reproduction of data. Note also that the second mechanism of formation of the long-lived stimulated photon echo is realized in the case of coincidence of levels d and b.

The third (polarization) mechanism of formation of the long-lived stimulated photon echo was proposed in [28]. It is featured by the formation of the signal of the stimulated photon echo not in three- or four-level systems but in a two-level one with degenerate energy levels (Fig. 5.3). The radiation decay of the upper excited state b is possible only to the lower

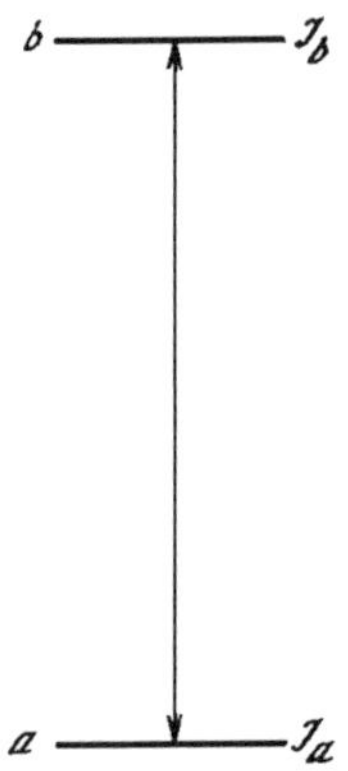

Fig. 5.3. The scheme of levels illustrating the third (polarization) mechanism of formation of the long-lived stimulated photon echo. J_a and J_b are the angular momenta of the resonance levels.

ground state *a*. It was mentioned several times that the degenerate level is characterized by the multipole moment of the *k*-th order ($0 \leq k \leq 2J$), where J is the value of the full angular momentum of the level; zero moment ($k = 0$) corresponds to the population of the level.

All multipole moments of the upper resonance level *b* decay down to the ground level *a* under the action of spontaneous radiation processes. The population is transferred completely, and the highest multipole moments ($k > 0$) are transferred not completely because a part of the angular momentum leaves the system with the photons. Consequently, after the time interval $1/\gamma_{ab}$ all the atoms (molecules, ions) can be found in the ground state, but the modulated nonequilibrium is preserved only for the multipole moments with $k \neq 0$. These are the moments that contribute to the stimulated photon echo after the time interval much longer than the lifetime $1/\gamma_{ab}$ of the excited state. Thus, the third mechanism of formation of the long-lived stimulated photon echo [28] can be realized in systems in which the angular momentum of the ground state is different from zero. The time of data storage with the use of the third mechanism is determined by the rates of transitions between the sublevels of the degenerate ground state, i.e., by the times of relaxation of the multipole moments of the ground state.

Let us note an interesting feature of the systems with the angular momentum of the ground state $J_a = 1/2$. In this case the ground state is characterized by two multipole moments: population ($k = 0$) and orientation ($k = 1$). The orientation of the ground state is formed only in the case of different polarizations of the first two ("recording") pulses. Hence, in the mentioned systems the data recorded by pulses polarized in one and the same plane are stored during a relatively short time $1/\gamma_{ab}$ whereas the data recorded by the pulses polarized in different planes can be stored much longer during the time interval equal to the time of relaxation of orientation of the ground state. Thus, there is a possibility of control of the time of data storage by means of variation of polarizations of the recording pulses in the system with angular momentum of the ground state $J_a = 1/2$. Note that this effect can be observed not only in gases but in the solid state as well (e.g., in ruby for the transition $^4A_2 - {}^2E\,(\bar{E}\,)$ of ^{52}Cr ions).

An example of the gas medium consisting of atoms with zero spin of nucleus is an illustration for the third (polarization) mechanism of formation of the long-lived stimulated photon echo. In contrast to Section 4.7 we take into account the radiation transition to the lower resonance level due to spontaneous emission from the upper one. It was assumed in [28] that the state a is the ground one and only the radiation decay of the state b to state a takes place. Such a situation was realized for example in the studies of the stimulated photon echo in a vapor of ^{174}Yb with zero spin of nucleus [11] and in a vapor of ^{23}Na with the spin of nucleus different from zero [26].

For gases the time of data storage with the use of the signal of the stimulated photon echo is limited by the time of flight T_{tr} and the times T_a and T_b of thermalization of the distributions over the velocities of atoms at the resonance levels a and b, respectively, due to their interactions with each other or with the atoms of the buffer gases. In reality (in experiments on stimulated photon echo) the time of data storage τ_2 is limited by much shorter times: the times of relaxation due to spontaneous relaxation transitions and due to elastic depolarizing collisions. That is why we consider below only the last two relaxation processes.

Recall that the relaxation of each of the multipole moments of the resonance level due to elastic depolarizing collisions does not depend on relaxation of the others and is characterized by the time $1/\Gamma_{\alpha}^{(k)}$ ($\alpha = a, b$ and $\Gamma_{\alpha}^{(0)} = 0$) for the collision integral in the model of elastic depolarizing collisions averaged over both the direction and the absolute value of the velocity $\mathbf{v}$ of the resonance atoms. All these multipole moments of the upper level decay in one and the same time $1/\gamma_{ab}$ and are transferred to the lower level under the action of the spontaneous radiation processes. The population is transferred completely and the highest multipole moments ($k > 0$) are transferred partly because some fraction of the angular momentum leaves the system with the photons. As a result all the atoms can be found in the ground state after the time interval $1/\gamma_{ab}$ after the arrival of the second excitation pulse. The nonequilibrium in the distribution of the atoms over the velocities is preserved only for the states with $k \neq 0$. Only these states contribute to the signal of the stimulated photon echo. Note that as the transition $b \rightarrow a$ is optically allowed, the time $1/\gamma_{ab}$ is usually not large.

In order to prove the aforesaid we present an expression for the intensity of the electric field $\boldsymbol{E}^{se}$ of the signal of the stimulated photon echo for the case $\tau_2 \gg 1/\gamma_{ab}$. The signal is formed by the excitation pulses of small area at the optically allowed transition $b \rightarrow a$ with arbitrary angular momenta J_a and J_b of the resonance levels. It follows from the solution of the system of equations (2.3.8)–(2.3.11) for $\tau_2 \gg 1/\gamma_{ab}$ that the value of $\boldsymbol{E}^{se}$ is given by the formula (4.8.1) in which the substitution $\boldsymbol{e}^{se} \rightarrow \boldsymbol{e}^{se}(\tau_2)$ must be made. The vector $\boldsymbol{e}^{se}(\tau_2)$ that characterizes the polarization properties of the signal of the stimulated photon echo has the following nonzero components:

$$e_x^{se}(\tau_2) = b_1\sin(\psi_2 - \psi_1) + b_2\sin(\psi_1 + \psi_2), \tag{5.2.1}$$

$$e_z^{se}(\tau_2) = (1/3)(2b_0 + b_2)\cos(\psi_2 - \psi_1) + b_2\cos(\psi_1 + \psi_2). \tag{5.2.2}$$

Here

$$b_k = \exp(-\Gamma_a^{(k)}\tau_2)\begin{Bmatrix} k & 1 & 1 \\ J_b & J_a & J_a \end{Bmatrix}\Bigg[\begin{Bmatrix} k & 1 & 1 \\ J_b & J_a & J_a \end{Bmatrix} +$$

$$+ (-1)^{J_a+J_b}\gamma_{ab}(2J_b+1)(\gamma_{ab} + \Gamma_b^{(k)} - \Gamma_a^{(k)})^{-1} \times$$

$$\times \begin{Bmatrix} k & 1 & 1 \\ J_a & J_b & J_b \end{Bmatrix}\begin{Bmatrix} k & J_a & J_a \\ 1 & J_b & J_b \end{Bmatrix}\Bigg], \tag{5.2.3}$$

ψ_1 and ψ_2 are the angles between the polarization vectors of the first and the second excitation pulses and the polarization vector of the third excitation pulse. Note that the formulas (5.2.1)–(5.2.3) for the vector $\boldsymbol{e}^{se}(\tau_2)$ for $J_a = 0$ or $J_a = 1/2$ (the most typical values in practice) are valid also for the case of arbitrary areas of the excitation pulses.

It follows from (5.2.1)–(5.2.3) that all the transitions $b \rightarrow a$ can be divided into three groups. The transition with $J_a = 0$ ($J_b = 1$) belongs to the first group. In this case the ground state a is characterized only by population and the time of data storage with the use of the stimulated photon echo at this transition is determined by short radiation lifetime of the upper resonance level. Such a situation is typical, for example, for the experiments on the stimulated photon echo in ytterbium vapor [11]. The transitions with $J_a = 1/2$ ($J_b = 1/2$ or $J_b = 3/2$) belong to the second group. Now the ground state is characterized by both population (k = 0) and orientation (k = 1). However, the elastic depolarizing collisions at J = 1/2 do not increase the rate of the decay of orientation: $\Gamma_a^{(1)} = 0$ (see, for example, [5]). Consider the transition $J_b = 1/2 \rightarrow J_a = 1/2$ at which the signal of the stimulated photon echo is formed in sodium vapor [26]. It follows from (5.2.1)–(5.2.3) that:

$$e_x^{se}(\tau_2) = (2/27)\sin(\psi_2 - \psi_1), \qquad e_z^{se}(\tau_2) = 0. \tag{5.2.4}$$

It follows from (5.2.4) that the time of data storage depends in this case on the polarization of the excitation pulses. If the "recording" pulses are polarized in one and the same plane ($\psi_1 = \psi_2$), then $\boldsymbol{e}^{se}(\tau_2) = 0$ and the time of data storage is limited by the radiation lifetime of the upper resonance level. If the "recording" pulses are polarized in different planes, then $e_x^{se}(\tau_2) \neq 0$ and does not depend on τ_2 within the frames of the considered relaxation scheme. It means that the time of data storage with the use of the signal of the stimulated photon echo is limited in this case only by the times T_{tr} and T_a that are large in comparison with $1/\gamma_{ab}$. Therefore, the formula (5.2.4) allows one to account for the experimental results [26]: it was demonstrated that the time of data storage with the use of signal of the stimulated photon echo is much larger than the radiation lifetime of the upper resonance level.

The transitions with $J_a > 1/2$ belong to the third group. In this case the ground state is characterized by a series of multipole moments that decay under the action of elastic depolarizing collisions. The time of data storage at these transitions depends on the relationship between the rates of the radiation relaxation and the relaxation under the action of the elastic depolarizing collisions. If the pressures in gases are rather high so that $\Gamma_{a,b}^{(k)} > \gamma_{ab}$, then the time of data storage is determined by the time $1/\gamma_{ab}$, and the signal of the stimulated photon echo is polarized along the vector of polarization of the third excitation pulse. If the gas is rarefied $\Gamma_{a,b}^{(k)} < \gamma_{ab}$, then it follows from (5.2.1)–(5.2.3) that the time of data storage is determined by the maximal relaxation time $1/\Gamma_a^{(k)}$ ($k > 0$) of the multipole moments of the ground state.

Thus, based on the results presented in [28], one can increase substantially the time of the data storage (relative to $1/\gamma_{ab}$) with the use of the signal of the stimulated photon echo in the gas medium by means of the proper choice of the resonance atoms of the medium and polarizations of the excitation pulses.

Another possibility controlling the time of data storage in the systems with $J_a = 1/2$ is related with application of the longitudinal magnetic field to the medium in which the signal of the stimulated photon echo is formed. It can be demonstrated [31] that the time of data storage with the use of the stimulated photon echo depends in such systems on the value of the external magnetic field intensity $\boldsymbol{H}$. Under certain values of H the time of data storage does not exceed the radiation lifetime $1/\gamma_{ab}$ of the excited state and under other values it is much larger than $1/\gamma_{ab}$ and is comparable with the relaxation time of the orientation of the ground state.

Note that none of the three considered mechanisms can explain the formation of the long-lived stimulated photon echo in experiments [24–26] where the hyperfine splitting of the resonance levels was observed. The formation of the stimulated photon echo at the levels with hyperfine structure depends on the relationship between the characteristic frequencies Δ_a and Δ_b of the hyperfine splitting of the resonance levels, spectral width δ of the excitation pulses and inhomogeneous width $1/T_2^*$ of the spectral line of the resonance transition [32]. If $\Delta_b >> \delta,\ 1/T_2^*$, then the first mechanism of formation of the long-lived stimulated photon echo is realized [29]. If $1/\Delta_a$ is much larger than all the characteristic times of the system then one can neglect the hyperfine structure of the resonance levels. In this case the third (polarization) mechanism of formation of the long-lived stimulated photon echo is realized provided the total angular electronic momentum of the ground state is different from zero [29]. In the case of arbitrary relationship between $1/\Delta_a$ and the characteristic times of the system the time of data storage with the use of the signal of the stimulated photon echo can be of the order of the lifetime $1/\gamma_{ab}$ of the excited state or much larger than this quantity. In particular, under certain conditions [29] the time of data storage τ_2 depends on the time of reproduction of the data τ_1. For certain values of τ_1 the time of data storage is either about $1/\gamma_{ab}$ or much larger than this quantity. This can be used for selection of even or odd elements from the sequence of the laser pulses.

Therefore, the formation of the long-lived stimulated photon echo in the systems with the hyperfine splitting of the ground state has the features of both the first and the third mechanisms. In the general case it can be reduced to none of them which allows one to propose the fourth mechanism. Its typical feature is the existence of quantum beats of intensity of signals of the stimulated photon echo as functions of time intervals between the excitation pulses and modulation oscillations of their shapes [29, 30, 32].

Modulation oscillations mentioned above impose on SPE signal which results in distortions of the prerecorded data. Mitsunaga *et al.* [33]

demonstrated that in the case of formation of the SPE signal at the levels with hyperfine structure the time of storage of undistorted data is limited by the lifetime $1/\gamma_{ab}$ of the upper resonance level. This is the main disadvantage of the fourth mechanism. Note that the first three mechanisms are free of this disadvantage. At the same time this is the fourth mechanism that provided the longest time of data storage with the use of the stimulated photon echo [24–26, 34–37].

Note that Mitsunaga *et al.* [33] did not study the polarization properties of the stimulated photon echo. Yevseyev and Reshetov [38, 39] demonstrated that the time of storage of undistorted data with the use of the stimulated photon echo formed at the levels with hyperfine structure can be much larger than $1/\gamma_{ab}$ under certain choice of polarization vectors of the excitation pulses. This result makes the systems with hyperfine structure of the resonance levels promising for the application in optical echo processors.

At the end of this section, we should note that the maximal time of data storage with the use of SPE was attained in the case of its formation in $Eu^{3+} : Y_2O_3$ crystals under helium temperatures [37]. Such a long time of storage seem to be impossible for the gaseous media. However, in some cases it is not necessary to provide such a long storage of data, for example, in the case of development of RAM modules or for the operation of the echo processor in real time. Gaseous media possess an additional advantage besides those mentioned above: they can be used under room or higher temperatures. All this provides wide perspectives for the application of the gaseous media in optical echo processors [40].

5.3 Optical Data Processing Based on the Photon Echo in Gaseous Media

As mentioned above, the correlation of the time shape of the echo signal with the time shape of certain radio pulses was demonstrated in the first works on the spin echo [2, 41, 42]. The practical outcome of these works is evident: they can be used for the development of the delay lines and memory devices. The studies in this direction went on and resulted in the construction of the spin echo processors that solve different problems of processing of the radio signals [43, 44].

Similar studies in the optical range could be expected after the discovery of the photon echo [45]. The correlation of the time shape of the primary photon echo (PPE) with symmetrical two-maxima excitation pulse was reported in [46, 47]. However, the conditions and reasons of its realization

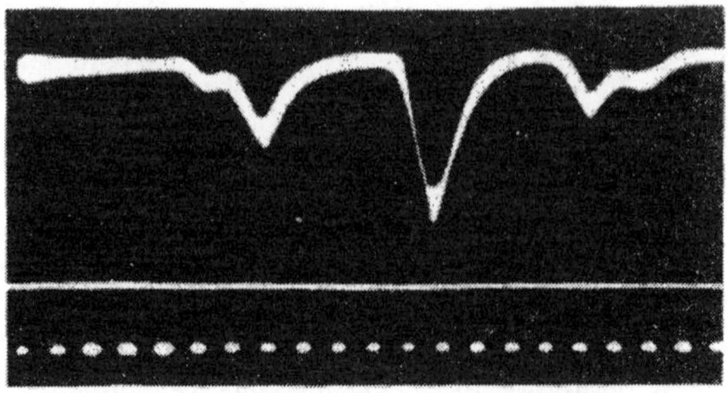

Fig. 5.4. Oscilloscope trace illustrating the effect of correlation of the time shape of the primary photon (light) echo with the shape of the first pulse in the saturated vapor of molecular iodine under the temperature of 24°C at the wavelength of 571.5 nm [56]. PPE signal is the first from the right. Markers correspond to 20 ns time interval.

were not studied. The first thorough theoretical analysis of the effect of correlation of the PPE shape with arbitrary shape of the first or second excitation laser pulse was presented in [3]. The experimental observation of this effect (hereafter – correlation effect) was reported in [6] for ruby crystal (see also [48–51]). In addition, the coincidence of the time shape of the stimulated photon echo with that of the second excitation pulse was observed for the first time under certain experimental conditions [6]. Later on the theoretical aspects of the effect of correlation of SPE signal were analyzed in [4, 10]. The effect of correlation was observed for the first time in gases (Yb vapor) by Mossberg *et al.* [8, 52, 53]. In 1984 this effect became the basis of the patent [54] on the development of the optical memory. Later the experiments on realization of the effect of correlation in different gases were reported in [55, 56]. Demonstrate some of the experimental results using the saturated vapor of molecular iodine (wavelength 571.5 nm, temperature 24°C) as an example. Figure 5.4 shows an oscilloscope trace demonstrating the effect of correlation of the PPE signal. It is seen that the time shape of PPE is reversed in time relative to the shape of the first pulse (i.e., the reversal of shape and time delay of the information signal is realized).

The analysis shows that the effect takes place if the area θ_1 of the code pulse meets the condition as follows: $\sin\theta_1 \leq \theta_1$, whereas the power of the second pulse must be rather high. The coincidence of the time shape of SPE with that of the second pulse (with the delay relative to this pulse) is demonstrated in Fig. 5.5. The limitation on the area of the code pulse (i.e., the

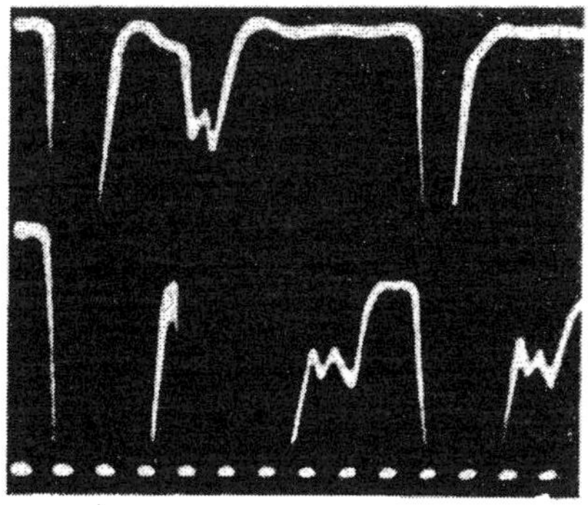

Fig. 5.5. Oscillograms illustrating the effect of correlation of the time shape of the stimulated photon echo with that of the second excitation pulse in the vapor of molecular iodine. Wavelength λ = 571 nm. Markers correspond to 20 ns time interval.

second pulse) is given by: $\sin\theta_2 \leq \theta_2$. It is evident that the code pulse can be a sum of several signals (data set). That is why it is worthwhile to assign an associative attribute to each fragment of the code pulse. Polarization was used as such an attribute in the experimental work [57]. This study was aimed at determination of the associative relationship between two different presentations of one and the same fragment of the code pulse when a predetermined shape of the signal of the primary photon echo is generated by means of setting of a certain direction of the polarization vector of the reading (second) pulse. The aforesaid was proved by experiments in which the vapor of I_2 were excited according to the scheme presented in Fig. 5.6. The pulse code was formed by two laser pulses (*a* and *b*) of different polarization and time shape. The sum time shape could be unrecognizable. The linear polarizations of these pulses were chosen to be orthogonal. In this case, one can expect (according to the general physical principles of formation of the photon echo) that reading by pulses of this or that polarization results in formation of the echo signal of the same polarization with maintaining of the time shape (Figs. 5.6, *d* and *e*). The experimental results were in complete agreement with these expectations. A single associative feature (polarization) was used in the procedure of reading of the PPE signal. In response, the gas medium (vapor of molecular iodine at the wavelength 571.5 nm that corresponds to the $X\,{}^1\Sigma_g^+ - B^3\Pi_{0u}^+$ band under the pressure of 20–70 mTorr) produced the PPE signal of a specific time shape. This result is illustrated by Fig. 5.6 and the comments to it.

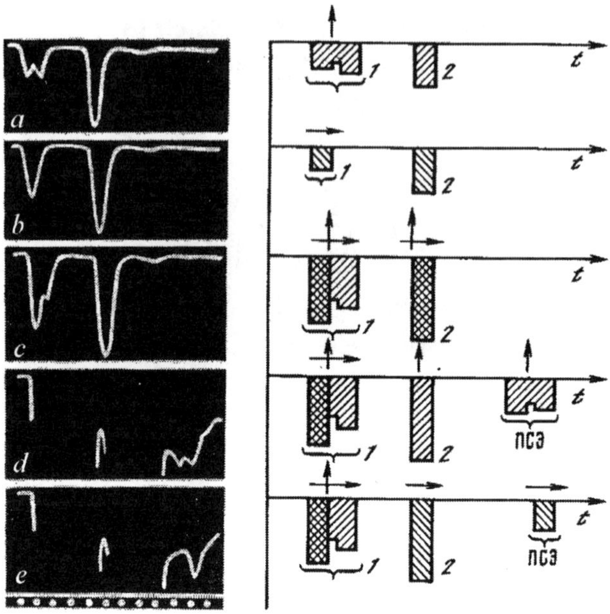

Fig. 5.6. The effect of the associative photon (light) echo [57]. The oscillograms in the left part illustrate the detection of the effect of the associative nature of PPE; the schematic presentation of the pulses of the corresponding oscillograms is given in the right part of the figure (orthogonally polarized polarization vectors of the light pulses are shown with arrows above the pulses and by means of different hatching of the pulses): (*a* – *c*) excitation pulses; (*d* and *e*) the results of the associative selection of a certain fragment of information according to the corresponding key. Markers correspond to 20-ns time interval.

Note that several papers [58–62] and a detailed review by Manykin *et al.* [63] are devoted to the problems of optical coherent data processing (OCDP). In these papers the physical principles and theoretical aspects of OCDP are discussed but the corresponding experiments are not considered. That is why we try to compensate here the lack of information. Mossberg *et al.* (see, for example, [64, 65]) obtained the most significant results in experimental realization of different regimes of OCDP. Consider in detail one of the experiments [64] devoted to the coherent compression of signals in the case of excitation of the photon echo by the pulses with a linear chirp

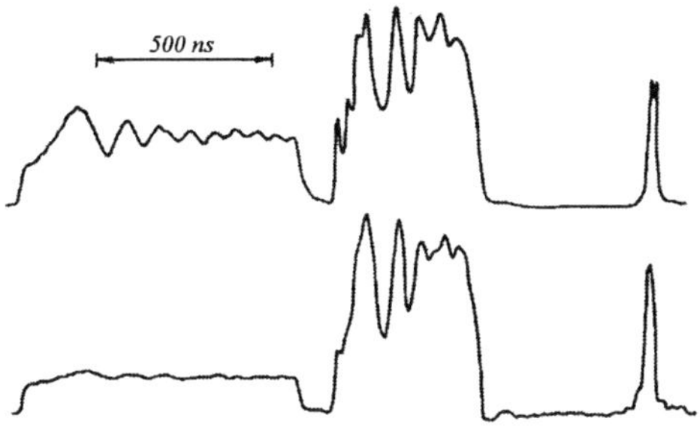

Fig. 5.7. Oscilloscope trace illustrating the compression of the information signal in the case of excitation of the vapor of atomic ytterbium by two laser LC pulses at the wavelength 555.6 nm [64]. The durations of the first and second LC pulses are 800 and 400 ns, respectively; the duration of the PPE signal is 27 ns.

(LC pulses). The physical reason of this compression is well known from the theory of the spin echo processes [66]. The following model is used in this case. Let the frequency of the excitation pulses change linearly in time:

$$\omega(t) = \omega_s + (\omega_f - \omega_s)(t - t_s)/\Delta t,$$

where ω_s and ω_f are the starting and final values of the frequency of the pulse with the duration Δt, and t_s is the beginning of the pulse. Assume, for example, that the first pulse is twice as long as the second one. Each pulse can be presented within the frames of the proposed model as a sequence of the monochromatic pulses of the duration δt_s, but the frequencies of the neighbor pulses differ from each other by the quantity Δ. Each of the subpulses excites its own spectral fragment of the inhomogeneously broadened band. As the second pulse is twice shorter than the first one, its series of the monochromatic pulses has the same frequencies but each of them has the duration $\delta t_i/2$. As a result, each pair of the subpulses of the same frequencies generates its own PPE signal at one and the same instant $t_e = 2\Delta t_1 + 2\tau$, where Δt_1 is the duration of the first pulse; τ is the time interval between the end of the first rectangular pulse and the beginning of the second one. The duration δt of such PPE signals is determined by the inhomogeneous broadening of the band fragment $\delta\omega$ that is excited by one of the subpulses

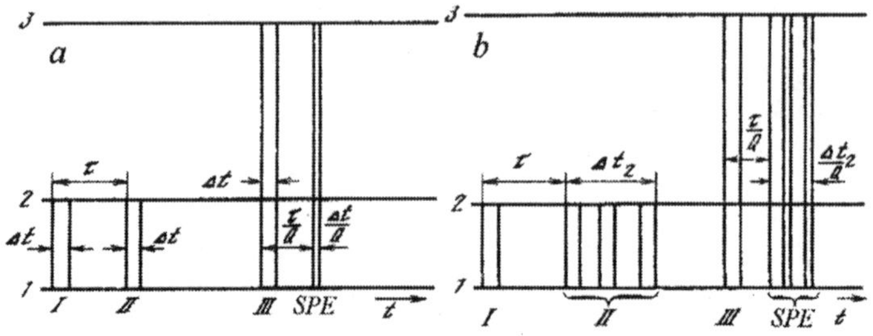

Fig. 5.8. The compression of (*a*) a single information signal, and (*b*) a series of information signals in a three-level system ("time telescope").

($\delta t = 1/\delta\omega$). As $\delta\omega = 1/\delta t_i$, the duration of the compressed echo signal is given by

$$\delta t_e = 1/\delta\omega = \delta t_i, \tag{5.3.1}$$

and $\Delta \geq 1/\delta t_i$.

The compression coefficient of the information signal (the first pulse) is determined as:

$$K = \sqrt{\Delta t_1(\omega_f - \omega_s)}\,. \tag{5.3.2}$$

It is evident that the span of the frequency modulation cannot be larger than the inhomogeneous bandwidth, i.e.,

$$K^p_{max} = (T_2/T_2{}^*)^{1/2}, \tag{5.3.3}$$

where T_2 and $T_2{}^*$ are the times of the reversible and irreversible cross relaxation, respectively.

Thus, in order to compress the information signal one has to approach the gas medium with two LC pulses with $\Delta t_1 > \Delta t_2$. Go back to the experiment described in [64]. It was performed in the ytterbium vapor (^{174}Yb) at the wavelength 555.6 nm. The durations of the excitation pulses were 0.8 and 0.4 μs, respectively, and the relaxation times for the medium were: $T_2 \approx$ 1.4 μs and $T_2^* = 5 \cdot 10^{-10}$ s. In spite of the fact that the estimates of K_{max} according to the formula (5.3.3) yield the value of about 50, in experiments the compression factor was as high as 30. The oscilloscope trace (Fig. 5.7) from [64] demonstrates the compression of the information signal in the Yb vapor with the use of the technique of the primary photon echo (echo signal is the first from the right).

The compression can be performed also with the use of the method of the stimulated photon echo. In this case the maximal compression coefficient is given by:

$$K^{s}_{max} = (T_1/T_2^*)^{1/2}, \tag{5.3.4}$$

where T_1 is the time of the longitudinal irreversible relaxation. A three-pulse experiment where the first and the third exciting signals are LC pulses and the second (information) one is a series of signals was also performed in ytterbium vapor ($\lambda = 555.6$ nm) [65].

In conclusion of this Section consider the possibilities of application of the regime of compression of the information signal in three-level media with the use of the method of the stimulated echo [67, 68]. For the details see Fig. 5.8. If the first two pulses work at the transition between the levels *1* and *2* and the third (read-out) one works at the transition *1* → *3*, SPE signal is generated at the transition *3* → *1* at $t = \tau_{13} + \tau_{12}/Q$ where $Q = \omega_{13}/\omega_{12}$, $\omega_{\alpha\beta}$ is the frequency of the transition between the levels α and β, τ_{lm} is the time interval between the l-th and the m-th pulses. The SPE signal appears to be compressed Q times (i.e., $\Delta t_e = \Delta t_p/Q$), in other words, under this regime the gas medium works as a "time telescope" [67]. Moreover, if the second excitation pulse is a series of coherent information signals of the nanosecond duration, each of them can be compressed Q times thus giving rise to a series of picosecond information signals [68].

Note, finally, that the description of different convolutions of the signals with the use of the SPE technique can be found in monographs [69, 70] and the reviews [43, 44, 63]. As the corresponding optical experiments have not been accomplished yet, we do not go into a detailed discussion.

5.4 Optical Echo Holography in Gas Media

A new and promising branch of holography–dynamic holography–emerged recently (see the review of Denisyuk [71] and the references therein). The structure of the dynamic holograms appears to be the function of not only the spatial coordinates but also of time and the very process of the dynamic holography is considered as the process of scattering of the probe light beam at the quasiperiodic inhomogeneities of the medium caused by the action of the writing waves (and of the probe wave in the case of high powers). Gerritsen was the first to use the resonance media for recording of the dynamic holograms [72] (many references on the resonance dynamic holography can be found in the monograph [73]). A simultaneous action of the object and reference signals on the resonance medium resulting in the formation of the interferogram appears to be an important condition of formation of such holograms. Since 1975 several works in which the recording of the resonance dynamic holograms with the use of the time-shifted object and reference coherent pulses was discussed were published [74–77]. In these works the new branch of holography was called echo holography which is a generally accepted term nowadays.

The theoretical possibility of recording of echo holograms in gas media was demonstrated in 1982 in [78] (see also a later work [79]). Note that many scientists expressed doubts on the very idea of recording of echo holograms in the media with intense internal motion, because in the conventional holography special efforts were aimed at making the recording medium more stable. As the idea is nontrivial, it is worthwhile to devote this Section to the description of the physics of recording of the echo holograms in the gas media and to the description of the first experiment [11] in this field and to omit the theoretical problems.

Note that in contrast to the conventional regime of excitation of the signals of the primary and stimulated photon echo by the pulses with plane wave fronts, in echo holography the wave front of one of the pulses carries the information on the object. In calculations the wave front of the object wave is expanded in Fourier series over the plane waves. After that the time evolution of each component is traced at all the stages of calculation of the nonequilibrium polarization and the difference of population of levels under the action of the next pulses. The calculations [74–77] show that in the gas medium the wave front of the PPE and SPE signals appears to be phase conjugated relative to the wave front of the first and second object pulses, respectively, if the condition $\sin\theta_c \leq \theta_c$ is met. Note that in the gas medium the Doppler effect appears to be the reason of the inhomogeneous

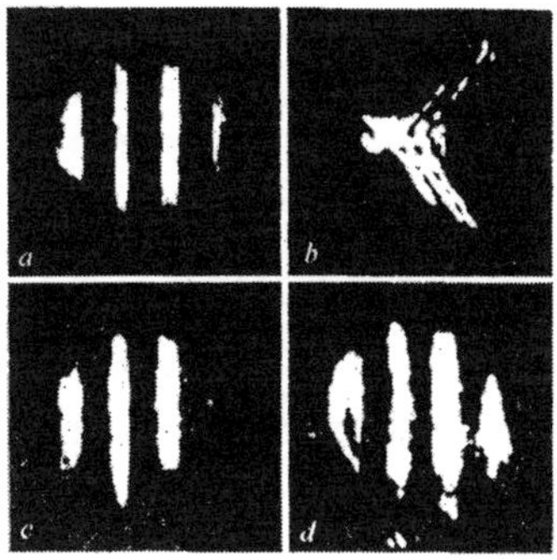

Fig. 5.9. Phase conjugation of the wave front of the signal of the stimulated photon (light) echo relative to the wave front of the second (information) pulse in a vapor of the atomic ytterbium. $T = 400$ K at the wavelength 555.6 nm [79] (see text for details).

broadening. In this case the condition of the spatial synchronism for the SPE signal for each η-th component of the expansion in the Fourier series (this is the signal we consider below) is given by $\mathbf{k}_{s\eta} = \mathbf{k}_1 - \mathbf{k}_{2\eta} + \mathbf{k}_3$, where $\mathbf{k}_1$ and $\mathbf{k}_3$ are the wave vectors of the reference pulses; $\mathbf{k}_{s\eta}$ is the wave vector of the η-th component of the expansion of the SPE signal. If $\mathbf{k}_1 = -\mathbf{k}_3$ then $\mathbf{k}_{s\eta} = -\mathbf{k}_{2\eta}$ and the corresponding condition for the phases is $\varphi_s(\mathbf{r}) = -\varphi_2(\mathbf{r})$, where $\mathbf{r}$ is the radius-vector that characterizes the wave front. In experiments the phase conjugation of the wave front of the SPE signal relative to the wave front of the second (object) pulse was detected [11]. Recall (see Appendix 3) that the frequencies of the energy transitions of gas particles depend on the velocities of these particles: $\omega_i = \omega_0 - (\omega_0/c)\mathbf{v}_i\boldsymbol{n}$, where ω_0 is the frequency of the transition between the energy levels of the particle at rest, c is the phase speed of light in the medium, $\mathbf{v}_i$ is the velocity of the i-th particle, and $\boldsymbol{n}$ is a unit vector in the direction of observation.

After expansion of the object wave in the Fourier series over the plane waves the problem of theoretical description of the echo holography in the SPE regime is reduced to the stages as follows: 1) in the beginning a plane wave (the first pulse) acts on each elementary volume; 2) then a multitude of the plane waves with the wave vectors $\mathbf{k}_{2\eta}$ (object pulse) acts on the same volume after the time interval τ_{12}; 3) at $t = \tau_{12} + \tau_{23}$ (where $\tau_{\alpha\beta}$ is the interval between the α-th and β-th pulses) another reference pulse arrives in the gas medium thus performing the read-out of the SPE signal with phase-conjugated wave front. Therefore, the complication of calculation of echo holograms in comparison with the description of the conventional SPE is related with simultaneous action (at the second stage) of a multitude of plane waves that propagate in different directions and have different amplitudes. However, the calculations are simplified due to the fact that in the gas medium each of these plane Fourier components can transform the nonequilibrium electric dipole moment and the nonequilibrium population (created by the first pulse) of only those gas particles that have the corresponding projections of velocities on the directions of the wave vectors $\mathbf{k}_{2\eta}$. In the opposite case these particles do not belong to one and the same spectral fragment of the inhomogeneously broadened band. During the time intervals τ_{12} and τ_{23} between the pulses the gas particles travel in different directions at different velocities $\mathbf{v}_i$. However, it is known from the physical principles of formation of the photon echo in gases (see Appendix 3) that such a "dispersion" of the particles is not an obstacle for the formation of the echo signals. Each η-th component of the expansion of the object wave forms together with the reference waves the η-th component of the expansion of the SPE wave with the spectral "weight" that is equal to the spectral weight of the corresponding component of the object wave. The first reference pulse must be a broad-band one in order to be able to excite a large number of gas particles into the superposition state (see Appendix 1). Moreover, it is worthwhile to make its wave front spherical rather than a plane in order to involve particles moving in opposite directions. The action of the object wave at $t = \tau_{12}$ results in introduction of different fragments of its complex wave front (as phase and amplitude information) into the shape of the electronic envelope of certain gas particles. As a consequence, certain dynamic distributions of the inhomogeneous electric polarizability and inhomogeneous difference of populations are formed in the gas medium. The subsequent action of the broad-band reading reference pulse (preferably with the spherical wave front) leads to generation of the SPE signal at $t = \tau_{23} + 2\tau_{12}$. The wave front of this SPE signal was demonstrated [11] to be phase-conjugated relative to the wave front of the object (second) pulse. Consider the results of the three-pulse echo experiment [11] carried out by Mossberg *et al.* in the vapor of atomic ytterbium (at the temperature of 400 K) at the wavelength 555.6 nm. In this experiment the duration of the excitation pulses was 7 ns and the powers of the first, second, and third pulses were 340, 170, and 200 W, respectively. Carlson *et al.* [11] try to determine the shape of the SPE wave front under the given wave front of the second pulse. In Fig. 5.9 we present the photographs that illustrate the results of this study. In photograph *a* one can see the trace of the second pulse that passed through the slide. Photograph *b* demonstrates the deformation of the wave front of the same pulse after passing through the phase plate. Two bottom photographs *c* and *d* show the traces of the SPE signals in two cases when both the second pulse and the SPE signal propagated outside the phase plate and through it, respectively. As the SPE wave front was phase-conjugated relative to the wave front of the second pulse, its passing through the phase plate resulted in restoration (correction) of the SPE wave front.

In conclusion of this Section consider the problem of writing and reading of the color echo holograms [80, 81] in the gas medium. Notice, that the essence of the color echo holography lies in the fact that several black-and-white holograms are recorded simultaneously in the medium. In particular, this operation can be performed with the use of femtosecond pulses. Wide spectrum of these pulses can "cover" several transitions at a time. If the transitions of a certain gas cannot reproduce this or that color of the object, one can use for recording a mixture of gases thus reproducing all the colors of the object. And, finally, consider the case of the object pulse with time-dependent wave front. It was demonstrated in [82] that the reconstructed phase conjugated stimulated echo wave is featured by the similar dynamics of variation of the wave front.

REFERENCES

1. E.L. Hahn, *Phys. Rev.* **77**: 746; **80**: 580–594 (1950).
2. S. Fernbach, W.G. Proctor, *J. Appl. Phys.* **26**: 170–181 (1955).
3. S.O. Elyutin, S.M. Zakharov, E.A. Manykin, *Zh. Eksp. Teor. Fiz.* **76**: 835–845 (1979).
4. I.V. Yevseyev, V.A. Reshetov, *Opt. Spektrosk.* **53**: 796–799 (1982).
5. I.V. Yevseyev, *The Theory of Polarization Echo Spectroscopy of Atoms and Molecules Interacting through Elastic Depolarizing Collisions* (Moscow: Doctor of Phys. and Math. Science Dissertation, 1987) (in Russian).
6. V.A. Zuikov, V.V. Samartsev, R.G. Usmanov, *Pis'ma Zh. Eksp. Teor. Fiz.* **32**: 293–297 (1980).
7. N.N. Rubtsova, *The Study of Relaxation of Populations and Dipole Moment of Colliding Molecules by Means of Stationary and Nonstationary Spectroscopy* (Novosibirsk, Cand. Phys. and Math. Sci. Dissertation, 1983) (in Russian).
8. N.W. Carlson, L.J. Rothberg, A.G. Yodh, *et al.*, *Opt. Lett.* **8**: 483–485 (1983).
9. L.S. Vasilenko, N.N. Rubtsova, *Opt. Spektrosk.* **59**: 52–56 (1985).
10. S.O. Elyutin, S.M. Zakharov, E.A. Manykin, *Opt. Spektrosk.* **52**: 577–579 (1982).
11. N.W. Carlson, W.R. Babbitt, T.W. Mossberg, *Opt. Lett.* **8**: 623–625 (1983).
12. A.V. Yevseyev, I.V. Yevseyev, V.A. Reshetov, *Fotonnoe ekho v gazakh: trekhurovnevye sistemy* (*Photon Echo in Gases: Three-level Systems*) (Moscow: IAE, Preprint No. 3849/1, 1983) (in Russian).
13. I.V. Yevseyev, V.M. Yermachenko, V.A. Reshetov, *J. Phys. B* **19**: 185–198 (1986).

14. I.V. Yevseyev, V.A. Reshetov, in *Phase Conjugation of the Wave Front of Laser Radiation in Nonlinear Media*, (Minsk: Inst. Phys. BSSR Acad. Sci., 1987) (in Russian), p. 143–148.
15. I.V. Yevseyev, V.M. Yermachenko, V.A. Reshetov, *Opt. Acta* **30**: 817–829 (1983).
16. I.V. Yevseyev, Yu.V. Men'shikova, V.N. Tsikunov, *Opt. Spektrosk.* **63**: 47–52 (1987).
17. I.V. Yevseyev, Yu.V. Men'shikova, V.N. Tsikunov, *J. Phys. B* **22**: 1863–1883 (1989).
18. I.V. Yevseyev, V.A. Reshetov, *Opt. Spektrosk.* **61**: 1053–1057 (1986).
19. I.V. Yevseyev, M.V. Fedorov, V.N. Tsikunov, *J. Opt. Soc. Amer. B* **6**: 1314–1316 (1989).
20. I.V. Yevseyev, A.V. Korovushkin, V.N. Tsikunov, *Opt. Spektrosk.* **67**: 666–670 (1989).
21. I.V. Yevseyev, Yu.V. Men'shikova, V.N. Tsikunov, *J. Mod. Opt.* **37**: 1185–1198 (1990).
22. I.V. Yevseyev, V.A. Reshetov, *J. Mod. Opt.* **38**: 143–150 (1991).
23. I.V. Yevseyev, A.V. Korovushkin, V.N. Tsikunov, *Laser Phys.* **1**: 79–80 (1991).
24. Y.C. Chen, K. Chiang, S.R. Hatrmann, *Opt. Commun.* **29**: 181–185 (1979).
25. J.B.W. Morsink, D.A. Wiersma, *Chem. Phys. Lett.* **65**: 105–108 (1979).
26. R. Kachru, T.W. Mossberg, S.R. Hartmann, *Opt. Commun.* **30**: 57–62 (1979).
27. J.B.W. Morsink, W.H. Hesselink, D.A. Wiersma, *Chem. Phys.* **71**: 289–294 (1982).
28. I.V. Yevseyev, V.A. Reshetov, *Pis'ma Zh. Eksp. Teor. Fiz.* **44**: 160–162 (1986).
29. I.V. Yevseyev, V.A. Reshetov, *Opt. Commun.* **67**: 144–148 (1988).
30. N.N. Akhmediev, I.V. Mel'nikov, *Opt. Spektrosk.* **62**: 969–975 (1987).
31. I.V. Yevseyev, V.M. Yermachenko, V.A. Reshetov, *J. Opt. Soc. Amer.* ***B*** **7**: 125–126 (1990).
32. A.V. Yevseyev, I.V. Yevseyev, V.A. Reshetov, *Opt. Spektrosk.* **58**: 518–523 (1985).
33. M. Mitsunaga, K. Kubodera, H. Kanbe, *Opt. Lett.* **11**: 339–341 (1986).
34. N.N. Akhmediev, B.S. Borisov, V.A. Zuikov, *et al.*, *Pis'ma Zh. Eksp. Teor. Fiz.* **45**: 122–125 (1987).
35. N.N. Akhmediev, B.S. Borisov, V.A. Zuikov, *et al.*, *Izv. AN SSSR, Ser. Fiz.* **52**: 1106–1112 (1988).
36. N.N. Akhmediev, B.S. Borisov, V.A. Zuikov, *et al.*, *Pis'ma Zh. Eksp. Teor. Fiz.* **48**: 585–587 (1988).
37. W.R. Babbitt, T.W. Mossberg, *Opt. Commun.* **65**: 185–188 (1988).
38. I.V. Yevseyev, V.A. Reshetov, *Opt. Spektrosk.* **67**: 226–228 (1989).
39. I.V. Yevseyev, V.A. Reshetov, *Opt. Commun.* **72**: 377–380 (1989).
40. I.V. Yevseyev, V.A. Reshetov, *Izv. AN SSSR, Ser. Fiz.* **53**: 2311–2316 (1989).
41. A.G. Anderson, R.L. Garwin, E.L. Hahn, *J. Appl. Phys.* **26**: 1324–1338 (1955).
42. M.P. Petrov, S.I. Stepanov, *Review on Electronics: UHF Electronics* no. **10(385)**: 3–28 (1976).
43. V.B. Ustinov, *Kvantovye ustroistva obrabotki signalov* (*Quantum Devices for Signal Processing*) (Leningrad: LETI, 1984) (in Russian).
44. V.S. Barsukov, *Ekho-protsessory* (*Echo processors*) (Moscow: Znanie, 1987) (in Russian).

45. N.A. Kurnit, I.D. Abella, S.R. Hartmann, *Phys. Rev. Lett.* **13**: 567–568 (1964).
46. U.Kh. Kopvillem, V.R. Nagibarov, V.A. Pirozhkov, *et al.*, *JETP Lett.* **20**: 139–144 (1974).
47. V.V. Samartsev, R.G. Usmanov, I.Kh. Khadyev, *JETP Lett.* **22**: 32–36 (1975).
48. V.A. Zuikov, V.V. Samartsev, E.A. Turiyanskii, *Sov. Phys. JETP* **81**: 653–663 (1981).
49. V.A. Zuikov, V.V. Samartsev, *Phys. Status Solidi (a)* **73**: 625–632 (1982).
50. V.A. Zuikov, V.V. Samartsev, R.G. Usmanov, *Izv. Akad. Nauk SSSR, Ser. Fiz.* **46**: 600–604 (1982).
51. S.O. Elyutin, S.M. Zakharov, V.A. Zuikov, *Sov. Phys. JETP* **88**: 401–16 (1985).
52. T.W. Mossberg, *Opt. Lett.* **7**: 77–79 (1982).
53. N.W. Carlson, W.R. Babbitt, Y.S. Bai, *et al.*, *Opt. Lett.* **9**: 232–234 (1984).
54. T.W. Mossberg, US Patent 4 459 682 (1984).
55. L.S. Vasilenko, *Coherent Doppler-free Laser Spectroscopy of Gas Media* (Novosibirsk: Doctor of Phys. Math. Science Dissertation, 1987) (in Russian).
56. I.I. Popov, I.S. Bikbov, V.V. Samartsev, *Izv. Akad. Nauk SSSR, Ser. Fiz.* **53**: 2334–2339 (1989).
57. I.S. Bikbov, I.I. Popov, V.V. Samartsev, *Zh. Prikl. Spektrosk.* **54**: 535–537 (1991).
58. V.B. Ustinov, M.M. Kovalevskii, S.A. Baruzdin, *Izv. Akad. Nauk SSSR, Ser. Fiz.* **50**: 1495–1499 (1986).
59. V.V. Samartsev, L.A. Nefed'ev, D.L. Chubakov, in: *Golografiya: teoreticheskie i prikladnye voprosy* (Holography: Theoretical and Applied Problems) (Leningrad: LFTI, 1988) (in Russian), p. 145.
60. V.A. Katulin, A.A. Biryukov, V.V. Samartsev, *Izv. Akad. Nauk SSSR, Ser. Fiz.* **53**: 2274–2280 (1989).
61. S.M. Zakharov, E.A. Manykin, *Sov. Phys. JETP* **95**: 1587–1597 (1989).
62. E.A. Manykin, S.M. Zakharov, *Izv. Akad. Nauk SSSR, Ser. Fiz.* **53**: 2281–2286 (1989).
63. E.A. Manykin, S.M. Zakharov, E.V. Onishchenko, *Funktsional'nye metody opticheskoi obrabotki informatsii na osnove fotonnogo ekha* (*Functional Methods of Optical Data Processing on the Basis of the Photon Echo*) (Moscow: TsNII Atominform, 1989) (in Russian).
64. Y.C. Bai, T.W. Mossberg, *Opt. Lett.* **11**: 30–32 (1986).
65. Y.C. Bai, W.R. Babbit, T.W. Mossberg, *Opt. Lett.* **11**: 724–726 (1986).
66. M.I. Kurkin, E.A. Turov, *YaMR v magnitouporyadochennykh veshchestvakh i ego primenenie* (*NMR in Magentically-ordered Substances and Its Applications*) (Moscow: Nauka, 1990) (in Russian).
67. Yu.V. Naboikin, V.V. Samartsev, *Izv. AN SSSR, Ser. Fiz.* **50**: 1458–1467 (1986).
68. I.I. Popov, I.S. Bikbov, V.V. Samartsev, in: *Svetovoe ekho i problemy kogerentnoi optiki* (*Light Echo and the Problems of Coherent Optics*) (Samara: Samara University, 1990) (in Russian), p. 32.
69. V.A. Golenishchev-Kutuzov, V.V. Samartsev, B.M. Khabibullin, *Impul'snaya opticheskaya i akusticheskaya kogerentnaya spektroskopiya* (*Pulse Optical and Acoustical Coherent Spectroscopy*) (Moscow: Nauka, 1988) (in Russian).
70. N.N. Akhmediev, V.V. Samartsev, *Novye fizicheskie printsipy opticheskoi obrabotki informatsii* (*New Physical Principles of Optical Data Processing*) (Moscow: Nauka, 1990) (in Russian), p. 326.
71. Yu.N. Denisyuk, *Zh. Prikl. Spektrosk.* **33**: 397–414 (1980).

72. H. Gerritsen, *Appl. Phys. Lett.* **10**: 239–241 (1967).
73. H.J. Eichler, P. Gunter, D.W. Pohl, *Laser Induced Dynamic Gratings* (Berlin: Springer, 1985).
74. E.I. Shtyrkov, V.V. Samartsev, *Rezonansnaya dinamicheskaya golografiya i opticheskoe sverkhizluchenie* (*Resonance Dynamical Holography and the Optical Super-raditation*) (Kazan: KF USSR Acad. Sci., 1975) (in Russian), p. 398.
75. E.I. Shtyrkov, V.V. Samartsev, *Opt. Spektrosk.* **40**: 392–393 (1976).
76. E.I. Shtyrcov, V.V. Samartsev, *Phys. Status Solidi A* **45**: 647–655 (1978).
77. V.V. Samartsev, E.I. Shtyrkov, in: *Spektroskopiya kristallov* (*Spectroscopy of Crystals*) (Leningrad: Nauka, 1978) (in Russian), p. 108.
78. V.A. Zuikov, L.A. Nefed'ev, V.V. Samartsev, in: *Prikladnye voprosy golografii* (*Applied Holography*) (Leningrad: Leningrad Inst. Nucl. Phys., 1982) (in Russian), p. 175.
79. L.A. Nefed'ev, V.V. Samartsev, *Primenenie metodov golografii v nauke i tekhnike* (*Application of the Holographic Methods in Science and Technology*) (Leningrad: Leningrad Inst. Nucl. Phys., 1987) (in Russian), p. 70.
80. L.A. Nefed'ev, V.V. Samartsev, *Opt. Spektrosk.* **62**: 701–703 (1987).
81. L.A. Nefed'ev, V.V. Samartsev, *Zh. Prikl. Spektrosk.* **47**: 640–648 (1987).
82. S.N. Andrianov, M.K. Gafiev, *Proceedings of Conference of Young Scientists, KFTI-90* 101 (1990).

Chapter 6

DOUBLE-MODE LASING IN STANDING-WAVE GAS LASERS WITH ALLOWANCE FOR DEPOLARIZING COLLISIONS

6.1 Theoretical Description of Double-Mode Lasing in Gas Lasers

At present gas lasers are widely used in science and technology due to some specific parameters of their radiation: monochromaticity, low divergence, small beam diameter, etc.

Single-mode lasers are used, for example, in laser interferometry, optical communication networks, and devices for plasma diagnostics. The theory of single-mode regime of lasing has been addressed in numerous studies. See, for example, [1–10] where different aspects of single-mode lasing were studied: lasing at low and high powers, in the presence of an absorbing cell, lasing in the external magnetic field, etc.

In the case of single-mode lasing the interaction of modes via the active medium plays an important role. The study of the properties of radiation of such lasers provides new perspectives of their application for the purposes of technology and science. The general approach to the theoretical consideration of the problems of multimode generation in the gas laser at low powers was put forward by Lamb [1]. The further studies were carried out within the frames of the proposed model, some of them were aimed at generalization of the model with regard to degeneration of the levels. In connection with this the works [11–13] are referenced. The studies of the stability of generation of two opposite traveling waves in a ring laser were reported in [14–17]. In this monograph we present the results of the studies of the double-mode lasing in a linear standing-wave gas laser. The main attention is paid to the regime of generation of two modes with linear polarizations orthogonal to each other. The first studies of the properties of radiation of such a laser made it possible to develop a quantum oscillator with high stability of the frequency [18] and a novel method of measurement of the low optical densities [19]. In general, the double-mode gas lasers with the possibility of a wide-range tuning of the intermode distance can be readily applied for both fundamental physical studies and the solution of many applied problems. It is demonstrated in the monograph that the understanding of the work of such lasers is impossible without consideration of the depolarizing collisions.

Electromagnetic field in the laser is considered on the basis of the Maxwell equations. Assume that z-axis is directed along the axis of the cavity, and write down the equation for the strength of the electric field in a way as follows:

$$-\frac{\partial^2 \boldsymbol{E}}{\partial z^2}+\frac{\partial^2 \boldsymbol{E}}{c^2 \partial t^2}+\frac{4\pi\tilde{\sigma}\partial \boldsymbol{E}}{c^2 \partial t}=-\frac{4\pi\partial^2 \boldsymbol{P}}{c^2 \partial t^2}, \tag{6.1.1}$$

where $\boldsymbol{P}$ is the polarization vector of the gas medium that includes both linear and nonlinear parts of the polarization. The calculation of the dependence of $\boldsymbol{P}$ on $\boldsymbol{E}$ is a complex problem that cannot be solved in the general case. $\tilde{\sigma}$ stands for the tensor that describes the losses of the electromagnetic field inside the cavity. Although the gas medium itself is isotropic, the presence of the anisotropic elements inside the cavity or the imposing of the external magnetic field results in the anisotropy.

Electric field strength $\boldsymbol{E}(z, t)$ can be presented in a standard way [1, 13] as an expansion over the eigenfunctions of the empty (without the active medium) cavity with slowly varying amplitudes and phases:

$$\boldsymbol{E}(z, t) = (½)\Sigma \boldsymbol{e}_n E_n(t)\sin(k_n z)\exp(-i\omega_n t - i\phi_n) + \text{c.c.}, \tag{6.1.2}$$

where $\boldsymbol{e}_n$, is a unit vector of polarization of the n-th mode that belongs to the xy plane; $E_n(t)$ and $\phi_n(t)$ are the slowly varying amplitudes and phases. In a similar way $\boldsymbol{P}(z, t)$ can be presented as:

$$\boldsymbol{P}(z, t) = (½)\Sigma \boldsymbol{D}_n(z, t)\exp(-i\omega_n t - i\phi_n) + \text{c.c.}, \tag{6.1.3}$$

where $\boldsymbol{D}_n(z, t)$ is a slowly varying time function (complex amplitude of polarization at the frequency ω_n). Substitute (6.1.2) and (6.1.3) in (6.1.1) and take into account the slowness of variation of the functions $E_n(t)$, $\phi_n(t)$, and $\boldsymbol{D}_n(z, t)$. Select the terms that oscillate at the optical frequency ω_n thus obtaining the equations for slowly varying envelopes that are well-known in nonlinear electrodynamics [20, 21]:

$$\dot{E}_n + \omega_0 E_n/(2Q_n) = -\, 2\pi\omega_0 \mathrm{Im} P_n\,, \tag{6.1.4}$$

$$(\dot{\phi}_n + \omega_n - \Omega_n)E_n = -\, 2\pi\omega_0 \mathrm{Re} P_n\,. \tag{6.1.5}$$

A dot over the quantity in (6.1.4) and (6.1.5) stands for time differentiation. ω_0 is the central frequency of the considered working transition. As in the case of lasing the difference between the values of ω_n and ω_0 is not larger then the width of the spectral emission band of the transition, the following relationship holds for helium-neon laser: $|\omega_n - \omega_0| << \omega_0$, which is used for derivation of (6.1.4) and (6.1.5). Q_n is the quality of the cavity for mode number n that is defined according to the expression $\omega_0/(2Q_n) = 2\pi\mathrm{Re}T_n$ [13], and Ω_n is the eigenfrequency of the empty cavity:

$$\Omega_n = ck_n + 2\pi\mathrm{Im}T_n$$

where

$$T_n = \boldsymbol{e}_n \tilde{\sigma}\, \boldsymbol{e}_n.$$

The quantity $P_n(t)$ is the part of the complex polarization of the medium at frequency ω_n that has the same spatial structure as the mode number n and that depends on $\boldsymbol{D}_n(z, t)$ in a way as follows:

$$P_n(t) = (2/L) \int dz\, \sin(k_n z)(\boldsymbol{D}_n(z, t), \boldsymbol{e}_n)\,, \tag{6.1.6}$$

where integration is from 0 to L, and L is the length of the cavity.

We consider here the double-mode lasing. Let the paramcters of the first and sccond mode be E_1, ϕ_1, ω_1, E_2, ϕ_2, and ω_2. Then the equations (6.1.4) and (6.1.5) can be rewritten in a way they are used below:

$$\dot{E}_1 = -\, \omega_0 E_1/(2Q_1) - 2\pi\omega_0 \mathrm{Im} P_1, \tag{6.1.7}$$

$$\dot{E}_2 = -\, \omega_0 E_2/(2Q_2) - 2\pi\omega_0 \mathrm{Im} P_2, \tag{6.1.8}$$

$$(\dot{\phi}_1 + \omega_1 - \Omega_1)E_1 = -2\pi\omega_0 \mathrm{Re}P_1, \tag{6.1.9}$$

$$(\dot{\phi}_2 + \omega_2 - \Omega_2)E_2 = -2\pi\omega_0 \mathrm{Re}P_2. \tag{6.1.10}$$

6.2 Polarization of the Gas Medium in the Case of Double-Mode Lasing

Let j_a and j_b be the angular momentum of the lower and upper working levels of the considered transition. The states of the atoms excited at these levels are described by means of the elements of the density-matrix introduced in Chapter 1. The atomic collisions can be taken into account within the frames of the model of depolarizing collisions under the approximation of the relaxation parameters averaged over the direction of the velocity of atoms. Under this approximation it is convenient to use the irreducible components of the density matrix that are introduced by the relationships (1.3.1). The notions are: $f_q^{(\kappa)}(\boldsymbol{r}, \mathrm{v}, t)$ for the upper level, $\phi_q^{(\kappa)}(\boldsymbol{r}, \mathrm{v}, t)$ for the lower level, and $\psi_q^{(\kappa)}(\boldsymbol{r}, \mathrm{v}, t)$ for the transition between them.

Circular components of the polarization vector $\boldsymbol{P}(z, t)$ related with the Cartesian components :

$$P_0 = P_z, \qquad P_{\pm 1} = -(\pm P_x + iP_y)/\sqrt{2}$$

can be expressed via $\psi_q^{(1)}(\boldsymbol{r},\mathrm{v},t)$. In the considered case of double-mode lasing the components of the matrix of optical coherence can be presented as:

$$\psi_q^{(1)}(\boldsymbol{r},\mathrm{v},t) = \psi_{1q}^{(1)}(\boldsymbol{r},\mathrm{v},t)\exp(i\omega_1 t + i\phi_1) + \psi_{2q}^{(1)}(\boldsymbol{r},\mathrm{v},t)\exp(i\omega_2 t + i\phi_2). \tag{6.2.1}$$

The quantities P_1 and P_2 in (6.1.7)–(6.1.10) are defined as the functions of the circular components of the vectors $\boldsymbol{D}_1(z,t)$ and $\boldsymbol{D}_2(z,t)$ and the latter can be expressed via the quantities $\psi_{1q}^{(1)}(\boldsymbol{r},\mathrm{v},t)$ and $\psi_{2q}^{(1)}(\boldsymbol{r},\mathrm{v},t)$. For example:

$$D_{1q}(z, t) = (-1)^{j_b - j_a + q}(2j_a + 1)^{-1/2} d^* \int d\mathrm{v}\, \psi_{1q}^{(1)}(\boldsymbol{r},\mathrm{v},t), \tag{6.2.2}$$

where d is the reduced matrix element of the operator of the dipole moment of the atom for the transition between the working levels.

The quantities $\psi_{1q}^{(1)}(\boldsymbol{r},\mathrm{v},t)$ and $\psi_{2q}^{(1)}(\boldsymbol{r},\mathrm{v},t)$ via which the polarization vector of the medium is expressed can be found from the solution of the system of equations (2.2.5)–(2.2.7) for the irreducible components of the density matrix in which the following alterations are made. The terms

$$\gamma_b^{(0)} N_b(2j_b+1)F(\mathbf{v})\delta_{\kappa 0}\delta_{q0}$$

and

$$\gamma_a^{(0)} N_a(2j_a+1)F(\mathbf{v})\delta_{\kappa 0}\delta_{q0}$$

are added to the right-hand sides of the equations (2.2.6) and (2.2.7), respectively, in order to take into account the continuous pumping that is assumed to be homogeneous and isotropic and results in the excitation of atoms to the working levels. The atoms obey Maxwell distribution over the velocities. Then in the right-hand side of the equation (2.2.7) we omit the term that takes into account the radiation transition from the upper working level to the lower one. For the considered helium–neon lasers this does not cause qualitative changes of the results but simplifies the solution substantially. An approximation described in Section 1.5 is used for the collision integrals that take into account the influence of the elastic depolarizing collisions. Under this approximation the relaxation parameters

$$\gamma_a^{(\kappa)} = \gamma_a^{(0)} + \Gamma_a(\mathrm{v})^{(\kappa)},\ \gamma_b^{(\kappa)} = \gamma_b^{(0)} + \Gamma_b(\mathrm{v})^{(\kappa)},$$

$$\gamma^{(\kappa)} = (\gamma_a^{(0)} + \gamma_b^{(0)})/2 + \mathrm{Re}\Gamma(\mathrm{v})^{(\kappa)} \tag{6.2.3}$$

are introduced into final solutions. The imaginary part of the quantity $\Gamma(\mathrm{v})^{(\kappa)}$ is added to the central frequency of the working transition ω_0. The detunings of the mode frequencies from this frequency with regard to its shift due to elastic collisions are introduced into the results:

$$\omega_{10} = \omega_1 - \omega_0 + \mathrm{Im}\Gamma(\mathrm{v})^{(1)}, \qquad \omega_{20} = \omega_2 - \omega_0 + \mathrm{Im}\Gamma(\mathrm{v})^{(1)}. \tag{6.2.4}$$

Intermode distance is an important parameter for the double-mode lasing:

$$\omega_{12} = \omega_1 - \omega_2 = \omega_{10} - \omega_{20}. \tag{6.2.5}$$

We used the expression (6.1.2) with two terms corresponding to the considered modes (n = 1, 2) for the circular field components in the right-hand sides of the equations (2.2.5)–(2.2.7). For simplification of the notation the factor 1/4 that appears in (6.1.2) after replacement of $\sin(k_n z)$ by two exponential functions is omitted in the further calculations which is equivalent to redefining of the amplitudes E_1 and E_2. The direction of the polarization of modes was assumed to be predetermined according to the experimental conditions [22].

The mentioned system of equations was solved under resonance approximation because the detunings of the mode frequencies (6.2.4) are small in comparison with ω_0. The calculations were performed by means of iterations. The interaction with the field was considered as the disturbance of the atomic system correct to the terms cubic over the field intensity. The relaxation parameters of the active medium (6.2.3) were considered to be large in comparison with the cavity bandwidth $\omega_0/(2Q_n)$ (n = 1, 2). Such an approximation is rather good for helium-neon lasers and results in the dependence of the polarization of the medium at a certain instant on the value of the electric field at the same instant.

Finally, the quantities P_1 and P_2 in the right-hand sides of the equations (6.1.7)–(6.1.10) can be presented as the sums of linear and nonlinear terms:

$$P_1 = P_1^{(l)} + P_1^{(nl)}, \quad P_2 = P_2^{(l)} + P_2^{(nl)}. \tag{6.2.6}$$

The linear part is given by:

$$P_1^{(l)} = -(2/3)in|d|^2 E_1 \int d\mathrm{v}\, F(\mathrm{v})[\gamma - i(\omega_{10} - k\mathrm{v}_z)]^{-1}, \tag{6.2.7}$$

where n is the mean value of the inverse population of the working levels:

$$n = l(N_b - N_a)/L$$

(l is the length of the tube with the active medium). For simplification reasons we used the notation

$$\gamma = \gamma^{(\kappa = 1)} \tag{6.2.8}$$

in (6.2.7) and everywhere below.

The dependence of the polarization of the medium on the relaxation parameters $\gamma^{(\kappa)}$ with $\kappa \neq 1$ (if they are different from zero for the considered working transition) can be found in the terms that are proportional to the fifth and higher powers of the electric field strength. In addition, we omitted subscripts for the wave vectors k_1 and k_2 in (6.2.7) and similar expressions, i.e., in integration over $\mathbf{v}$ we neglect the difference between k_1 and k_2 because for the value of v_z that are important for calculations of the integrals we have:

$$|(k_1 - k_2)v_z| \approx |\omega_{12}v_z|/c << |\omega_{12}|.$$

The following expression is obtained for the nonlinear part of polarization at the frequency of the first mode:

$$P_1^{(nl)} = in|d|^4 E_1 \int d\mathbf{v}\, F(\mathbf{v})[\gamma - i(\omega_{10} - kv_z)]^{-1}\times$$

$$\times \sum_{j\kappa} \{E_1^2\, [2\gamma(a_j^{(\kappa)}/\gamma_j^{(\kappa)})(\frac{1}{\gamma^2 + (\omega_{10} + kv_z)^2} + \frac{1}{\gamma^2 + (\omega_{10} - kv_z)^2}) +$$

$$+ a_j^{(\kappa)}(\gamma_j^{(\kappa)} + 2ikv_z)^{-1}(\frac{1}{\gamma - i(\omega_{10} - kv_z)} + \frac{1}{\gamma + i(\omega_{10} + kv_z)})] +$$

$$+ E_2^2\, [2\gamma(b_j^{(\kappa)}/\gamma_j^{(\kappa)})(\frac{1}{\gamma^2 + (\omega_{20} + kv_z)^2} + \frac{1}{\gamma^2 + (\omega_{20} - kv_z)^2}) +$$

$$+ c_j^{(\kappa)}(\gamma_j^{(\kappa)} - i\omega_{12})^{-1}(\frac{1}{\gamma - i(\omega_{10} - kv_z)} + \frac{1}{\gamma + i(\omega_{20} - kv_z)}) +$$

$$+ c_j^{(\kappa)} (\gamma_j^{(\kappa)} - i\omega_{12} + 2ikv_z)^{-1} \Big(\frac{1}{\gamma - i(\omega_{10} - kv_z)} +$$

$$+ \frac{1}{\gamma + i(\omega_{20} + kv_z)} \Big) + (n_2/n)\, b_j^{(\kappa)} (\gamma_j^{(\kappa)} + 2ikv_z)^{-1} \times$$

$$\times \Big(\frac{1}{\gamma - i(\omega_{20} - kv_z)} + \frac{1}{\gamma + i(\omega_{20} + kv_z)} \Big) +$$

$$+ (n_2/n)\, c_j^{(\kappa)} (\gamma_j^{(\kappa)} - i\omega_{12})^{-1} \Big(\frac{1}{\gamma - i(\omega_{10} + kv_z)} +$$

$$+ \frac{1}{\gamma + i(\omega_{20} + kv_z)} \Big) \Big] \Big\}. \qquad (6.2.9)$$

Here by n_2/n stands for the ratio of the spatial harmonic of the difference of populations of the working levels to its mean value [1]:

$$n_2/n = (\omega_{12} l/c)^{-1} \sin(\omega_{12} l/c) \cos(\omega_{12}(2z_0 + l)/c), \qquad (6.2.10)$$

where z_0 is the distance from the tube with the active medium to the nearest mirror. The summation in (6.2.9) is over $\kappa = 0, 1, 2$ and $j = a, b$.

The coefficients $a_j^{(\kappa)}$, $b_j^{(\kappa)}$, $c_j^{(\kappa)}$ depend on the polarization of modes and the values of momenta of the working levels. The values of the quantities $a_j^{(\kappa)}$ and $c_j^{(\kappa)}$ for the case of linear mutually orthogonal polarizations of the modes and three types of transitions with the angular momenta $j_b = 1$, $j_a = 0, 1, 2$ are given in Tables 6.1, 6.2, and 6.3. Note that $a_j^{(1)} = c_j^{(0)} = 0$ ($j = a, b$). The quantities $b_j^{(\kappa)}$ are defined in a way as follows:

$$b_j^{(\alpha)} = a_j^{(\alpha)},\ b_j^{(2)} = -\, a_j^{(2)}/2,\ (j = a, b, \alpha = 0, 1).$$

Table 6.1 (transition $j_b = 1, j_a = 2$)

$a_b^{(0)}$	$a_b^{(2)}$	$a_a^{(0)}$	$a_a^{(2)}$	$c_b^{(1)}$	$c_b^{(2)}$	$c_a^{(1)}$	$c_a^{(2)}$
1/27	1/1350	1/45	7/450	1/72	1/1800	1/40	7/600

Table 6.2 (transition $j_b = 1, j_a = 1$)

$a_b^{(0)}$	$a_b^{(2)}$	$a_a^{(0)}$	$a_a^{(2)}$	$c_b^{(1)}$	$c_b^{(2)}$	$c_a^{(1)}$	$c_a^{(2)}$
1/27	1/54	1/27	1/54	1/72	1/72	1/72	1/72

Table 6.3 (transition $j_b = 1, j_a = 0$)

$a_b^{(0)}$	$a_b^{(2)}$	$a_a^{(0)}$	$a_a^{(2)}$	$c_b^{(1)}$	$c_b^{(2)}$	$c_a^{(1)}$	$c_a^{(2)}$
1/27	2/27	1/9	0	1/18	1/18	0	0

These transitions correspond, for example, to the generation of helium-neon lasers at the wavelengths 0.63, 1.15, and 3.3922 μm (working transitions with the angular momenta of levels $j_b = 1, j_a = 2$); 3.3912 μm (transition $j_b = j_a = 1$); and 1.52 μm (transition $j_b = 1, j_a = 0$).

For the linear parallel polarizations of the modes the values of the coefficient $a_j^{(\kappa)}$ ($j = a, b$) coincide with the corresponding values of these quantities for the orthogonal polarizations. For parallel polarizations the values of coefficients $b_j^{(\kappa)}$ and $c_j^{(\kappa)}$ are equal and coincide with the values $a_j^{(\kappa)}$.

For the optically allowed transitions with angular momenta $j_b = j_a = j$ and $j_b = j, j_a = j + 1$ for $j > 1$ the values of these coefficients can be determined with the use of the data presented in Table 6.1 and with the use of the fact that their dependence on the value of the angular momenta of the levels is determined by the factors that can be presented as the squares of 6j-symbols [23]

$$g_b^{(\kappa)} = \begin{Bmatrix} 1 & 1 & \kappa \\ j_b & j_b & j_a \end{Bmatrix}^2, \quad g_a^{(\kappa)} = \begin{Bmatrix} 1 & 1 & \kappa \\ j_a & j_a & j_b \end{Bmatrix}^2.$$

Using the expressions for 6j-symbols we arrive at the expressions as follows:

$$g_a^{(0)} = 1/[3(2j_a + 1)],$$

$$g_a^{(1)} = [j_b(j_b+1) - j_a(j_a+1) - 2]^2[12j_a(2j_a+1)(2j_a+2)]^{-1},$$

$$g_a^{(2)} = \{3[j_b(j_b+1) - j_a(j_a+1)]^2 - 9j_b(j_b+1) +$$
$$+ j_a(j_a+1) + 6\}^2[120j_a(j_a+1)(2j_a-1)(2j_a+1)(2j_a+3)]^{-1}. \qquad (6.2.11)$$

The expressions for $g_b^{(\kappa)}$ ($\kappa = 0, 1, 2$) can be obtained from the corresponding quantities $g_a^{(\kappa)}$ by means of replacement of j_a by j_b and of j_b by j_a.

The quantities $P_2^{(l)}$ and $P_2^{(nl)}$ in the right-hand side of the equations (6.1.8) and (6.1.10) can be obtained from (6.2.7) and (6.2.9) by means of the replacements ω_{10} by ω_{20}, ω_{20} by ω_{10}, and ω_{12} by $-\omega_{12}$.

When calculating the integrals over velocity in (6.2.7) and (6.2.9) we neglected the dependence of the relaxation parameters on the velocity of the atom on the basis of consideration presented in Chapter 1. They were presented as the functions of the gas pressure p in a way as follows:

$$\gamma_a^{(\kappa)} = \gamma_a^{(0)} + A_a^{(\kappa)} p, \ \gamma_b^{(\kappa)} = \gamma_b^{(0)} + A_b^{(\kappa)} p,$$

$$\gamma = (\gamma_a^{(0)} + \gamma_b^{(0)})/2 + Ap. \qquad (6.2.12)$$

The values of quantities $\gamma_a^{(0)}$, $\gamma_b^{(0)}$, A, $A_a^{(2)}$, and $A_b^{(2)}$ were taken from the experimental works. For the ratio $A_{a(b)}^{(1)} / A_{a(b)}^{(2)}$ we used the theoretical results of Chapter 1.

For the real and imaginary parts of P_1 and P_2 on the right-hand sides of equations (6.1.7)–(6.1.10) we introduce the standard notation used for the description of the double-mode lasing:

$$-2\pi\omega_0 \mathrm{Im}\, P_1^{(l)} - \omega_0 E_1/(2Q_1) = \alpha_1(\omega_{10})E_1,$$

$$-2\pi\omega_0 \mathrm{Re}\, P_1^{(l)} = \sigma_1(\omega_{10})E_1, \qquad (6.2.13)$$

$$-2\pi\omega_0 \mathrm{Im}\, P_1^{(nl)} = [\beta_1(\omega_{10})E_1^2 + \theta_{12}(\omega_{10}, \omega_{20})E_2^2]E_1,$$

$$-2\pi\omega_0 \mathrm{Re}\, P_1^{(nl)} = [\rho_1(\omega_{10})E_1^2 + \tau_{12}(\omega_{10}, \omega_{20})E_2^2]E_1. \qquad (6.2.14)$$

The coefficients $\alpha_2(\omega_{20})$, $\beta_2(\omega_{20})$, $\theta_{21}(\omega_{20}, \omega_{10})$, $\sigma_2(\omega_{20})$, $\rho_2(\omega_{20})$, and $\tau_{21}(\omega_{20}, \omega_{10})$ are introduced in a similar way.

With the use of these notations the equations (6.1.7) and (6.1.8) that will be used for further transformations can be rewritten as:

$$\dot{E}_1 = [\alpha_1(\omega_{10}) - \beta_1(\omega_{10})E_1^2 - \theta_{12}(\omega_{10}, \omega_{20})E_2^2]E_1 , \qquad (6.2.15)$$

$$\dot{E}_2 = [\alpha_2(\omega_{20}) - \beta_2(\omega_{20})E_2^2 - \theta_{21}(\omega_{20}, \omega_{10})E_1^2]E_2 . \qquad (6.2.16)$$

The quantities $\alpha_1(\omega_{10})$ and $\alpha_2(\omega_{20})$ have the meaning of the linear coefficients of amplification of modes with regard to losses. The integrals with the help of which they are defined can be calculated analytically in the limiting cases of either inhomogeneously broadened band when

$$\gamma << ku \qquad (6.2.17)$$

and with the thermal velocity of the atom u defined in Chapter 1 or homogeneously broadened band when $\gamma >> ku$.

Consider, for example, an expression for $\alpha_1(\omega_{10})$ under $\gamma << ku$:

$$\alpha_1(\omega_{10}) = (\omega_0/2Q_1)[\eta_1\exp(-(\omega_{10}/ku)^2) - 1], \qquad (6.2.18)$$

where $\eta_1 = n/n_{1t}$, is the parameter that characterizes the excess of the difference of populations of the working levels over the threshold value n_{1t}. It is defined in such a way that lasing takes place in the case $\eta_1 > 1$.

The coefficients $\beta_1(\omega_{10})$ and $\beta_2(\omega_{20})$ determine the value of saturation of amplification by the field of the mode itself and, therefore, are called the coefficients of self-saturation. For example, under the limit of inhomogeneously broadened band we obtain:

$$\beta_1(\omega_{10}) = 2\pi\omega_0 n|d|^4\sqrt{\pi}\,[1 + \gamma^2/(\gamma^2 + \omega_{10}^2)]/(ku\gamma\gamma_{ab}), \qquad (6.2.19)$$

where γ_{ab} stands for the following combination of the relaxation parameters of the working levels:

$$1/\gamma_{ab} = \sum a_j^{(\kappa)} / \gamma_j^{(\kappa)}. \tag{6.2.20}$$

The summation in (6.2.20) is over $\kappa = 0, 1, 2$ and $j = a, b$.

The coefficients $\theta_{12}(\omega_{10}, \omega_{20})$ and $\theta_{21}(\omega_{20}, \omega_{10})$ are called the coefficients of cross saturation. For example, under the limit of the inhomogeneously broadened band we have:

$$\theta_{12}(\omega_{10}, \omega_{20}) = 2\pi\omega_0 n|d|^4 \sqrt{\pi}\,(ku\gamma\gamma'_{ab})^{-1}\Big[\frac{4\gamma^2}{4\gamma^2+\omega_{12}^2} +$$

$$+\frac{4\gamma^2}{4\gamma^2+(\omega_{10}+\omega_{20})^2}+\frac{2\gamma\gamma'_{ab}}{(4\gamma^2+\omega_{12}^2)}\sum c_j^{(\kappa)}\frac{(2\gamma_j^{(\kappa)}\gamma-\omega_{12}^2)}{(\gamma_j^{(\kappa)2}+\omega_{12}^2)}+$$

$$+(n_2/n)\frac{\gamma\gamma'_{ab}}{(\gamma^2+\omega_{10}^2)}\sum c_j^{(\kappa)}\frac{(\gamma_j^{(\kappa)}\gamma-\omega_{10}\omega_{12})}{(\gamma_j^{(\kappa)2}+\omega_{12}^2)}\Big], \tag{6.2.21}$$

where the summation is over $\kappa = 0, 1, 2$ and $j = a, b$ and the quantity γ'_{ab} is defined as:

$$1/\gamma'_{ab} = \sum b_j^{(\kappa)} / \gamma_j^{(\kappa)}. \tag{6.2.22}$$

The expressions for the coefficients in the equation (6.2.15) can be obtained from the expressions presented above by means of replacement of the index 1 by 2 and by inverse replacement in the mode frequencies.

6.3 Stability of the Stationary Double-Mode Regime of Lasing

For the case of the stationary lasing the left-hand sides of the equations (6.2.15) and (6.2.16) are equal to zero and solving them we arrive at:

$$E_1^2 = (\alpha_1\beta_2 - \alpha_2\theta_{12})/(\beta_1\beta_2 - \theta_{12}\theta_{21}),$$

$$E_2^2 = (\alpha_2\beta_1 - \alpha_1\theta_{21})/(\beta_1\beta_2 - \theta_{12}\theta_{21}), \quad (6.3.1)$$

where we omitted for the simplicity reasons the arguments of all the coefficients.

The conditions of the stability of the double-mode regime of lasing and the positiveness of E_1^2 and E_2^2 are given by the following inequalities:

$$\beta_1\beta_2 - \theta_{12}\theta_{21} > 0, \;\; \alpha_1\beta_2 - \alpha_2\theta_{12} > 0, \;\; \alpha_2\beta_1 - \alpha_1\theta_{21} > 0. \quad (6.3.2)$$

Here we assume that the linear coefficients of amplification are positive.

We consider the interaction of modes for the case of equal losses:

$$Q_1 = Q_2 = Q.$$

In experiments special measures are taken in order to meet this equality with sufficient accuracy. Assuming that the equality is met, we arrive at the equality of the parameters: $\eta_1 = \eta_2 = \eta$. That is why the range of the frequencies inside which the amplification is higher than losses is one and the same for both modes. It is equal to the double value of the quantity ω_{10m} that can be defined, e.g., from the equation

$$\alpha_1(\omega_{10m}) = 0. \quad (6.3.3)$$

Using (6.2.18) we obtain

$$\omega_{10m} = ku(\ln\eta)^{1/2}.$$

In the case of low relative excitation of the medium, i.e., for the case $\eta - 1 << 1$, we have

$$\omega_{10m} = ku(\eta - 1)^{1/2}.$$

Consider a symmetrical position (relative to the center of the band) of the modes in the frequency scale:

$$\omega_{10} = -\,\omega_{20} = \omega_{12}/2, \tag{6.3.4}$$

where we assume that $\omega_{12} > 0$. It is in the vicinity of such a position of the modes that their strong interaction takes place. Indeed, the first mode interacts effectively with the group of atoms of the medium for which the condition $|\omega_{10} \pm kv| \leq \gamma$ is met, where v is the projection of atom velocity on the laser axis. A similar condition for the second mode is given by $|\omega_{20} \pm kv| \leq \gamma$. Taking into account the fact that for the predetermined intermode distance ω_{12}, the position of the second mode is related with the position of the first one: $\omega_{20} = \omega_{10} - \omega_{12}$, we arrive at the conclusion that for the symmetrical position of modes (6.3.4) both modes interact effectively with the same atoms. Note, however, that as $k_1 \neq k_2$, the positions of the maximums of the field intensity for the first and the second modes do not coincide in the coordinate space.

In the case of the symmetrical position we have the following relationship between the coefficients that determine the intensity of the generated modes:

$$\alpha_1(\omega_{12}/2) = \alpha_2(-\,\omega_{12}/2) = \alpha(\omega_{12})\,,\ \beta_1(\omega_{12}/2) = \beta_2(-\,\omega_{12}/2) = \beta(\omega_{12})\,,$$

$$\theta_{12}(\omega_{12}/2,\ -\,\omega_{12}/2) = \theta_{21}(-\,\omega_{12}/2,\ \omega_{12}/2) = \theta(\omega_{12}). \tag{6.3.5}$$

Note that the value of the intermode distance ω_{12} has an upper-bound limit:

$$\omega_{12} < \omega_{12m} = 2\omega_{10m}\,. \tag{6.3.6}$$

Assuming that this inequality is met (in the given case it is equivalent to the positiveness of α), we see that all the inequalities (6.3.2) that are necessary for the stability of the double-mode regime of lasing are reduced to only one:

$$\beta(\omega_{12}) > \theta(\omega_{12}). \tag{6.3.7}$$

In the case of parallel polarizations of the modes when $\gamma'_{ab} = \gamma_{ab}$ the inequality (6.3.7) can be presented under the limit of the inhomogeneously broadened band as:

$$(1 + n_2/n)\Sigma c_j^{(\kappa)} \frac{(2\gamma_j^{(\kappa)}\gamma - \omega_{12}^2)}{(\gamma_j^{(\kappa)2} + \omega_{12}^2)} < 0. \tag{6.3.8}$$

If ω_{12} decreases, then $n_2/n \to 1$ (see (6.2.10)). That is why it is clear that the inequality (6.3.8) is not met for the values of ω_{12} that are smaller than a certain critical value ω_{12cr} regardless of the values of the coefficients $c_j^{(\kappa)}$ ($c_j^{(\kappa)} > 0$), i.e., for any values of the momenta of levels j_a and j_b. Thus, for the parallel polarizations of the modes the stationary double-mode regime of lasing is possible if the intermode distance meets the inequality:

$$\omega_{12cr} < \omega_{12} < \omega_{12m}\,, \tag{6.3.9}$$

where the upper-bound limit ω_{12m} is determined by the equation (6.3.3) and the lower ω_{12cr} is determined by the equation

$$\beta(\omega_{12}) = \theta(\omega_{12}). \tag{6.3.10}$$

In order to estimate ω_{12cr} consider the case of the inhomogeneously broadened band and neglect the difference of the relaxation parameters of the working levels:

$$\gamma_j^{(\kappa)} = \gamma_0 \;\; (j = a, b, \kappa = 0, 1, 2). \tag{6.3.11}$$

Then for ω_{12cr} we have

$$\omega_{12cr} = (2\gamma\gamma_0)^{1/2}\,. \tag{6.3.12}$$

The rejection of the approximation (6.3.11) makes the expression for

ω_{12cr} more complex but does not change the order of magnitude of this quantity. Therefore, for the parallel polarizations of the modes the regard to the depolarizing collisions that results in the difference of the relaxation parameters of the levels does not lead to qualitative changes of the results.

Consider now the case of the orthogonal polarizations of the modes. The analysis of the expressions for the coefficients $\beta(\omega_{12})$ and $\theta(\omega_{12})$ shows that the violation of the inequality (6.3.7) can be expected for the decreasing ω_{12}. That is why consider the limiting value of the difference $\beta(\omega_{12}) - \theta(\omega_{12})$ for $\omega_{12} = 0$. For the limiting cases of inhomogeneously and homogeneously broadened band and neglecting the insufficient factors we arrive at:

$$\beta(0) - \theta(0) \sim \Sigma(a_j^{(\kappa)} - b_j^{(\kappa)} - c_j^{(\kappa)})/ \gamma_j^{(\kappa)} \tag{6.3.13}$$

where the summation is over $\kappa = 0, 1, 2$ and $j = a, b$. It is evident that the relationship (6.3.13) holds also for the case of the parallel polarizations of the modes. The negative sign of the difference $\beta(0) - \theta(0)$ in this case becomes also evident if one takes into account the mentioned relationship between the coefficients: $a_j^{(\kappa)} = b_j^{(\kappa)} = c_j^{(\kappa)}$.

For the orthogonal polarizations of the modes consider at first the transition with the angular momenta of the levels $j_a = 2,\ j_b = 1$. Using the data of Table 6.1 and (6.3.13) we have:

$$\beta(0) - \theta(0) \sim (1/\gamma_b^{(2)} - 25/\gamma_b^{(1)})/3 + 7/\gamma_a^{(2)} - 15/\gamma_a^{(1)} . \tag{6.3.14}$$

Analyzing the right-hand side of (6.3.14) with the use of (6.2.3) and the relationship between the relaxation rates of orientations and alignment of the levels under the action of collisions (see Chapter 1), we arrive at the inequality $\beta(0) - \theta(0) < 0$.

It means that there is a certain critical intermode distance ω_{12cr} below which the double-mode regime becomes unstable. The value of ω_{12cr} can be determined from the equation (6.3.10) where β and θ are the corresponding coefficients.

Estimate the value of ω_{12cr} for the transition $j_a = 2, j_b = 1$ in the case of the inhomogeneously broadened band. For simplicity consider the situation when

$$\gamma_j^{(\kappa)} \ll \gamma \;\; (\kappa = 0, 1, 2, \; j = a, b). \tag{6.3.15}$$

This relationship holds for helium-neon lasers because for the considered transitions the band broadens much faster than the polarization parameters of the levels. Using the results of Section 6.2 we conclude that

$$a_j^{(0)} = b_j^{(0)}, \quad c_j^{(0)} = 0 \quad (j = a, b),$$

i.e., only the terms with $\kappa = 1$ and $\kappa = 2$ contribute to the result. The approximate equalities for the rates of relaxation of orientation and alignment are based on the results of Chapter 1:

$$\gamma_a^{(1)} \approx \gamma_a^{(2)}, \quad \gamma_b^{(1)} \approx \gamma_b^{(2)}. \tag{6.3.16}$$

Assuming, for simplicity, that $\gamma_a^{(1)} = \gamma_b^{(1)}$ and omitting a numerical factor that is close to unity, we arrive at:

$$\omega_{12cr} \approx \gamma_a^{(1)}. \tag{6.3.17}$$

Thus, the value of the critical intermode distance is determined first of all by the polarization parameters of the levels for the transition with angular momenta of the working levels $j_a = 2, j_b = 1$ and orthogonal polarizations of the modes. More detailed theoretical study of this quantity can be found in [24]. The results of experiments for helium-neon laser with the wavelength 3.3922 μm are presented in [25, 26].

Consider now the transition with the angular momenta of levels $j_a = j_b = 1$ for the orthogonal polarizations of the modes. Using the data of Table 6.2 and (6.3.13) we have:

$$\beta(0) - \theta(0) \sim 1/\gamma_b^{(2)} - 1/\gamma_b^{(1)} + 1/\gamma_a^{(2)} - 1/\gamma_a^{(1)}. \tag{6.3.18}$$

Going back to the results of Chapter 1 that give the ratio of the rates of relaxation of orientation and alignment of levels for $j = 1$ we can conclude that the expression (6.3.18) is positive. Hence, the double-mode lasing at the transitions with the angular momenta of the working levels $j_a = j_b = 1$ for orthogonal polarizations of modes must be stable even for the intermode distances that are smaller than the polarization parameters of the working levels in contrast to the transition with the angular momenta $j_a = 2, j_b = 1$.

Formally, under the considered approximation the double-mode lasing at the transition $j_a = j_b = 1$ with orthogonal polarization of the modes must be stable for infinitely small intermode distances. However, the input of the combination tones was not taken into account in calculation of polarization of the medium. It will be demonstrated in the next Section that they determine the lower boundary of the stability of double-mode lasing in such lasers. The same conclusions follow from the theoretical description of the double-mode lasing for the orthogonal polarization of the modes in lasers with the angular momenta of the working levels $j_a = 0, j_b = 1$. These theoretical results based on the model of the depolarizing collisions were verified in the experimental works [27, 28].

Consider the general case of the optically allowed transitions that can be of two types: without changes of the angular momentum and with its change by unity: $j_a = j_b = j$ and $j_a = j + 1, j_b = j$ for $j > 1$ (transitions $j_a = j, j_b = j + 1$ are considered in a similar way). Recall that we considered already the particular cases $j_a = 0, j_b = 1, j_a = j_b = 1$ and $j_a = 2, j_b = 1$.

For the linear orthogonal polarizations and arbitrary values of the angular momenta of the working levels from (6.3.13) we have:

$$\beta(0) - \theta(0) \sim g_b^{(2)} / \gamma_b^{(2)} - g_b^{(1)} / \gamma_b^{(1)} + g_a^{(2)} / \gamma_a^{(2)} - g_a^{(1)} / \gamma_a^{(1)}. \tag{6.3.19}$$

The relationships (6.3.18) and (6.3.14) follow from (6.3.19) in the particular cases $j_a = j_b = 1$ and $j_a = 2, j_b = 1$. It is seen that the sign of the difference $\beta(0) - \theta(0)$ that determines the stability of the regime of double-mode lasing is determined by the relationship between the rates of relaxation of orientations and alignment of the levels. The results presented in Chapter 1 allow one to determine the sign in (6.3.19) for the transition $j_a = j_b = 2$ also. In this case from (6.2.11) we obtain:

$$g_a^{(1)} = g_b^{(1)} = 1/180, \; g_a^{(2)} = g_b^{(2)} = 7/300.$$

Using this and the fact that the ratio $\gamma^{(1)}_{a(b)}/\gamma^{(2)}_{a(b)}$ changes from 1 to 0.78 with the increase of pressure we can conclude that the difference $\beta(0) - \theta(0)$ is positive. Therefore, the boundary of the intermode distances under which the double-mode lasing is possible is determined for this transition by the input of the combination tones into the polarization of the active medium like in the case of the transition $j_a = j_b = 1$.

In the general case for the transitions $j_a = j_b = j$ from (6.2.11) we have:

$$g^{(2)}_{a(b)}/g^{(1)}_{a(b)} = (2j-1)(2j+3)/5. \qquad (6.3.20)$$

As the ratio $\gamma^{(1)}_{a(b)}/\gamma^{(2)}_{a(b)}$ must be close to unity the difference $\beta(0) - \theta(0)$ is positive for these transitions and the lower boundary of the intermode distances under which the double-mode lasing is possible is determined by the input of the combination tones into the polarization of the active medium.

Consider now the transitions of the type $j_a = j + 1$, $j_b = j$ for $j > 1$. It follows from (6.2.11) that

$$g^{(2)}_{a}/g^{(1)}_{a} = (2j+5)/5(2j+1),\ g^{(2)}_{b}/g^{(1)}_{b} = (2j-1)/5(2j+3). \qquad (6.3.21)$$

In contrast to (6.3.20) these quantities tend to a finite value of 1/5 if j increases. That is why the sign of the difference $\beta(0) - \theta(0)$ depends on the ratios $\gamma^{(2)}_{a}/\gamma^{(1)}_{a}$ and $\gamma^{(2)}_{b}/\gamma^{(1)}_{b}$. One can expect these ratios to be close to unity. Then the critical intermode distance at these transitions is given by the equation (6.3.10). Note that the difference between the types of the transitions holds also for $j >> 1$.

6.4 The Influence of Combination Tones on the Stability of the Stationary Double-Mode Regime of Lasing

In the case of generation of two and more modes the nonlinear part of polarization of the medium contains the terms at the frequencies that do not coincide with the frequencies of the considered modes and that are known as combination tones [1]. In the case of double-mode generation of the orthogonally polarized modes an additional term from the combination tone at the frequency $2\omega_2 - \omega_1$ appears in the expression (6.2.9) that determines the nonlinear part of polarization of the medium at the frequency ω_1 of the first mode. In a similar way an input from the combination tone at the frequency $2\omega_1 - \omega_2$ is added to the nonlinear part of polarization of the medium at the frequency ω_2. Therefore, the equations for the slowly varying envelopes (6.2.15) and (6.2.16) can be rewritten with regard to these additional terms in a way as follows:

$$\dot{E}_1 = \{\alpha_1(\omega_{10}) - \beta_1(\omega_{10})E_1^2 - \theta_{12}(\omega_{10}, \omega_{20})E_2^2 - [\chi_{12}\cos(2\psi) + \\ + \xi_{12}\sin(2\psi)]E_2^2\}E_1, \tag{6.4.1}$$

$$\dot{E}_2 = \{\alpha_2(\omega_{20}) - \beta_2(\omega_{20})E_2^2 - \theta_{21}(\omega_{20}, \omega_{10})E_1^2 - [\chi_{21}\cos(2\psi) - \\ - \xi_{21}\sin(2\psi)]E_1^2\}E_2, \tag{6.4.2}$$

where ψ is the phase difference of the generated modes:

$$\psi = (\omega_2 - \omega_1)t + \phi_2 - \phi_1. \tag{6.4.3}$$

The coefficients χ_{12}, ξ_{12}, χ_{21}, and ξ_{21}, that characterize the input of the combination tones are determined in the following way:

$$\chi_{12}(\omega_{20}, \omega_{10}) + i\xi_{12}(\omega_{20}, \omega_{10}) = 2\pi\omega_0 n|d|^4 \int d\mathbf{v}\, F(\mathbf{v})[\gamma - i(\omega_{20} - \omega_{12} - kv_z)]^{-1} \times$$

$$\times \sum (-1)^\kappa\, c_j^{(\kappa)} \{(\gamma_j^{(\kappa)} + i\omega_{12})^{-1}(\frac{1}{\gamma - i(\omega_{20} + kv_z)} +$$

$$+ \frac{1}{\gamma + i(\omega_{10} + kv_z)}) + (\gamma_j^{(\kappa)} + i\omega_{12} + 2ikv_z)^{-1}(\frac{1}{\gamma - i(\omega_{20} - kv_z)} +$$

$$+ \frac{1}{\gamma + i(\omega_{10} + kv_z)}) + (n_2/n)(\gamma_j^{(\kappa)} + i\omega_{12})^{-1}(\frac{1}{\gamma - i(\omega_{20} - kv_z)} +$$

$$+ \frac{1}{\gamma + i(\omega_{10} - kv_z)})\}, \tag{6.4.4}$$

where the summation is over $\kappa = 0, 1, 2$ and $j = a, b$, the coefficients $c_j^{(\kappa)}$ are given in Table 1, and the ratio n_2/n is determined by the expression (6.2.10). The expression for $\chi_{21}(\omega_{20}, \omega_{10}) + i\xi_{21}(\omega_{20}, \omega_{10})$ is obtained from (6.4.4) by means of the replacement under the integral of ω_{10} by ω_{20}, ω_{20} by ω_{10}, and ω_{12} by $-\omega_{12}$. The equations (6.4.1) and (6.4.2) must be supplemented with the equation for the difference of phase of the modes:

$$\dot{\psi} = \Omega_{21} + \sigma_2 - \sigma_1 + (\rho_2 - \tau_{12})E_2^2 - (\rho_1 - \tau_{21})E_1^2 + (\chi_{21}E_1^2 + \\ + \chi_{12}E_2^2)\sin(2\psi) + (\xi_{21}E_1^2 - \xi_{12}E_2^2)\cos(2\psi), \tag{6.4.5}$$

where $\Omega_{21} = \Omega_2 - \Omega_1$ is the difference of the mode frequencies for the empty cavity. The coefficients σ_2, σ_1, ρ_2, ρ_1, τ_{21}, and τ_{12} are defined in Section 6.2.

Before going to the discussion of the system of equations (6.4.1), (6.4.2), and (6.4.5) note the following. Dienes [29] studied the combination tones and stated that the combination tones are absent at the transitions with $j_a = 0$, $j_b = 1$, and $j_a = j_b = 1$. This statement is not true because in [29] the depolarizing atomic collisions were not taken into account. Indeed, if one neglects the difference between $\gamma_j^{(1)}$ and $\gamma_j^{(2)}$ ($j = a, b$) that results from the depolarizing atomic collisions in the expression (6.4.4) then for the transitions with such angular momenta we obtain $\chi_{12} = \xi_{12} = 0$. Our results that take into account the depolarizing collisions show that the input of the combination tones into the polarization of the medium is different from zero. This input results at such transitions in the phenomenon of trapping of the modes (self-mode-locking). In this case the double-mode regime is changed by a single-frequency one for the intermode distances that are smaller than a certain value, and the difference of phases does not depend on time:

$$\dot{\psi} = 0. \tag{6.4.6}$$

Let Ω_{12cr} be the upper boundary of the intermode distances below which the trapping of the modes takes place (assume for definiteness that $\Omega_{12} = \Omega_1 - \Omega_2 > 0$). If $\Omega_{12} < \Omega_{12cr}$, then single-frequency lasing takes place, and the polarization of the laser light is elliptical in the general case. The quantities E_1 and E_2 lose the meaning of the amplitudes of modes. They must be interpreted now as the projections of the electric field intensity generated under the regime of trapping of the radiation on the directions determined by the anisotropic element of the cavity. Assuming that $\Omega_{12} < \Omega_{12cr}$, rewrite the system of equations (6.4.1), (6.4.2), and (6.4.5) for the case of the stationary lasing under the regime of trapping for E_1 and E_2 different from zero in a way as follows:

$$\beta_0 E_1^2 + [\theta_0 + \chi\cos(2\psi) + \xi\sin(2\psi)]E_2^2 = \alpha_1(\omega_{10}), \tag{6.4.7}$$

$$\beta_0 E_2^2 + [\theta_0 + \chi\cos(2\psi) - \xi\sin(2\psi)]E_1^2 = \alpha_2(\omega_{10}), \tag{6.4.8}$$

$$\begin{aligned}&\chi(E_1^2 + E_2^2)\sin(2\psi) + \xi(E_1^2 - E_2^2)\cos(2\psi) + \\ &+ (\tau - \rho)(E_1^2 - E_2^2) = \Omega_{12},\end{aligned} \tag{6.4.9}$$

where the following notations are introduced:

$$\beta_0 = \beta_1(\omega_{10}) = \beta_2(\omega_{10}), \quad \theta_0 = \theta_{12}(\omega_{10}, \omega_{10}) = \theta_{21}(\omega_{10}, \omega_{10}),$$

$$\chi = \chi_{12}(\omega_{10}, \omega_{10}) = \chi_{21}(\omega_{10}, \omega_{10}), \quad \xi = \xi_{12}(\omega_{10}, \omega_{10}) = \xi_{21}(\omega_{10}, \omega_{10}),$$

$$\tau = \tau_{12}(\omega_{10}, \omega_{10}) = \tau_{21}(\omega_{10}, \omega_{10}), \quad \rho = \rho_1(\omega_{10}) = \rho_2(\omega_{10}),$$

$$\alpha_2(\omega_{10}) = \alpha_1(\omega_{10}) + (\omega_0/2Q)(\Delta Q/Q). \tag{6.4.10}$$

Note that in the case of trapping single-frequency lasing takes place $\omega_{10} = \omega_{20}$, but the frequencies of the modes of the empty cavity are not equal to each other: $\Omega_1 \neq \Omega_2$. ΔQ stands for the difference of the quality factors of the cavity $Q_2 - Q_1$ where Q_2 and Q_1 are the quality factors for the modes with different polarizations. It is assumed that $\Delta Q \ll Q$ ($Q \approx Q_1 \approx Q_2$).

Consider only the case of tuning to the center of the transition band $\omega_{10} = 0$ and equal quality factors $\Delta Q = 0$. Then $\xi = 0$, and from (6.4.7) and (6.4.8) we obtain:

$$E_1^2 = E_2^2 = \alpha_1(\omega_{10})/[\beta_0 + \theta_0 + \chi\cos(2\psi)]. \tag{6.4.11}$$

Substituting (6.4.11) into (6.4.9) we arrive at:

$$2\chi\alpha_1\sin(2\psi)/[\beta_0 + \theta_0 + \chi\cos(2\psi)] = \Omega_{12}. \tag{6.4.12}$$

It follows that such a regime of lasing is possible if only the value of Ω_{12} is not larger than the value Ω_{12cr}, where

$$\Omega_{12cr} = 2\chi\alpha_1[(\beta_0 + \theta_0)^2 - \chi^2]^{-1/2}. \tag{6.4.13}$$

Estimate the value Ω_{12cr} for the transition with angular momenta of the working levels $j_a = 0$, $j_b = 1$. Under the extreme cases of both homogeneously and inhomogeneously broadened band we arrive at one and the same expression:

$$\Omega_{12cr} = (\omega_0/2Q)(\eta - 1)(1/\gamma_b^{(2)} - 1/\gamma_b^{(1)})[2/\gamma_a^{(0)} + 2/(3\gamma_b^{(0)}) + 5/(6\gamma_b^{(2)}) + $$

$$+ 1/(2\gamma_b^{(1)})]^{-1}. \tag{6.4.14}$$

Thus, Ω_{12cr} really equals zero if one neglects the depolarizing atomic collisions. The experimental results on the phenomenon of trapping for the case of generation of two modes with linear mutually orthogonal polarizations [27] are in good agreement with (6.4.13).

Discuss now the role of the combination tones in double-mode lasing of the gas lasers at the transitions with the angular momenta $j_a = 2$, $j_b = 1$. It was demonstrated in the previous Section that for such transitions and for orthogonal polarizations of the modes the double-mode regime is changed by the single-mode one due to the competition (interaction) of the modes if the intermode distance is smaller than ω_{12cr}. Here we estimate the influence of the combination tones for $\omega_{12} > \omega_{12cr}$, i.e., in the range of stability of the double-mode lasing.

With regard to the combination tones the equations for amplitudes of the modes appear to be related with the equation for the phase difference ψ. As the value of ψ is time-dependent out of the trapping regime, the equations (6.4.1) and (6.4.2) have the solution only for the time-dependent amplitudes E_1 and E_2. Here we search for an approximate solution that is valid under the limitations mentioned below. Introduce the notation as follows:

$$E_1^2 = I_{10} + \delta I_1(t), \quad E_2^2 = I_{20} + \delta I_2(t), \tag{6.4.15}$$

where I_{10} and I_{20} are the stationary solutions of the system when the combination tones are neglected (see (6.3.1)). Assuming that the influence of combination tones is a minor correction:

$$|\delta I_1(t)| << I_{10}, \qquad |\delta I_2(t)| << I_{20}, \tag{6.4.16}$$

we write down the linearized equations for determination of $\delta I_1(t)$ and $\delta I_2(t)$:

$$\tfrac{1}{2}\,\delta \dot{I}_1 = -\beta_1 I_{10}\delta I_1 - \theta_{12} I_{10}\delta I_2 - I_{10}I_{20}[\chi_{12}\cos(2\psi) + \xi_{12}\sin(2\psi)], \tag{6.4.17}$$

$$\tfrac{1}{2}\,\delta \dot{I}_2 = -\beta_2 I_{20}\delta I_2 - \theta_{21} I_{20}\delta I_1 - I_{10}I_{20}[\chi_{21}\cos(2\psi) - \xi_{21}\sin(2\psi)]. \tag{6.4.18}$$

The quantity ψ in these equations can be replaced by an approximate expression [24]:

$$\psi(t) \approx (\Omega_{21} + \sigma_2 - \sigma_1)t + \psi_0, \tag{6.4.19}$$

where ψ_0 is a constant. With regard to this expression the system of equations (6.4.17), (6.4.18) can be considered as a system of the standard differential equations the solution of which can be obtained by means of the conventional methods. We are looking for only a partial oscillating solution that is related with the combination tones because the solution that depends on the initial conditions decays in time due to the stability of the double-mode regime. It can be presented as:

$$\delta I_1 = A_1\cos(2\psi) + A_2\sin(2\psi), \quad \delta I_2 = A_3\cos(2\psi) + A_4\sin(2\psi),$$

where the A_j ($j = 1, 2, 3, 4$) are constants. We do not present here the explicit expressions because they are quite cumbersome and restrict ourselves only to the approximate expression for the amplitudes $\delta I_{1m} = (A_1^2 + A_2^2)^{1/2}$ that is obtained for the case of the symmetrical position of the modes when $\chi_{12} = \chi_{21} = \chi$, and $\xi_{12} = -\xi_{21} = \xi$, and the intermode distance is much larger than the value of the relaxation characteristics of the working levels. In this case we obtain:

$$\delta I_{1m} = I_0\alpha(\Omega_{21} + \sigma_2 - \sigma_1)^{-1}(\beta + \theta)^{-1}(\chi^2 + \xi^2)^{1/2}, \qquad (6.4.20)$$

where $I_{10} = I_{20} = I_0$ for the symmetrical tuning of the modes.

Thus, the influence of the combination tones on the intensity of a mode for the range of the stable double-mode lasing manifests itself as an addition to the stationary (mean) value that oscillates at a frequency equal to the double intermode distance. The amplitude of oscillations increases with the decrease of the intermode distance. This effect is similar to a certain extent to the noises upon double-mode lasing that get larger in the vicinity of the critical intermode distance.

The approach described above is valid, of course, for the transitions with other values of the angular momenta of the working levels. As in the case described above the combination tones in the range of the stable double-mode lasing always lead to oscillating additions to the intensities of the modes. The intensity of these additional inputs depend on the intermode distance.

REFERENCES

1. W.E. Lamb, *Phys. Rev.*, **134A**: 1429–1450 (1964).
2. S.G. Rautian, *Trudy FIAN*, **43**: 3–115 (1968).
3. M.I. D'yakonov, V.I. Perel', *Sov. Phys. JETP*, **58**: 1090–1097 (1970).
4. I.M. Beterov, Yu.A. Matyugin, S.G. Rautian, *et al.*, *Sov. Phys. JETP*, **58**: 1243–1258 (1970).
5. B.J. Feldman, M.S. Feld, *Phys. Rev.*, **A1**: 1375–1396 (1970).
6. H.K. Holt, *Phys. Rev.*, **A2**: 233–249 (1970).
7. H. Greenstein, *Phys. Rev.*, **175**: 438–452 (1968).
8. A.P. Kazantsev, S.G. Rautian, G.I. Surdutovich, *Sov. Phys. JETP*, **54**: 1409–1421 (1968).
9. M.I. D'yakonov, V. I. Perel', *Opt. Spektrosk.*, **20**: 472–480 (1966).
10. R. Salomaa, S. Stenholm, *Phys. Rev.*, **A8**: 2695–2726 (1973).
11. R.L. Fork, M.A. Pollack, *Phys. Rev.*, **139A**: 1408–1414 (1965).
12. S.G. Zeiger, E.E. Fradkin, *Opt. Spektrosk.*, **21**: 386–393 (1966).
13. M. Sargent, III, W.E. Lamb (Jr.), R.L. Fork, *Phys. Rev.*, **164**: 436–467 (1967).
14. Yu.L. Klimontovich, P.S. Landa, E.G. Lariontsev, *Sov. Phys. JETP*, **52**: 1616–1631 (1967).

15. B.L. Zhelnov, V.S. Smirnov, A.P. Fadeev, *Opt. Spektrosk.*, **28**: 744–756 (1970).
16. S.G. Zeiger, E.E. Fradkin, P.P. Filatov, Opt. *Spektrosk.*, **26**: 622–631 (1969).
17. F. Aranowitz, *Phys. Rev.*, **139A**: 635–646 (1965).
18. N.G. Basov, M.A. Gubin, V.V. Nikitin, *et al.*, *JETP Lett.*, **15**: 525–28 (1972).
19. G.I. Kozin, N.A. Konovalov, E.S. Nikulin, *et al.*, *Sov. Phys. JTP*, **43**: 1781–1782 (1973).
20. F. Zernike, J.E. Midwinter, *Applied Nonlinear Optics: Basics and Applications* (New York: Wiley, 1973).
21. M. Schubert, B. Wilhelmi, *Nonlinear Optics and Quantum Electronics* (New York: Wiley, 1986).
22. M.A. Gubin, G.I. Kozin, E.D. Protsenko, *Opt. Spektrosk.*, **36**: 567–571 (1974).
23. I.I. Sobel'man, *Atomic Spectra and Radiative Transitions* (Berlin: Springer, 1979).
24. V.M. Yermachenko, *Theory of Depolarizing Collisions in Gas Lasers and Amplifiers* (Moscow: Doctor of Phys. and Math. Sci. Dissertation, 1982) (in Russian).
25. Yu.A. Vdovin, S.A. Gonchukov, M.A. Gubin, *et al.*, *Vliyanie atomnykh stolknovenii i pleneniya rezonansnogo izlucheniya na kharakteristiki gazovykh lazerov* (*The influence of atomic collisions and imprisonment of the resonance radiation on the parameters of gas lasers*) (Moscow: FIAN, Preprint No. 116, 1972) (in Russian).
26. Yu.A. Vdovin, M.A. Gubin, V.M. Yermachenko, *et al.*, *Kvantovaya Elektronika*, No. 4 (16), 35 (1973).
27. S.A. Gonchukov, V.M. Yermachenko, V.N. Petrovskii, *et al.*, *Sov. Phys. JETP*, **73**: 462–464 (1977).
28. S.A. Gonchukov, V.M. Yermachenko, V.N. Petrovskii, *et al.*, *Zh. Prikl. Spektrosk.*, **26**: 626–632 (1977).
29. A. Dienes, *Phys. Rev.*, **174**: 400–426 (1968).

Chapter 7

INTERACTION OF STRONG AND WEAK RUNNING WAVES IN A RESONANT GAS MEDIUM

7.1 The Gain of a Weak Wave Passing through a Medium Saturated with a Strong Wave

Many methods of nonlinear laser spectroscopy employ the fact that the parameters of the detected nonlinear electromagnetic signals are determined by the homogeneous width of the relevant spectral line and the widths of the levels involved in the transition under study under conditions when the Doppler line width is much greater than the homogeneous line widths. An extensive bibliography on spectroscopic applications of lasers is provided by several review papers and monographs (e.g., see [1, 2]).

In this chapter, we consider the method of using a weak electromagnetic wave to probe a gas medium saturated with a strong field. Various aspects of this approach have been thoroughly investigated theoretically [3–7]. The main features of this method can be described in the following way. A sufficiently strong electromagnetic wave passes through a layer of an amplifying (absorbing) medium. The frequency of this wave ω_1 is stabilized, for example, against the central frequency ω_0 of the transition under study [7]. Simultaneously a weak wave propagates through this medium in the same direction as the strong wave or in the opposite direction. The frequency of the weak wave ω_2 is varied within some interval around the frequency of the strong field. The detected frequency dependence of the intensity of the transmitted weak field displays weak resonances whose parameters are related to the homogeneous width of the relevant spectral line and the widths of the levels of the considered transition.

We can qualitatively understand the appearance of such resonances in the following way. The strong wave changes the electrooptical characteristics of a medium at different frequencies, including the frequency of the weak probing wave. In other words, we deal with a high-frequency electrooptical effect of the second order [8, 9], when the dielectric constant of the medium at the frequency of the weak wave can be written as

$$\varepsilon(\omega_2) = \varepsilon^{(1)}(\omega_2) + 4\pi\,\tilde{\kappa}^{(3)}(\omega_2, \omega_1)\,|\boldsymbol{E}_1|^2\,, \tag{7.1.1}$$

where $\varepsilon^{(1)}(\omega_2)$ is the conventional dielectric constant associated with the linear susceptibility, $\tilde{\kappa}^{(3)}(\omega_2, \omega_1)$ are the relevant components of the tensor of the nonlinear susceptibility of the medium, and $\boldsymbol{E}_1$ is the slowly varying amplitude of the strong wave. Expressions of the form (7.1.1) can be derived in a natural way in nonlinear electrodynamics, when we include not only linear, but also nonlinear terms of low orders in the expansion of the polarization vector of a medium $\boldsymbol{P}$ in the strength of the electric field $\boldsymbol{E}$. Specifically, in the case of a gas medium, this relation can be formally represented as

$$\boldsymbol{P} = \tilde{\kappa}^{(1)}\boldsymbol{E} + \tilde{\kappa}^{(3)}E^2\boldsymbol{E},$$

where $\tilde{\kappa}^{(1)}$ and $\tilde{\kappa}^{(3)}$ are the tensors of the linear and nonlinear susceptibilities of the medium. Therefore, the gain (absorption coefficient) of a weak wave can be written as

$$g(\omega_2) = \alpha(\omega_2) - \theta_{21}(\omega_2, \omega_1)|\boldsymbol{E}_1|^2, \tag{7.1.2}$$

where $\alpha(\omega_2)$ is the conventional linear gain (for definiteness, we will consider the case of an amplifying medium), which is related to the imaginary part of $\tilde{\kappa}^{(1)}$. The second term describes the saturation effect, i.e., the decrease in additional populations in the levels involved in the transition under study due to the interaction with the field. Note that the coefficient $\theta_{21}(\omega_2, \omega_1)$ is proportional to the imaginary part of $\tilde{\kappa}^{(3)}$. The difference of the coefficients $\alpha(\omega_2)$ and $\theta_{21}(\omega_2, \omega_1)$ considered in this Chapter from analogous quantities introduced in Chapter 6 is associated with the fact that, while Chapter 6 considered standing electromagnetic waves, we deal with running electromagnetic waves in this chapter.

In the limiting case of an inhomogeneously broadened line, the linear gain α is subject to considerable changes when the frequency of the wave changes by a value on the order of the Doppler width ku, while the changes in the cross-saturation coefficient θ_{21} become noticeable when the frequency changes by a value on the order of γ which is small as compared with ku. Indeed, a high efficiency of interaction with the field is achieved for the atoms that meet the resonance condition

$$|\omega_j - \omega_0 - \boldsymbol{k}\mathbf{v}| \le \gamma, \quad (j = 1, 2). \tag{7.1.3}$$

Therefore, the strong field considerably changes the populations of the levels involved in the transition under study for those atoms whose velocities satisfy inequality (7.1.3). Consequently, the weak field may detect this change when it interacts with this group of atoms, i.e., when $|\omega_{12}| \le \gamma$ for copropagating waves and $|\omega_1 + \omega_2 - 2\omega_0| \le \gamma$ for counterpropagating waves. Because of this reason, the cross-saturation coefficient θ_{21} features a resonance structure with a half-width equal to γ.

In the case when the strong and weak waves propagate in the same direction, the function $\theta_{21}(\omega_2, \omega_1)$ displays an even narrower resonance, whose width is determined by the widths of the levels involved in the transition under study. The appearance of such a resonance is due to the interference action of the strong and weak fields on the populations in the working levels. Indeed, the change in the population (more rigorously, the change in the density matrix) of a level under the action of the field is proportional to the square of the matrix elements of the field–atom interaction operator,

$$|\boldsymbol{d}_{ab}\boldsymbol{E}|^2, \tag{7.1.4}$$

where

$$\boldsymbol{E} = \boldsymbol{E}_1 \exp(i\omega_1 t + ik_1 z) + \boldsymbol{E}_2 \exp(i\omega_2 t + ik_2 z) + \text{c.c.} \tag{7.1.5}$$

Expression (7.1.4) including the field (7.1.5) involves, along with time-independent terms of the form $(\boldsymbol{d}_{ab}\boldsymbol{E}_1)(\boldsymbol{d}_{ba}\boldsymbol{E}_1{}^*)$, oscillatory terms of the form

$$(\boldsymbol{d}_{ba}\boldsymbol{E}_1{}^*)(\boldsymbol{d}_{ab}\boldsymbol{E}_2)\exp(-\, i\omega_{12} t - i(k_1 - k_2)z) + \text{c.c.}$$

These terms provide a considerable contribution to θ_{21} if they do not change substantially within the lifetime of the excited state, i.e., if

$$|\omega_{12}| \le \gamma_a^{(0)}, \gamma_b^{(0)}. \tag{7.1.6}$$

We emphasize that, in the case of copropagating waves, we deal with a nearly exact compensation of Doppler shifts in the interaction of an atom with quanta of the first and second fields. Therefore, the interference resonance in such a situation is not masked by the thermal motion of atoms, as it would be the case with counterpropagating waves, when no shift compensation is observed.

The coefficients $\alpha(\omega_2)$ and $\theta_{21}(\omega_2, \omega_1)$ in the case of running waves are calculated in the same way as it was done in the previous chapter for standing waves. To illustrate this, we present the expression for the coefficient $\theta_{21}(\omega_2, \omega_1)$ derived for the case of an inhomogeneously broadened line and copropagating waves with allowance for depolarizing collisions:

$$\theta_{21}(\omega_2, \omega_1) = A|d|^4(N_b - N_a)\Big[\frac{4\gamma^2}{(4\gamma^2 + \omega_{12}^2)}\sum b_j^{(\kappa)}/\gamma_j^{(\kappa)} + \\ + \frac{2\gamma}{(4\gamma^2 + \omega_{12}^2)}\sum c_j^{(\kappa)}\frac{(2\gamma_j^{(\kappa)}\gamma - \omega_{12}^2)}{(\gamma_j^{(\kappa)2} + \omega_{12}^2)}\Big], \tag{7.1.7}$$

where we introduced A to denote such factors as ku, π, etc., which are of no importance for our further analysis. The numerical factors $b_j^{(\kappa)}$ and $c_j^{(\kappa)}$, which depend on the polarizations of waves and the angular momenta of the levels involved in the transition under consideration, have been defined in the previous chapter for the cases of orthogonal and parallel linear polarizations of pump waves. Tables 7.1, 7.2, and 7.3 summarize the coefficients $b_j^{(\kappa)}$ and $c_j^{(\kappa)}$ for circularly polarized waves where polarization vectors rotate in opposite directions and the angular momenta of the levels are equal to $j_b = 1, j_a = 0, 1, 2$ ($c_j^{(0)} = c_j^{(1)} = 0$). For circularly polarized waves where the polarization vectors rotate in the same direction, we have $b_j^{(\kappa)} = c_j^{(\kappa)}$ and the quantities $b_j^{(\kappa)}$ can be obtained from the values summarized in Tables 7.1, 7.2, and 7.3 by multiplying by $(-1)^\kappa$.

Table 7.1 (transition $j_b = 1, j_a = 2$)

$b_b^{(0)}$	$b_b^{(1)}$	$b_b^{(2)}$	$b_a^{(0)}$	$b_a^{(1)}$	$b_a^{(2)}$	$c_b^{(2)}$	$c_a^{(2)}$
1/27	– 1/72	1/540	1/45	– 1/40	7/1800	1/900	7/300

Table 7.2 (transition $j_b = 1, j_a = 1$)

$b_b^{(0)}$	$b_b^{(1)}$	$b_b^{(2)}$	$b_a^{(0)}$	$b_a^{(1)}$	$b_a^{(2)}$	$c_b^{(2)}$	$c_a^{(2)}$
1/27	– 1/72	1/216	1/27	– 1/72	1/216	1/36	1/36

Table 7.3 (transition $j_b = 1, j_a = 0$)

$b_b^{(0)}$	$b_b^{(1)}$	$b_b^{(2)}$	$b_a^{(0)}$	$b_a^{(1)}$	$b_a^{(2)}$	$c_b^{(2)}$	$c_a^{(2)}$
1/27	– 1/18	1/54	1/9	0	0	1/9	0

The first term in (7.1.7) features a resonance structure, which is due to the change in the distribution of atoms in the considered levels under the action of the strong field only. The width of this resonance is determined by the homogeneous width of the spectral line, and its magnitude depends on the relaxation parameters of the levels and polarizations of the fields. When inequality (6.3.15) is satisfied, the second term in (7.1.7), which is responsible for the interference action of the fields on the distribution of atoms in the working levels, displays even narrower resonances. A variation in the polarizations of the fields changes not only the magnitude, but also the width of such resonances. Note that, when the strong and weak waves have the same polarization, the term with $\kappa = 0$ provides a dominant contribution to the interference resonance, while when these waves have different polarizations, this contribution vanishes, because $c_j^{(0)} = 0$ ($j = a, b$). Since the quantity $\gamma_j^{(0)}$ displays only weak changes (only due to inelastic collisions) with the growth in the gas pressure and the quantities $\gamma_j^{(1)}$ and $\gamma_j^{(2)}$ ($j = a$, b) grow much faster due to depolarizing collisions (see Chapter 1), the width of the interference resonance in the case of pump waves with the same polarization remains virtually unchanged as the gas pressure grows. In the case of pump waves with different polarizations, the interference resonance broadens, and its magnitude decreases. These effects were observed in experiments [7] in the case when polarization vectors of the strong and weak case of pump waves with different polarizations, the interference resonance pump waves were parallel or orthogonal to each other (transition in neon with $\lambda = 3.3922$ μm).

In principle, studying the width of the interference resonance for different combinations of polarizations of the strong and weak waves, one can obtain information concerning the spectroscopic parameters $\gamma_j^{(\kappa)}$ ($j = a, b, \kappa = 0, 1, 2$) of the levels involved in the transition under investigation. The same information can be extracted from the investigation of the total magnitude of both resonances, which is determined by the formula

$$\theta_{21}(\omega_2 = \omega_1) = A|d|^4(N_b - N_a)\Sigma(b_j^{(\kappa)} + c_j^{(\kappa)})/\gamma_j^{(\kappa)}. \tag{7.1.8}$$

From the experimental point of view, such an approach may be more convenient, as it allows one to employ also counterpropagating waves to extract the information concerning the spectroscopic parameters $\gamma_j^{(\kappa)}$ ($j = a, b, \kappa = 0, 1, 2$). As mentioned above, in the case of counterpropagating waves, the interference resonance is masked by the thermal motion of atoms. Therefore, the depth of the dip in this case is described by expression (7.1.8), where we should set $c_j^{(\kappa)} = 0$.

The authors of [10] employed this technique to determine the spectroscopic parameters of the levels involved in $3s_2 - 3p_4$ and $3s_2 - 3p_2$ transitions of neon. The frequencies of the strong and weak waves were stabilized at the central frequencies of these transitions. The depth of the dip arising in the weak signal transmitted through the amplifying medium was measured for different polarizations of the waves. The parameters of the cell with an amplifying medium were chosen in such a way as to ensure the applicability of expression (7.1.2) for the gain of the weak wave and the proportionality of the output signal to this coefficient. The results of measurements were processed with the use of an expression more accurate than (7.1.8). The necessity of improving the accuracy of data processing in this case was associated with two circumstances. First, the expression for the depth of the dip was written for arbitrary γ/ku, since the spectral lines related to the transitions under consideration do not display a clearly pronounced inhomogeneous broadening. The second circumstance is related to resonant radiation trapping, since the $3s_2$ level is dipole-coupled with the ground state of neon. All the spectroscopic characteristics of $3s_2$, $3p_2$, and $3p_4$ levels in [10] were measured in experiments with counterpropagating waves. Since the amplifying medium contained a mixture of neon and helium, the authors of these experiments were able to separately determine parameters of depolarizing collisions of excited neon atoms with neon and neon with helium by independently varying the partial pressures of neon and helium. The experimental ratios of orientation and alignment relaxation rates were found to agree well with theoretical predictions of Chapter 1.

7.2 Amplification of a Weak Wave through a Transition Adjacent to a Strong Wave

Relaxation parameters of a single level can be obtained by the investigation of the amplification of a weak wave through a transition adjacent to a strong wave. Different methods of three-level spectroscopy were described in [2]. A comprehensive theory of a three-level quantum amplifier was presented in [11], where the nonlinear gain was calculated for a weak wave resonant to a transition $c - b$ with angular momenta of the levels equal to j_c and j_b, respectively, in the case when a strong wave resonant to a transition $a - b$ (with the angular momentum of level a equal to j_a) passes through this medium. Calculations performed with allowance for depolarizing atomic collisions for different polarizations of the strong and weak waves demonstrate the possibility of determining all the polarization parameters of the common level b, including the parameters defining the population, orientation, and alignment relaxation times of this level.

Along with the elements of the density matrix considered above, now, we should calculate the gain of the weak wave taking into account the elements related to the optically allowed transition $c - b$ $\rho^{(cc)}_{mm_1}(\boldsymbol{r},\mathrm{v},t)$ and $\rho^{(cb)}_{mm_1}(\boldsymbol{r},\mathrm{v},t)$, as well as the elements of the density matrix $\rho^{(ac)}_{mm_1}(\boldsymbol{r}, \mathrm{v}, t)$ for the optically forbidden transition $a - c$ transition. Equations governing these elements in the model of depolarizing collisions are presented in Chapter 2. We assume that the field $\boldsymbol{E}$ acting on the medium includes two plane running waves with frequencies ω_1 and ω_2 [see (7.1.5)]. The frequency ω_1 is assumed to be resonant to the transition $a - b$, while the frequency ω_2 is assumed to be resonant to the transition $c - b$. For definiteness, the wave with frequency ω_1 is assumed to be strong, while the wave with frequency ω_2 is assumed to be weak. The central frequencies of optically allowed transitions $a - b$ and $c - b$ are denoted as $\omega_0^{(ab)}$ and $\omega_0^{(cb)}$. To simplify our analysis, we assume that the central frequency of the optically forbidden transition, which is equal to the modulus of the difference of these frequencies, considerably exceeds all the spectroscopic parameters involved in the problem.

The gain of a weak wave can be still written in the form (7.1.2). The cross-saturation coefficient $\theta_{21}(\omega_2, \omega_1)$ involved in this formula can be represented as a sum of two terms:

$$\theta_{21}(\omega_2, \omega_1) = \theta_{21}^{(1)}(\omega_2, \omega_1) + \theta_{21}^{(2)}(\omega_2, \omega_1). \tag{7.2.1}$$

Here, the quantity $\theta_{21}^{(1)}$, which describes the contribution of optically allowed transitions, is written (with an accuracy up to an insignificant factor) in the following form:

$$\theta_{21}^{(1)}(\omega_2, \omega_1) = |d|^2|d_1|^2(N_b - N_a)\Sigma(2\, d_b^{(\kappa)}/\gamma_b^{(\kappa)}) \int dv\, F(v)[\gamma_{cb}^2 + (\omega_{20} + k_2 v)^2]^{-1}[\gamma_{ab}^2 + (\omega_{10} + k_1 v)^2]^{-1}, \tag{7.2.2}$$

where the summation is performed in the index $\kappa = 0, 1, 2$. The quantities γ_{ab} and γ_{cb} in this expression stand for the homogeneous half-widths of spectral lines for optically allowed transitions, d is the reduced matrix element of the dipole moment operator for the transition $a - b$, d_1 is a similar quantity for the transition $c - b$, and ω_{10} and ω_{20} are the frequency detunings of the strong and weak waves from the central frequencies of the corresponding resonant transitions with allowance for atomic collisions. The quantities $d_b^{(\kappa)}$ are introduced to denote numerical factors depending on the angular momenta of the levels j_a, j_b, and j_c and polarizations of the waves. For example, for linearly polarized waves with parallel polarization vectors and levels with angular momenta $j_a = 2, j_b = j_c = 1$ we find that $d_b^{(0)} = 1/27$, $d_b^{(1)} = 0$, and $d_b^{(2)} = 1/270$, while for angular momenta $j_a = 2$, $j_b = 1$, and $j_c = 0$, we have $d_b^{(0)} = 1/27$, $d_b^{(1)} = 0$, and $d_b^{(2)} = 1/135$.

In the case of linearly polarized waves with orthogonal polarization vectors and levels with angular momenta $j_a = 2, j_b = j_c = 1$, these coefficients are equal to $d_b^{(0)} = 1/27$, $d_b^{(1)} = 0$, and $d_b^{(2)} = 1/540$, while for angular momenta $j_a = 2$, $j_b = 1$, and $j_c = 0$, we have $d_b^{(0)} = 1/27$, $d_b^{(1)} = 0$, and $d_b^{(2)} = -1/270$. Expression (7.2.2) is written for the case of copropagating waves. In the case of counterpropagating waves, we should change the sign of the quantity k_2.

The coefficient $\theta_{21}^{(2)}(\omega_2, \omega_1)$ can be expressed in terms of the contribution of the optically forbidden transition $a - c$:

$$\theta_{21}^{(2)}(\omega_2, \omega_1) = |d|^2|d_1|^2 \mathrm{Re} \int dv\, F(v)(\gamma_{cb} + i\omega_{20} + ik_2 v)^{-1}[(N_b - N_a)(\gamma_{ab} - i\omega_{10} - ik_1 v)^{-1} + (N_b - N_c)(\gamma_{cb} + i\omega_{20} + ik_2 v)^{-1}]\Sigma f^{(\kappa)}[\gamma_{ac}^{(\kappa)} - i\omega_{10} + i\omega_{20} - i(k_1 - k_2)v]^{-1}. \tag{7.2.3}$$

Summation in κ in (7.2.3) is performed over the values of this index equal to 0, 1, and 2. The quantities $\gamma_{ac}^{(\kappa)}$ ($|j_a - j_c| \le \kappa \le j_a + j_c$) stand for the relaxation parameters of the optical coherence matrix for the optically forbidden transition $a - c$. These parameters can be expressed in terms of the radiative widths of the levels a and c and the terms including depolarizing atomic collisions (see Chapter 1). The numerical factors $f^{(\kappa)}$ depend on the angular momenta of the levels j_a, j_b, and j_c and polarizations of the waves.

Specifically, for transitions with angular momenta $j_a = 2$, $j_b = j_c = 1$, we find that $f^{(0)} = f^{(1)} = 0$, and $f^{(2)} = 1/30$ in the case of parallel linear polarizations and $f^{(0)} = 0$, $f^{(1)} = 1/72$, and $f^{(2)} = 1/40$ in the case of orthogonal linear polarizations.

For angular momenta $j_a = 2$, $j_b = 1$, and $j_c = 0$, the nonvanishing component $f^{(2)}$ is equal to 2/45 in the case of parallel polarizations and $f^{(2)} = 1/30$ in the case of orthogonal polarizations of the pump waves.

Suppose that the frequencies of the light waves are stabilized at the centers of the corresponding transitions, i.e., $\omega_{10} = \omega_{20} = 0$. Then, measuring the depth of the dip in the gain for different polarizations of the strong and weak waves, one can extract the information concerning the spectroscopic parameters $\gamma_b^{(\kappa)}$ of the common level and the parameters $\gamma_{ac}^{(\kappa)}$ of the forbidden transition.

Vdovin *et al.* [12] considered adjacent transitions $3s_2 - 3p_4$ ($\lambda = 3.3922$ μm) and $3s_2 - 3p_2$ ($\lambda = 3.3912$ μm) to experimentally demonstrate the potential of the three-level approach for determining the quantities $\gamma_b^{(\kappa)}$. The results of these investigations were processed with the formulas presented above, where resonant radiation trapping was additionally included. Measurements have been performed in counterpropagating waves, when the contribution $\theta_{21}^{(2)}$ due to the forbidden transition is masked by the thermal motion of atoms.

To extract the information concerning the relaxation parameters of an optically forbidden transition, one should detect the difference of signals measured for co- and counterpropagating waves. The coefficient $\theta_{21}^{(1)}$ does not contribute to such a difference, and we have

$$\theta_{21}^{(2)(+)} - \theta_{21}^{(2)(-)} = 2(N_b - N_a)|d|^2|d_1|^2\Sigma(f^{(\kappa)}/\gamma_{ac}^{(\kappa)})(2\gamma +$$

$$+ \gamma_{ac}^{(\kappa)})^2 \int dv\, F(v)(kv)^2[\gamma_{ac}^{(\kappa)2} + (2kv)^2]^{-1}[\gamma^2 + (kv)^2]^{-1}. \tag{7.2.4}$$

In deriving (7.2.4), we employed several approximate equalities derived experimentally [13, 14]: $k_1 \approx k_2 = k$, $\gamma_{ab} \approx \gamma_{bc} = \gamma$, $N_b - N_a \approx N_b - N_c$. The plus index in this formula corresponds to the case of copropagating waves, while the minus sign stands for counterpropagating waves. In the limiting case of an inhomogeneously broadened line, the contribution of $\theta_{21}^{(2)(-)}$ vanishes, and we arrive at

$$\theta_{21}^{(2)(+)} - \theta_{21}^{(2)(-)} = (N_b - N_a)|d|^2|d_1|^2 \Sigma f^{(\kappa)}/(\gamma\ \gamma_{ac}^{(\kappa)}\ ku).$$

For the considered transitions, only the quantity $f^{(2)}$ remains nonvanishing for the waves with parallel polarizations. Consequently, in this case, we can obtain the information on one of the relaxation parameters, $\gamma_{ac}^{(2)}$, of the optically forbidden transition under study. For orthogonal polarizations, the coefficients $f^{(1)}$ and $f^{(2)}$ differ from zero. Therefore, knowing $\gamma_{ac}^{(2)}$, we can determine $\gamma_{ac}^{(1)}$. The difference between these parameters is associated with depolarizing collisions.

REFERENCES

1. W. Demtroder, *Phys. Rep.* **7**: 223–277 (1973).
2. V.S. Letokhov, V.P. Chebotayev, *Nonlinear Laser Spectroscopy* (Berlin: Springer, 1977).
3. A.I. Alekseyev, *Zh. Eksp. Teor. Fiz.*, **58**: 2064–2071 (1970).
4. A.P. Kol'chenko, A.A. Pukhov, S.G. Rautian, *et al.*, *Zh. Eksp. Teor. Fiz.*, **63**: 1173–1182 (1972).
5. S.G. Rautian, G.I. Smirnov, A.M. Shalagin, *Zh. Eksp. Teor. Fiz.*, **62**: 2097–2110 (1972).
6. Yu.A. Vdovin, V.M. Yermachenko, V.K. Matskevich, *Kvantovaya Elektronika*, **2**: 902–911 (1975).
7. Yu.A. Vdovin, V.M. Yermachenko, A.I. Popov. *et al.*, *Pis'ma Zh. Eksp. Teor. Fiz.*, **15**: 401–404 (1972).
8. F. Zernike, J.E. Midwinter, *Applied Nonlinear Optics: Basics and Applications* (New York: Wiley, 1973).
9. M. Schubert, B. Wilhelmi, *Nonlinear Optics and Quantum Electronics* (New York: Wiley, 1986).
10. I.P. Konovalov, E.D. Protsenko, *Kvantovaya Elektronika*, **3**: 1991–2004 (1976).
11. Th. Hansch, P. Toschek, *Z. Phys.* **236**: 213–244 (1970).
12. Yu.A. Vdovin, V.M. Yermachenko, I.P. Konovalov, *Kvantovaya Elektronika*, **4**: 1867–1872 (1977).
13. V.A. Balakin, I.P. Konovalov, E.D. Protsenko, *Kvantovaya Elektronika*, **2**: 1064–1068 (1975).
14. P.W. Murphy, *J. Opt. Soc. Am.*, **58**: 1200–1209 (1968).

Appendices

As mentioned in Chapter 3, some nonlinear coherent phenomena will be considered in the Appendices. Here, we will use an approach that differs from the one employed in this book. This approach is based on optical Bloch equations. We will consider two coherent phenomena. One of these phenomena arises in the case of stationary laser excitation, the second one is caused by a multipulse excitation of a resonant medium. These nonlinear phenomena (which are often referred to as transient effects) are characterized by spatial features associated with large sizes of a resonant medium as compared with the wavelength (λ) of incident light.

A.1 Optical Bloch Equations

In 1957, Feynmann *et al.* [1] proposed a theoretical approach that later allowed Pao [2] to derive optical Bloch equations. Recall that earlier Bloch equations [3] have been widely employed for the description of transient phenomena in radio-frequency spectroscopy [4, 5].

Let us briefly reproduce the analysis of [1]. Consider an ensemble of two-level particles, assuming that each particle is characterized by an energy splitting $\hbar\omega_0^j$ (j = 1, 2, ..., N, N is the number of active particles in the ensemble). Resonant electromagnetic radiation with a carrier frequency ω may transfer these particles into a superposition state $\Psi(t)$, which is related to the resonant stationary states Ψ_1 and Ψ_2 by the following expression:

$$\Psi(t) = a_1(t)\Psi_1 + a_2(t)\Psi_2, \tag{A.1.1}$$

where $a_1(t)$ and $a_2(t)$ are the coefficients whose moduli squared represent the probabilities to find a particle in the lower and upper energy states, respectively.

Following [1], we employ $a_1(t)$ and $a_2(t)$ to construct three functions (R_1, R_2, R_3), which will be considered as components of the vector $\boldsymbol{R}$ having the same functions in optics as the magnetic dipole in radio-frequency spectroscopy. These components are written as

$$R_1 = a_1a_2^* + a_2a_1^*, \quad R_2 = i(a_1a_2^* - a_2a_1^*), \quad R_3 = a_1a_1^* - a_2a_2^*. \qquad \text{(A.1.2)}$$

We can easily derive the equation governing the motion of this vector ***R*** by using the differential equations for the relevant coefficients that follow from the Schrödinger equation [6]:

$$i\hbar\frac{da_m}{dt} = (\hbar\omega_0/2)a_m + \sum_{k=1,2} V_{mk}\, a_k, \qquad \text{(A.1.3)}$$

where V_{mk} is the matrix element related to the operator of the interaction of particles with the external field for the transition between sublevels m and k (m, k = 1, 2), $V = -\boldsymbol{pE}$, $\boldsymbol{p}$ is the dipole moment operator of a particle, and $\boldsymbol{E}$ is the vector of the electric field in the electromagnetic wave. Note that the condition $V_{11} = V_{22} = 0$ is usually satisfied in optics. The equation for a_m^* can be written in a form similar to (A.1.3). Summing and subtracting these expressions, we derive the equations for the components R_η and the vector ***R***:

$$d\boldsymbol{R}/dt = [\omega_1, \boldsymbol{R}], \qquad \text{(A.1.4)}$$

where the vector ω_1 has the following components: $\omega_1 = (V_{12} + V_{21})/\hbar$, $\omega_2 = i(V_{12} - V_{21})/\hbar$, and $\omega_3 = \omega_0$. Note that the equation for the sum $(a_1a_1^* + a_2a_2^*)$, which represents the modulus of the vector ***R***, brings us to a physically clear result: the dependence on the time t is absent (obviously, the total probability to find a particle in one of the two states under consideration at any moment of time is equal to unity and is independent of time).

Following [7] and introducing new notations, we can rewrite differential equation (A.1.4) as

$$\boldsymbol{P}^j = p[\boldsymbol{x}\sigma_x + \boldsymbol{y}\sigma_y + \boldsymbol{z}\sigma_z]/\sqrt{2}\,, \qquad \text{(A.1.5)}$$

$$\boldsymbol{E}^j = \boldsymbol{x}\,E_x^j + \boldsymbol{y}\,E_y^j - \boldsymbol{z}\hbar\,\omega_0^j/\sqrt{2}\,p, \qquad \text{(A.1.6)}$$

where $\boldsymbol{x}$, $\boldsymbol{y}$, and $\boldsymbol{z}$ are the unit vectors of the laboratory system of coordinates; p is the modulus of the electric dipole moment related to the resonant transition;

$$E_x^j = \varepsilon_0 \sin(\omega t - kz - \varphi),\ E_y^j = \varepsilon_0 \cos(\omega t - kz - \varphi);$$

ε_0, ω, k, and φ are the amplitude of the electric field strength, the carrier frequency, the wave vector, and the phase of the pump electromagnetic wave, respectively; and σ_x, σ_y, and σ_z are the Pauli operators (which are related to the energy spin through a factor of 1/2: $R_1 = ½\sigma_x$, $R_2 = ½\sigma_y$, and $R_3 = ½\sigma_z$). The vectors $\boldsymbol{P}^j$ and $\boldsymbol{E}^j$ are referred to as the pseudoelectric dipole and the pseudoelectric field (called pseudoelectric due to the z component). As a result, equation (A.1.4) is reduced to

$$d\boldsymbol{P}^j/dt = \gamma_e[\boldsymbol{P}^j, \boldsymbol{E}^j], \tag{A.1.7}$$

where $\gamma_e = \sqrt{2}\, p/\hbar$ is the gyroelectric ratio. Note that equation (A.1.7) is similar to the equation of motion of a magnetic dipole $\boldsymbol{M}$ in a magnetic field $\boldsymbol{H}(t)$,

$$d\boldsymbol{M}^j/dt = \gamma_m[\boldsymbol{M}, \boldsymbol{H}], \tag{A.1.8}$$

where γ_m is the gyromagnetic ratio. Equation (A.1.8) is well known in radio-frequency spectroscopy. This analogy allows vector representations of transient processes developed in the radio-frequency range to be transferred to the optical range. Pao [2] employed this analogy to introduce, following Bloch [3], relaxation terms into equation (A.1.7):

$$d\boldsymbol{P}^j/dt = \gamma_e[\boldsymbol{P}^j, \boldsymbol{E}^j] + \boldsymbol{x}\, P_x^j/T_2 + \boldsymbol{y}\, P_y^j/T_2 + \boldsymbol{z}(P_z^0 - P_z^j)/T_1, \tag{A.1.9}$$

where $P_z^0 = \sqrt{2}\, p\,\mathrm{th}[\hbar\omega_0/(2k_BT)]$; k_B is the Boltzmann constant; T is the absolute temperature of the resonant medium; T_1 and T_2 are the longitudinal and transverse irreversible relaxation times, respectively; and $\boldsymbol{x}$, $\boldsymbol{y}$, and $\boldsymbol{z}$ are the unit vectors of the laboratory system of coordinates.

Equation (A.1.9) is usually written for three components of the pseudoelectric dipole. As a result, we derive a set of three differential equations, referred to as the optical Bloch equations. In the system of coordinates rotating with the frequency ω with respect to the z axis of the laboratory system of coordinates, these equations are written as

$$dU/dt = V\Delta\omega - U/T_2, \tag{A.1.10}$$

$$dV/dt = -U\Delta\omega - k^2\varepsilon_0 W/\omega - V/T_2, \tag{A.1.11}$$

$$dW/dt = V\varepsilon_0\omega - (W - W_0)/T_1, \tag{A.1.12}$$

where U, V, and W are the components of the pseudoelectric dipole moment,

$$\boldsymbol{P} = \boldsymbol{x}U + \boldsymbol{y}V - \boldsymbol{z}Wk/\omega,$$

$$U = \tfrac{1}{2}Np\sigma_x,\quad V = \tfrac{1}{2}Np\sigma_y,\quad W = \tfrac{1}{2}N\hbar\omega_0\sigma_z,\quad W_0 = -N\hbar\omega_0/2,$$

$$\Delta\omega = \omega_0 - \omega,\quad k = 2p/\hbar.$$

Solutions to equations of the form (A.1.10)–(A.1.12) were investigated in [8,9]. Introducing F to denote any of the components (U, V, W) of the pseudodipole, we can represent the solution in the following general form:

$$F = A\exp(-at) + B\exp(-bt)\cos st + s^{-1}C\exp(-bt)\sin st + D, \tag{A.1.13}$$

where the coefficients A, B, C, D, a, b, and s are the functions of the detuning and relaxation times T_1 and T_2. These coefficients depend on the experimental conditions and are well known for some important cases [10]. In particular, in the case of an exact resonance ($\Delta\omega = 0$), we have

$$A_U = A_V = A_W = 0,\quad B_U = 0,\quad B_V = \alpha/(1+\alpha\beta),\quad B_W = W_i/W_0 - \alpha\beta/(1+\alpha\beta),$$

$$C_U = 0,\quad C_V = \tfrac{1}{2}\alpha(\alpha+\beta)/(1+\alpha\beta) - W_i/W_0,$$

$$C_W = W_i(\beta-\alpha)/W_0 + \alpha[1 - \tfrac{1}{2}\beta(\alpha+\beta)/(1+\alpha\beta)],$$

$$D_U = \alpha\delta/R,\quad D_V = -\,\alpha\beta/R,\quad D_W = \alpha(\beta^2+\delta^2)/R,$$

$$a = \beta\omega_1,\quad b = \tfrac{1}{2}\omega_1(\alpha+\beta),\quad s = [1 - \tfrac{1}{4}(\beta-\alpha)^2]^{1/2}\omega_1,$$

$$\alpha = (\omega_1 T_1)^{-1},\quad \beta = (\omega_1 T_2)^{-1},\quad R = \alpha(\beta^2+\delta^2)+\beta,\quad \delta = \Delta\omega/\omega_1, \tag{A.1.14}$$

where $\omega_1 = p\varepsilon_0/\hbar$ is the Rabi frequency, and W_i is the initial condition (for an equilibrium system, $W_i = W_0$).

If the components of the pseudoelectric dipole are known, then we can find the optical electric field reemitted by a medium using the wave equation

$$\frac{\partial^2 E_+(z,t)}{\partial z^2} = \frac{\eta^2}{c^2}\frac{\partial^2 E_+(z,t)}{\partial t^2} + \frac{4\pi}{c^2}\frac{\partial^2 P_+(z,t)}{\partial t^2}, \tag{A.1.15}$$

where

$$E_+(z,t) = \varepsilon_0(z,t)\exp\{i[\omega t - kz - \varphi(z)]\},$$

$$P_+(z,t) = \int_{-\infty}^{\infty} d(\Delta\omega)\, g(\Delta\omega)[U + iV]\exp\{i[\omega t - kz - \varphi(z)]\},$$

$g(\Delta\omega)$ is the distribution function of the detuning parameter, η is the refractive index of light in the medium, and c is the speed of light in a vacuum.

Many physical situations can be adequately described within the framework of approximation assuming that the variation in the amplitude of the electric field is small,

$$|\partial\varepsilon_0/\partial z| << |\varepsilon_0/\lambda|, \quad |\partial\varepsilon_0/\partial t| << \omega|\varepsilon_0|, \tag{A.1.16}$$

where λ is the wavelength of light propagating along the z axis. When inequalities (A.1.16) are satisfied, differential equation (A.1.15) yields the following equations:

$$\partial\varepsilon_0/\partial z = -(\eta/c)\partial\varepsilon_0/\partial t + (2\pi\omega/\eta c)\int_{-\infty}^{\infty} d(\Delta\omega)\, g(\Delta\omega)V(\Delta\omega), \tag{A.1.17}$$

$$\varepsilon_0\partial\varphi/\partial z = (2\pi\omega/\eta c)\int_{-\infty}^{\infty} d(\Delta\omega)\, g(\Delta\omega)U(\Delta\omega). \tag{A.1.18}$$

Thus, wc derived a set of coupled integrodifferential equations (A.1.10)–(A.1.12), (A.1.17), and (A.1.18) governing the behavior of a light wave in a two-level resonant medium taking into account the response of this medium to the action of this light wave.

A.2 Stimulated Photon Induction

Following papers [11–13], we will discuss the specific features of coherent optical emission from a resonant medium excited with laser radiation in a stationary regime. Obviously, the reemission of light in this case is accompanied by active relaxation processes in a medium. In radio-frequency spectroscopy, such an emission was demonstrated by Bloch [3]. Therefore, this phenomenon is sometimes referred to as the Bloch induction.

Let us consider the behavior of an ensemble of two-level particles with the splitting of levels $\hbar\omega_0$ in the stationary field of resonant laser radiation

$$E(z, t) = \varepsilon_0(z)\exp\{i[\omega t - kz - \varphi(z)]\} + c.c., \tag{A.2.1}$$

where ε_0, k, ω, and φ are the amplitude, the wave vector, the carrier frequency, and the phase of the light wave propagating along the z axis.

The stationary solutions to the optical Bloch equations (A.1.10)–(A.1.12) are written as

$$U = A^{-1}\omega_1 \Delta\omega T_2^2 P_z^0,$$

$$V = A^{-1}\omega_1 T_2 P_z^0, \qquad \text{(A.2.2)}$$

$$W = (\omega/k)A^{-1}(1 + \Delta\omega T_2) P_z^0,$$

where $A = 1 + (\Delta\omega T_2)^2 + \omega_1^2 T_1 T_2$.

Let us rewrite the wave equation (A.1.15) in the following form:

$$\left(\frac{\partial}{\partial z} - \frac{\eta}{c}\frac{\partial}{\partial t}\right)\left(\frac{\partial}{\partial z} + \frac{\eta}{c}\frac{\partial}{\partial t}\right)E(z,t) =$$

$$= -\frac{4\pi\omega^2}{c^2} P_+ \exp[i(\omega t - kz - \varphi)]. \qquad \text{(A.2.3)}$$

Introducing new variables $\xi = t - z\eta/c$ and $\sigma = t + z\eta/c$, we can reduce this equation to

$$\frac{\partial^2}{\partial\xi\partial\sigma} E = \frac{4\pi\omega^2}{c^2} P_+ \exp(i\xi\omega). \qquad \text{(A.2.4)}$$

Solutions to equations of the form (A.2.4) are well known from field theory:

$$E(\xi) = [\varepsilon_0 \exp(-\alpha z) - i\frac{4\pi\omega^2}{c^2} P_+]\exp(i\xi\omega), \qquad \text{(A.2.5)}$$

where the second term in the square brackets describes the field of the coherent response of a medium – stimulated photon induction (SPI). The SPI intensity emitted within unit solid angle $\Delta\Omega$ in the direction of the wave vector $\boldsymbol{k}_l$ can be calculated with the use of the following formula:

$$\Delta I_{SPI}(\boldsymbol{k}) = (c/8\pi)EE^*\Delta\Omega. \tag{A.2.6}$$

Expressing E in terms of the transverse components of the pseudoelectric dipole (A.2.2) and substituting the result into formula (A.2.6), we find that

$$\Delta I_{SPI}(\boldsymbol{k}) = I_0(3\pi/8)(\omega_1 T_2)^2 f\Delta\Omega, \tag{A.2.7}$$

where $I_0 = 4p^2\omega^4/(3c^3)$ is the intensity of spontaneous radiation of single particle;

$$f = \left| \int_{-\infty}^{\infty} d(\Delta\omega)\, g(\Delta\omega)(i + \Delta\omega T_2)/[1 + (\Delta\omega T_2)^2 + \omega_1^2 T_1 T_2] \right|^2. \tag{A.2.8}$$

Integration over the solid angles is well known [14] and leads to appearance of the factor $8\pi/3$. Thus, we arrive at the following expression for the total SPI intensity:

$$I_{SPI} = I_0(\pi N\omega_1 T_2)^2 f. \tag{A.2.9}$$

In the case of stationary excitation of a resonant medium, the pump laser field may have a rather narrow bandwidth. Therefore, we can assume that the detuning is equal to zero, which brings us to the following formula:

$$I_{SPI} = I_0(\pi N\omega_1 T_2)^2/[1 + \omega_1^2 T_1 T_2]^2. \tag{A.2.10}$$

It is of interest to consider two regimes of excitation. Under conditions of saturation, when $\omega_1^2 T_1 T_2 >> 1$, we have

$$I_{SPI} = I_0(\pi N)^2/(\omega_1 T_1)^2. \quad \text{(A.2.11)}$$

In the regime of weak excitation of a medium, when the opposite condition is satisfied, we find that

$$I_{SPI} = I_0(\pi N \omega_1 T_2)^2. \quad \text{(A.2.12)}$$

Note that saturation can be achieved in a CO_2-laser – SF_6-gas system with a resonance at the wavelength of 10.6 μ with $p = 10^{-20}$ CGSE, ε_0 being on the order of several CGSE units, $T_1 = 10^{-5}$ s, and $T_1 = 10^{-6}$ s at 0.01 Torr.

Numerical analysis performed in [14] with the use of the joint set of differential equations (A.1.10)–(A.1.12), (A.1.17), and (A.1.18) has shown that the growth in the intensity of a stationary laser wave propagating through a resonant medium may give rise to fluctuations in the electric field strength. If

$$\omega_1^6 > T_2^2/(T_2^3 T_1^3 T_2^{*2}), \quad \text{(A.2.13)}$$

where ω_1 is the Rabi frequency and T_1, T_2, and T_2^* are the longitudinal, transverse irreversible, and transverse reversible relaxation times, respectively, then the build-up of these fluctuations may transfer stationary optical emission into pulsed emission.

Finally, let us consider the detection of SPI. In accordance with [15], the detection of this effect is considerably simplified if an ensemble of particles is excited in the two-quantum regime at frequencies ω_1' and ω_2' and the SPI signal is detected in the single-quantum regime at the frequency $\omega = \omega_1' + \omega_2'$. In this case, the Rabi frequency in the expressions presented above is defined as

$$\omega_1 = \hbar^{-2}[\langle 2|\varepsilon_0^{(1)} p|\gamma\rangle\langle\gamma|\varepsilon_0^{(2)} p|1\rangle/(\omega_1' - \omega_\gamma + i\Gamma_\gamma) +$$

$$+ \langle 2|\varepsilon_0^{(2)}p|\gamma\rangle\langle\gamma|\varepsilon_0^{(1)}p|1\rangle/(\omega_2' - \omega_\gamma + i\Gamma_\gamma)], \tag{A.2.14}$$

where ω_γ and Γ_γ are the frequency and the width of the intermediate level γ, $\varepsilon_0^{(1)}$ and $\varepsilon_0^{(2)}$ are the amplitudes of the electric field strengths in the pump waves, and $\langle\alpha|p|\beta\rangle$ is the matrix element of the electric dipole moment related to the $\alpha - \beta$ transition. The SPI wave vector satisfies the following spatial phase-matching condition:

$$\boldsymbol{k}_{SPI} = \boldsymbol{k}_1 \pm \boldsymbol{k}_2. \tag{A.2.15}$$

Thus, the SPI signal in this regime is separated from the pump waves both in frequency and in space.

A.3 The Photon (Optical) Echo in Gas Media

An echo signal is a coherent response of a resonant medium to the action of two or more pump pulses delayed with respect to each other. In the case of two-pulse excitation, this response is called the primary echo. The primary-echo signal is generated by a medium at the moment of time equal to twice the delay time τ_{12} of the second pulse with respect to the first pulse. This process is illustrated in Fig. A.1. The action of the third pump pulse (obviously, we assume that the conditions of nonlinear coherent interaction specified in the previous chapters of this book are satisfied) gives rise to the generation of several echo signals. The regime of stimulated echo, when the echo signal are produced with a delay time τ_{12} after the third pulse, is especially promising for various applications. Other echo signals are multipulse analogs of the primary echo. The restored echo, where the primary echo is involved as the first pump pulse (see Fig. A.1), is also of considerable interest. Multiple echo signals are also very promising. One of the models of the multiple echo employs the primary echo signal as the second pump pulse.

Echo signals were observed for the first time by Hahn [16] in the radio-frequency range and were called spin echo signals. The possibility of observing echo signals in optics was not obvious because of several reasons. In particular, optical echo signals are excited through electric-dipole transitions, which implies the absence of the intrinsic electric dipole moment in each of the resonant stationary states in a nondegenerate system, which eventually means that the vector model of the spin echo cannot be employed

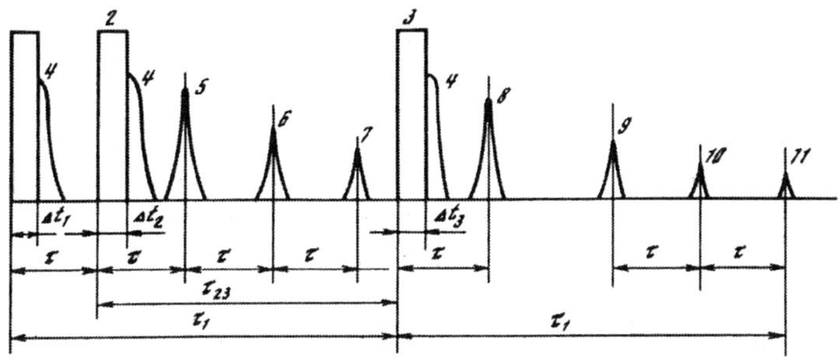

Fig. A.1. Sequence of pump pulses and the relevant coherent optical responses in a two-level resonant medium: (1–3) pulses, (4) signals of free optical induction, (5) primary echo, (6,7) multiple echo signals, (8) stimulated echo, (9) restored echo, (10, 11) three-pulse analogs of the primary echo.

in its conventional form. This problem was solved by Kopvillem and Nagibarov [17], who predicted the phenomenon of optical echo in 1963. This effect was experimentally detected by Kurnit, Abella, and Hartmann in a ruby crystal [18], who called this phenomenon the photon echo. Currently, both names of this effect are widely used. Experiments [18] (and especially subsequent studies [19]) stimulated the analysis of the possibility of observing the photon (optical) echo in media with intense inner motion and, first of all, in gas media [20–22]. Among the studies performed in this direction, special attention should be given to the paper [20], which gave the answer to the question of why the excitation of the photon echo in a rarefied gas does not require the use of pico- or femtosecond laser pulses. The first experiment on the photon echo in a gas medium (in SF_6) was carried out by Patel and Slusher [23], who revealed considerable differences of the polarization properties of the photon echo in a gas from the polarization properties of the photon echo produced in a solid. These properties were explained in theoretical papers [24, 25], which, in fact, open the direction of polarization echo spectroscopy. Among the first experimental papers on the optical echo in gases, we should also focus a special attention on the study by Alimpiev and Karlov [26].

Let us consider in greater detail the physics of the photon echo. We should mention, first of all, that this phenomenon can be observed only when conditions of nonlinear coherent interaction are satisfied. In the case of the primary photon echo, this implies that the duration of pump pulses

(Δt_η) and the time interval τ_{12} between these pulses should meet the inequalities

$$\Delta t_\eta, \tau_{12} < T_1, T_2. \tag{A.3.1}$$

Along with the inequality (A.3.1), the observation of the stimulated echo with $\eta = 1, 2, 3$ requires the fulfillment of the additional condition

$$\tau_{23} < T_1, \tag{A.3.2}$$

where τ_{23} is the time interval between the second and third pump pulses.

Thus, the formation of the photon echo is based on the nonlinear coherent interaction of two and more laser pulses with unpaired electrons of active particles. The nonlinear character of the interaction implies that the field of laser pulses distorts the electronic shells of these particles in such a way that these particles are transferred to a superposition state (A.1.1) rather than to an upper resonant state. The coherence of interaction implies that this distortion provides the information concerning the phase of the pump light waves. Each of the unpaired electrons in the superposition state (A.1.1) is characterized by a nonequilibrium electric dipole moment, which consists [as can be seen from expression (A.1.5)] of the electric dipole moments of the resonant transition. The first coherent pulse (it would be appropriate to choose the area of this pulse equal to $\pi/2$) phases electric dipoles in each infinitely thin layer. Suppose that a broadband pump pulse is circularly polarized. Then, due to the coherent interaction of this pulse with electrons, the instantaneous polarization phase of the pulse is transferred to the phase of induced nonequilibrium electric dipoles. After the action of the pump pulse, these dipoles are involved in the rotation (precession) with the natural frequency equal to the frequency of the resonant transition. Due to the Doppler effect, this frequency differs for different gas particles, and nonequilibrium electric dipoles cease to be in phase within the time T_2^* ($\sim 1/\Delta\omega_D$, where $\Delta\omega_D$ is the Doppler line width). However, this dephasing has a reversible character (the time T_2^* is called the transverse reversible relaxation time), and the gas particles can be phased again under the action of the second laser pulse.

Suppose that the area θ of the second pump pulse is equal to π (although this is not necessary). Such a pulse transfers the entire electron density to

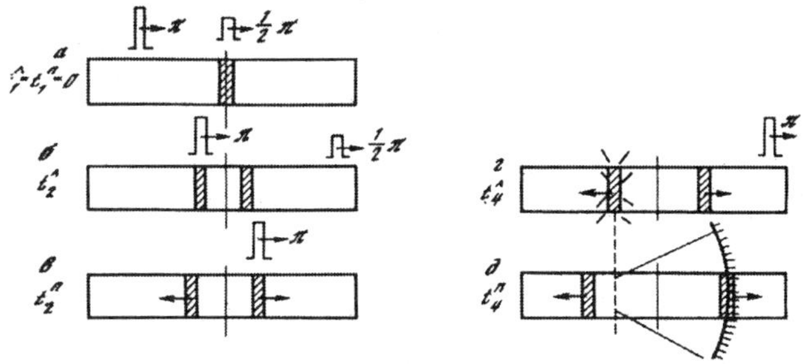

Fig. A.2. Illustration of photon-echo formation in a gas medium [27].

the excited state, leading to the total population inversion. The states 1 and 2 switch their places under these conditions, reversing the direction of the velocity ω_j (recall that $\hbar\omega_j$ is the splitting of resonant levels of the j-th particle) of angular rotation (precession). As a result, nonequilibrium electric dipoles start rotating (precessing) in opposite direction, and the dipoles become phased again at the moment of time equal twice the time interval between the laser pulses. This moment of time corresponds to the maximum of the primary photon echo signal. Formation of the stimulated photon echo and other echo signals can be explained in a similar way.

We still have to understand how the photon echo is produced in a gas with an intense motion of gas particles. Let us employ the physical model substantiated in [20]. The system under study is illustrated in Fig. A.2, which shows a cell with a gas irradiated from the left with short laser pulses with a resonant frequency. We choose a thin gas layer where particles may, in principle, have different velocities and different displacement directions. First, all the particles are irradiated with the first laser pulse. Then, within a certain time interval τ_{12}, a certain group of particles move to the left, while the remaining particles move to the right. Obviously, the second pump pulse interacts with these groups of particles at different moments of time. The particles moving to the left produce an echo signal earlier than particles moving to the right, because the time required for the second pump pulse to cover the distance between these groups of particles can be estimated as $t \approx \tau_{12}(v^r + v^l)/c$, where v^r and v^l are the velocities of particles moving to the right and to the left, respectively. The same delay arises for the echo signal, and the total time required to recover the phase is equal to $2\tau_{12}(v^r + v^l)/c$. On the other hand, during this time interval, a light pulse (first, the second pump pulse and then the echo signal) covers the distance between these two groups of particles twice. An important point is that the second pump pulse and the echo signal have equal velocities. Thus, regardless of the velocities of particles moving in opposite directions, echo signals emitted by these groups of particles reach the end of the cell having the same phases. This is also true for particles moving at arbitrary angles with respect to the direction of pulse propagation. As mentioned above, the fact that the echo signals produced by particles with different velocities reach the end of the cell having the same phase implies that the transverse shift of a particle does not lead to the dephasing of echo signals. Such echo signals may reach a photodetector at somewhat different points, but echo signals are usually focused, and the resulting spot produced by the photon echo signal is very small anyway.

We should address yet another issue associated with the Doppler broadening of spectral lines under conditions of the thermal scatter of velocities of gas particles. In the vector representation of echo-signal formation in gases, this source of inhomogeneous broadening gives rise to dephasing and subsequent phasing of pseudoelectric dipoles in the pseudoelectric field. For the i-th moving particle, the frequency of absorption and reemission of light is equal to

$$\omega_i = \omega_0 - (\omega_0/c)\mathbf{v}_i\boldsymbol{n}, \tag{A.3.3}$$

where ω_0 is the frequency of the energy transition for a stationary particle, c is the speed of light, $\mathbf{v}_i$ is the velocity of the i-th particle, and $\boldsymbol{n}$ is the unit vector characterizing the direction of observation. Particles moving to the left and to the right are characterized by different angles between the velocity $\mathbf{v}_i$ and the unit vector $\boldsymbol{n}$. Therefore, echo signals emitted by particles with different velocities have different carrier frequencies. We should note, however, that, if the spectrum of pump pulses could be monochromatic, then only definite groups of particles would absorb energy from these pulses. In reality, the spectrum of pump pulses is rather broad. Therefore, the spectrum of the echo signal is similar to the spectrum of pump pulses (although various convolutions of signals can also be obtained).

To calculate optical echo signals produced in a gas medium, we should modify equations (A.1.10)–(A.1.12) to include the motion of particles. Such a modification was done in paper [26], where these equations were also solved for different echo signals. By now, various methods have been used to perform such calculations. Since we have already described such calculations in this book, we are not going to discuss the details of this procedure here. However, we should note that optical echo signals (in contrast to their radio-frequency analogs) are emitted in definite directions meeting the spatial phase-matching conditions. For the primary photon echo, this condition is written in the same way for solids and for gas media:

$$\boldsymbol{k}_p = 2\boldsymbol{k}_2 - \boldsymbol{k}_1, \tag{A.3.4}$$

where $\boldsymbol{k}_1$ and $\boldsymbol{k}_2$ are the wave vectors of the primary photon echo and the first and second pump pulses, respectively.

For the stimulated photon echo (SPE) in gas media, the condition of spatial phase matching differs from the spatial phase-matching condition for this process in solids [27]. This circumstance was pointed out for the first time in [27], where it was demonstrated, that the spatial phase-matching condition for the SPE signal in a gas medium is written as

$$\boldsymbol{k}_s = \boldsymbol{k}_1 - \boldsymbol{k}_2 + \boldsymbol{k}_3. \tag{A.3.5}$$

With $\boldsymbol{k}_1 = -\boldsymbol{k}_3$, we have $\boldsymbol{k}_s = -\boldsymbol{k}_2$. Such a regime of SPE excitation is called the backward echo.

In extended and dense gas media, the response of the resonant medium may play an important role in the formation of the photon echo. Among the theoretical studies devoted to this problem, special attention should be given to the paper [28].

REFERENCES

1. R.P. Feynmann, F.L. Vernon, K.W. Hellwarth, *J. Appl. Phys.* **28**: 49–52 (1957).
2. Y.H. Pao, *J. Opt. Soc. Am.* **52**: 871–878 (1962).
3. F. Bloch, *Phys. Rev.* **70**: 460–473 (1946).
4. A. Abraham, *The Principles of Nuclear Magnetism* (London: Oxford Univ. Press, 1961).
5. K.M. Salikhov, A.G. Semenov, Yu.D. Tsvetkov, *Elektronnoe spinovoe ekho i ego primenenie* (*Electron Spin Echo and Its Applications*) (Novosibirsk: Nauka, 1976) (in Russian).
6. F. Berten, *Osnovy kvantovoi elektroniki* (*Fundamentals of Quantum Electronics*) (Moscow: Mir, 1971) (in Russian).
7. I.D. Abella, *Prog. Opt.* **7**: 141–168 (1969).
8. R.K. Wangsness, F. Bloch, *Phys. Rev.* **89**: 728–740 (1954).

9. H.C. Torrey, *Phys. Rev.* **76**: 1059–1069 (1949).
10. A. Löshe, *Kerninduktion* (Berlin: Wissenschaften, 1957) (in German).
11. V.R. Nagibarov, V.A. Pirozhkov, V.V. Samartsev, *Ukr. Fiz. Zh.* **16**: 767–768 (1971).
12. V.V. Samartsev, N.N. Kurkin, *Opt. Spektrosk.* **32**: 413–415 (1972).
13. V.R. Nagibarov, V.V. Samartsev, *Opt. Spektrosk.* **27**: 467–472 (1969).
14. E.A. Manykin, V.V. Samartsev, *Opticheskaya ekho-spektroskopiya* (*Optical Echo Spectroscopy*) (Moscow: Nauka, 1984) (in Russian).
15. U.Kh. Kopvillem, V.R. Nagibarov, V.V. Samartsev, *Zh. Prikl. Spektrosk.* **12**: 888–892 (1970).
16. E.L. Hahn, *Phys. Rev.* **80**: 580–594 (1950).
17. U.Kh. Kopvillem, V.R. Nagibarov, *Fiz. Metallov Metallovedenie* **15**: 313–315 (1963).
18. N.A. Kurnit, I.D. Abella, S.R. Hartmann, *Phys. Rev. Lett.* **13**: 567–570 (1964).
19. I.D. Abella, N.A. Kurnit, S.R. Hartmann, *Phys. Rev.* **141**: 391–411 (1966).
20. M. Scully, M.J. Stephen, D.C. Burnham, *Phys. Rev.* **171**: 171–173 (1969).
21. V.V. Samartsev, *Ukr. Fiz. Zh.* **14**: 1045–1046 (1969).
22. V.R. Nagibarov, V.V. Samartsev, *Opt. Spektrosk.* **27**: 467–472 (1969).
23. C.K.N. Patel, R.E. Slusher, *Phys. Rev. Lett.* **20:** 1087–1089 (1968).
24. A.I. Alekseyev, I.V. Yevseyev, *Zh. Eksp. Teor. Fiz.* **56**: 2118–2128 (1969).
25. J.P. Gordon, C.H. Wang, C.K.N. Patel, *et al.*, *Phys. Rev.* **179**: 294–309 (1969).
26. S.S. Alimpiev, N.V. Karlov, *Zh. Eksp. Teor. Fiz.* **63**: 482–490 (1972).
27. L. Allen, J. Eberly, *Optical Resonance and Two-Level Atoms* (New York: Wiley, 1975).
28. V.R. Nagibarov, V.V. Samartsev, *Svetovoe ekho v gazakh, zhidkotyakh i steklakh* (*Optical Echo in Gases, Liquids, and Glasses*) (Kazan: dep. VINITI No. 1100–1169, 1969) (in Russian).

Subject Index